SEAGRASSES

Monitoring, Ecology, Physiology, and Management

The CRC Marine Science Series is dedicated to providing state-of-the-art coverage of important topics in marine biology, marine chemistry, marine geology, and physical oceanography. The series includes volumes that focus on the synthesis of recent advances in marine science.

CRC MARINE SCIENCE SERIES

SERIES EDITOR

Michael J. Kennish, Ph.D.

PUBLISHED TITLES

Artificial Reef Evaluation with Application to Natural Marine Habitats, William Seaman, Jr.

The Biology of Sea Turtles, Volume I, Peter L. Lutz and John A. Musick

Chemical Oceanography, Second Edition, Frank J. Millero

Coastal Ecosystem Processes, Daniel M. Alongi

Ecology of Estuaries: Anthropogenic Effects, Michael J. Kennish

Ecology of Marine Bivalves: An Ecosystem Approach, Richard F. Dame

Ecology of Marine Invertebrate Larvae, Larry McEdward

Ecology of Seashores, George A. Knox

Environmental Oceanography, Second Edition, Tom Beer

Estuary Restoration and Maintenance: The National Estuary Program, Michael J. Kennish

Eutrophication Processes in Coastal Systems: Origin and Succession of Plankton Blooms and Effects on Secondary Production in Gulf Coast Estuaries, Robert J. Livingston

Handbook of Marine Mineral Deposits, David S. Cronan

Handbook for Restoring Tidal Wetlands, Joy B. Zedler

Intertidal Deposits: River Mouths, Tidal Flats, and Coastal Lagoons, Doeke Eisma

Marine Chemical Ecology, James B. McClintock and Bill J. Baker

Morphodynamics of Inner Continental Shelves, L. Donelson Wright

Ocean Pollution: Effects on Living Resources and Humans, Carl J. Sindermann

Physical Oceanographic Processes of the Great Barrier Reef, Eric Wolanski

The Physiology of Fishes, Second Edition, David H. Evans

Pollution Impacts on Marine Biotic Communities, Michael J. Kennish

Practical Handbook of Estuarine and Marine Pollution, Michael J. Kennish

Practical Handbook of Marine Science, Third Edition, Michael J. Kennish

Seagrasses: Monitoring, Ecology, Physiology, and Management, Stephen A. Bortone

SEAGRASSES

Monitoring, Ecology, Physiology, and Management

edited by
Stephen A. Bortone, Ph.D.

CRC Press is an imprint of the
Taylor & Francis Group, an **informa** business

CRC Press
Taylor & Francis Group
6000 Broken Sound Parkway NW, Suite 300
Boca Raton, FL 33487-2742

First issued in paperback 2019

© 2000 by Taylor & Francis Group, LLC
CRC Press is an imprint of Taylor & Francis Group, an Informa business

No claim to original U.S. Government works

ISBN-13: 978-0-8493-2045-3 (hbk)
ISBN-13: 978-0-367-39914-6 (pbk)
Library of Congress Card Number 99-39826

This book contains information obtained from authentic and highly regarded sources. Reasonable efforts have been made to publish reliable data and information, but the author and publisher cannot assume responsibility for the validity of all materials or the consequences of their use. The authors and publishers have attempted to trace the copyright holders of all material reproduced in this publication and apologize to copyright holders if permission to publish in this form has not been obtained. If any copyright material has not been acknowledged please write and let us know so we may rectify in any future reprint.

Except as permitted under U.S. Copyright Law, no part of this book may be reprinted, reproduced, transmitted, or utilized in any form by any electronic, mechanical, or other means, now known or hereafter invented, including photocopying, microfilming, and recording, or in any information storage or retrieval system, without written permission from the publishers.

For permission to photocopy or use material electronically from this work, please access www.copyright.com (http://www.copyright.com/) or contact the Copyright Clearance Center, Inc. (CCC), 222 Rosewood Drive, Danvers, MA 01923, 978-750-8400. CCC is a not-for-profit organization that provides licenses and registration for a variety of users. For organizations that have been granted a photocopy license by the CCC, a separate system of payment has been arranged.

Trademark Notice: Product or corporate names may be trademarks or registered trademarks, and are used only for identification and explanation without intent to infringe.

Library of Congress Cataloging-in-Publication Data

Seagrasses : monitoring, ecology, physiology, and management /
edited by Stephen A. Bortone.
p. cm. — (Marine science series)
Proceedings of a workshop "Subtropical and Tropical Seagrass Management Ecology: Responses to Environmental Stress" held in Fort Myers, Florida (USA), 14–16 October 1998.
Includes bibliographical references.
ISBN 0-8493-2045-3 (alk. paper)
1. Seagrasses Congresses. 2. Seagrasses—Ecology Congresses. I. Bortone, Stephen A. II. Subtropical and Tropical Seagrass Management Ecology: Responses to Environmental Stress (1998 : Fort Myers, Fla.) III. Series.
QK495.A14 S438 1999
584′.9—dc21

99-39826
CIP

Visit the Taylor & Francis Web site at
http://www.taylorandfrancis.com

and the CRC Press Web site at
http://www.crcpress.com

Preface

Seagrasses are becoming widely used as *in situ* indicators of the relative health and condition of subtropical and tropical estuarine ecosystems. To permit meaningful management of our estuaries, there is clearly a need to develop and refine ways of effectively monitoring and assessing seagrasses. Moreover, data on their status allows us the vision of being able to observe trends as well as assess the effectiveness of management actions, especially with regard to restoration.

The state of our knowledge on seagrasses in the subtropics and tropics was summarized in a symposium held in 1985. Since that time, however, considerable new research has greatly enhanced our ability to monitor and evaluate their ecological conditions. Additionally, local, state, regional, and federal resource management agencies are placing increasing emphasis on the development and use of biological indicators of ecological health. For example, several states have begun to develop programs to assess estuarine ecosystem health from a vantage point similar to the *Index of Biological Integrity* (IBI), which has been used successfully in freshwater ecosystems.

An overall objective is to use seagrasses effectively as a biological marker to measure the relative fitness of the estuarine ecosystem. To meet this objective, it is essential to bring the environmental management and research communities together so that a common level of understanding can occur. It is also essential to develop an appreciation of the problems each group faces when confronting seagrass management. To this end a workshop, Subtropical and Tropical Seagrass Management Ecology: Responses to Environmental Stress, was held in Fort Myers, Florida, October 14–16, 1998. A total of 28 original research and review presentations were offered during the workshop. The workshop was organized around several subtopics with complementary themes: Ecology and Physiology, Monitoring and Trends, Management, and Restoration.

The timing of this workshop was appropriate as new assessment methods and data analyses techniques had become available and ecosystem managers were being pressed to apply these techniques to their jurisdictional programs. Just as critical, the feedback and interaction between the scientific community and user groups was long overdue.

Several needs were met by this workshop. It provided a forum for researchers, administrators, and environmental managers interested in knowing the latest about seagrass assessment technology and how it can be adapted to assess estuarine health. It provided an atmosphere of cooperative interaction that permitted the experience and concerns of ecosystem managers to be relayed to an audience of active researchers. The published proceedings, based on the workshop presentations, will serve as a "Bible" to which estuarine resource managers will refer when seeking guidance or clarification on management/monitoring issues regarding seagrasses.

This publication is essentially the written results of many of the workshop presentations following peer review. A few additional research study results, not completed at the time of the workshop, are also included as they are directly related to the topic of seagrass management ecology. It is sincerely hoped that this work will serve as a foundation for the continued improvement of our ability to effectively monitor the overall health and condition of our near coastal waters. Further, it is anticipated that this work will serve as a reference that will be central to the premise that effective and efficient assessment of seagrasses will aid in estuarine ecosystem management.

Stephen A. Bortone, Ph.D.

Acknowledgments

Many people and agencies were responsible for the initial workshop upon which this work is based. The steering committee helped guide the early program and theme of the workshop. Steering committee members included: Steve Bortone (Florida Center for Environmental Studies, Chair), Peter Doering, Bob Chamberlain, Dave Rudnick and Chris Madden (South Florida Water Management District), Rob Mattson (Suwannee River Water Management District), Bob Virnstein (St. Johns River Water Management District), Paul Thorpe (Northwest Florida Water Management District), and Dave Tomasko (Southwest Florida Water Management District). Support for the workshop was gratefully provided by the South Florida Water Management District, Florida Center for Environmental Studies, Florida Sea Grant College, and Mississippi-Alabama Sea Grant College. The workshop was the idea of Pat Gostel (South Florida Water Management District), whose continuous support made the event possible. Registration was conducted under the auspices of Florida Atlantic University. Administrative support was provided by Doreen DiCarlo, Florida Center for Environmental Studies. Numerous other individuals associated with the Florida Center for Environmental Studies and the South Florida Water Management District were of great assistance in preparing and organizing the workshop. Shelby Curry helped in editing earlier drafts of abstracts and manuscripts. Many individuals who are actively involved with seagrass research and management gave their time and lent their expertise to help with the editorial review process. Among these were R.H. Chamberlain and P.H. Doering (South Florida Water Management District), R.W. Virnstein (St. Johns River Water Management District), C.L. Gallegos (Smithsonian Environmental Research Center), F.G. Lewis (Northwest Florida Water Management District), W.M. Kemp (University of Maryland), K. Moore (Virginia Institute of Marine Science), A. Squires (Pinellas County Department of Environmental Management), H. Neckles (U.S. Geological Survey), T. Ries (Scheda Ecological Associates, Inc.), J.O.R. Johansson (City of Tampa, Bay Study Group), L.K. Dixon (Mote Marine Laboratory), and R.R. Lewis (Lewis Environmental Services, Inc.). Lastly, thanks are due John Sulzycki, who foresaw the need to disseminate the workshop proceedings, and the editorial staff at CRC Press for making this publication a reality.

Editor

Stephen A. Bortone, Ph.D., is Director of Environmental Science at the Conservancy of Southwest Florida. He is Research Professor at Florida Atlantic University, Boca Raton, and has a Courtesy Faculty appointment at Florida Gulf Coast University, Fort Myers. In addition, he has served as Visiting Scientist with the Florida Center for Environmental Studies, and as Director of the Institute for Coastal and Estuarine Research. For over 25 years, he was Professor of Biology at the University of West Florida, Pensacola.

Dr. Bortone received a B.S. degree from Albright College, Reading, PA; a M.S. degree from Florida State University, Tallahassee; and a Ph.D. from University of North Carolina, Chapel Hill.

For the past 30 years, he has conducted research on the life history of estuarine organisms, including fishes and seagrasses, chiefly in the southeastern U.S. and the Gulf of Mexico. He has published over 100 scientific articles on the broadest aspects of Biology, including such diverse fields as Anatomy, Physiology, Endocrinology, Systematics and Taxonomy, Evolution, Reproductive Biology, Biogeography, Behavior, Histology, Ecology, Oceanography, and Sociobiology.

Widely traveled in conducting his research and teaching activities, Dr. Bortone was a Visiting Scientist at the Johann Gutenberg University, Mainz, Germany and has conducted extensive collaborative work with colleagues at the University of La Laguna in the Canary Islands. He was a Mary Ball Washington Scholar at University College, Dublin, Ireland. He has received several other teaching and research awards, including the title "Fellow" from the American Institute of Fishery Research Biologists.

Dr. Bortone has served as scientific editor and reviewer for numerous organizations, such as the National Science Foundation, the Environmental Protection Agency, and the U.S. Fish and Wildlife Service, and for several journals, including the *Bulletin of Marine Science, Copeia, Estuaries,* and *Transactions of the American Fisheries Society.*

Contributors

Walter Avery
Bay Study Group
City of Tampa
Tampa, Florida

Richard D. Bartleson
Center for Environmental Science
Horn Point Laboratory
University of Maryland
Cambridge, Maryland

Jeffrey L. Beal
Office of Coastal and Aquatic Managed Areas
Florida Department of Environmental Protection
Port St. Lucie, Florida

Norman J. Blake
Department of Marine Science
University of South Florida
St. Petersburg, Florida

Stephen A. Bortone
The Conservancy of Southwest Florida
Naples, Florida

Diana Burdick
Southwest Florida Water Management District
Tampa, Florida

Robert H. Chamberlain
Ecosystem Restoration Department
South Florida Water Management District
West Palm Beach, Florida

Kenneth W. Cummins
South Florida Water Management District
Fort Myers, Florida

Michelle R. Davis
Gonzalez, Florida

William P. Davis
Gulf Ecology Division
National Health and Environmental Effects Laboratory
United States Environmental Protection Agency
Gulf Breeze, Florida

Donald R. Deis
PBS&J, Inc.
Jacksonville, Florida

L. Kellie Dixon
Mote Marine Laboratory
Sarasota, Florida

Peter H. Doering
Ecosystem Restoration Department
South Florida Water Management District
West Palm Beach, Florida

Craig W. Dye
Southwest Florida Water Management District
Brooksville, Florida

Ernest D. Estevez
Mote Marine Laboratory
Sarasota, Florida

Robert Finck
Geonex Corporation
St. Petersburg, Florida

David A. Flemer
Gulf Ecology Division
National Health and Environmental Effects Laboratory
United States Environmental Protection Agency
Gulf Breeze, Florida

Nelly Gómez
CINVESTAV-IPN, Unidad Mérida
Yucatán, México

Holly S. Greening
Tampa Bay Estuary Program
St. Petersburg, Florida

Lauren M. Hall
St. Johns River Water Management District
Melbourne, Florida

Mark A. Hammond
Southwest Florida Water Management District
Tampa, Florida

M. Dennis Hanisak
Harbor Branch Oceanographic Institution
Fort Pierce, Florida

W. Schlese Heidelbaugh
St. Johns River Water Management District
Melbourne, Florida

Jorge A. Herrera-Silveira
CINVESTAV-IPN, Unidad Mérida
Yucatán, México

J. O. Roger Johansson
Bay Study Group
City of Tampa
Tampa, Florida

W. Michael Kemp
Center for Environmental Science
Horn Point Laboratory
University of Maryland
Cambridge, Maryland

W. Judson Kenworthy
Beaufort Laboratory
National Oceanic and Atmospheric Administration
Beaufort, North Carolina

Raymond C. Kurz
Scheda Ecological Associates, Inc.
Tampa, Florida

James Locascio
Sanibel-Captiva Conservation Foundation
Sanibel, Florida

Robert A. Mattson
Suwannee River Water Management District
Live Oak, Florida

Janice D. Miller
St. Johns River Water Management District
Palatka, Florida

Robbyn Miller-Myers
St. Johns River Water Management District
Jacksonville, Florida

Lori J. Morris
St. Johns River Water Management District
Palatka, Florida

Laura Murray
Center for Environmental Science
Horn Point Laboratory
University of Maryland
Cambridge, Maryland

Merrie Beth Neely
Department of Marine Science
University of South Florida
St. Petersburg, Florida

Keith Patterson
Geonex Corporation
St. Petersburg, Florida

Jane A. Provancha
Dynamac Corporation
Kennedy Space Center, Florida

Javier Ramírez-Ramírez
CINVESTAV-IPN, Unidad Mérida
Yucatán, México

Thomas F. Ries
Scheda Ecological Associates, Inc.
Tampa, Florida

Lourdes M. Rojas
Florida Center for Environmental Studies
Fort Myers, Florida

Paul J. Rudershausen
Florida Center for Environmental Studies
Fort Myers, Florida

Douglas M. Scheidt
Dynamac Corporation
Kennedy Space Center, Florida

Brandon S. Schmit
Office of Coastal and Aquatic Managed Areas
Florida Department of Environmental Protection
Port St. Lucie, Florida

William Severn
Center for Environmental Science
Horn Point Laboratory
University of Maryland
Cambridge, Maryland

R. Brian Sturgis
Center for Environmental Science
Horn Point Laboratory
University of Maryland
Cambridge, Maryland

David A. Tomasko
Southwest Florida Water Management District
Tampa, Florida

Robert K. Turpin
Florida Center for Environmental Studies
Fort Myers, Florida

Robert W. Virnstein
St. Johns River Water Management District
Palatka, Florida

Paula Whitfield
Beaufort Laboratory
National Oceanic and Atmospheric Administration
Beaufort, North Carolina

Margaret A. Wilzbach
Florida Center for Environmental Studies
Fort Myers, Florida

Arturo Zaldivar-Jimenez
CINVESTAV-IPN, Unidad Mérida
Yucatán, México

Table of Contents

1 Seagrass Ecology and Management: An Introduction

W. Michael Kemp

Anyone who spends time in and around tropical marine waters cannot help but notice the vast meadows of underwater plants that typically occupy much of the shoal area. Viewed from above the water, the variegated patterns of green reflect the plant community's sweeping systematic structure, but looking below the water surface with mask and snorkel, one is struck by the fine-scale complexity of this habitat. Waves generated by tropical breezes cause the ribbon-like leaves to undulate in oblique oscillating motions. Against this pulsating emerald backdrop, the darting movements of fish and spineless creatures within and around the plants create the sense of a well-choreographed dance ensemble. Indeed, there is an ageless allure that has always drawn fishermen, scientists, and sunburned visitors to these seagrass meadows. While perhaps initially drawn by the beauty of these vegetated environs, it does not take an expert's eye to see that these are also important habitats, which nourish, protect, and congregate their abundant tropical animal resources.

During the past 50 years, and particularly in the last decade, there has been ever increasing interest in the ecology of seagrasses in both tropical and temperate coastal waters. While the characteristically clear water of tropical coastal regions makes seagrass more conspicuous, its presence is also recognized in the more turbid temperate waters. Regardless of latitudinal differences, these vegetated ecosystems are widely valued as habitats for diverse fish and invertebrate populations, as subjects for scientific inquiry, as sites for human recreation, and as sensitive indicators of human pollution and perturbations (Kemp 1983). However, these presumed values were, perhaps, not fully appreciated until natural and anthropogenic disturbances caused local, regional, and global patterns of seagrass decline. The widespread losses of eelgrass in coastal regions of the North Atlantic certainly focused international attention on seagrasses in the 1930s (den Hartog and Polderman 1975). Although the exact circumstances underlying this so-called "wasting disease" are not fully understood even today, they set the stage for a present day vigilance in monitoring seagrass distribution and abundance in coastal regions worldwide (e.g., Short et al. 1988).

Scientific studies have provided detailed quantitative understanding of how seagrass populations influence local and regional ecosystem functions. The high biomass and productivity of seagrasses — compared to adjacent plankton-dominated waters — provide food and shelter for animals, sites for enhanced biogeochemical processes, and structures that modify physical circulation and sediment transport. It is well established that abundance and production of fish and invertebrate populations tend to be higher in seagrass habitats than in adjacent unvegetated areas (e.g., Stoner 1983; Orth and van Montfrans 1990; Lubbers et al. 1990; Heck et al. 1995); however, it is less clear what are the mechanisms which effect habitat enrichment (e.g., Bell and Westoby 1986; Petersen 1986) or how animal interactions might in turn influence plant growth and survival (e.g., Reusch et al. 1994). A substantial fraction of seagrass production may be directly consumed by herbivores in tropical habitats (e.g., Zieman et al. 1984), while in temperate coastal waters most of the grazing occurs only on senescent plants late in the growing season (Thayer et al. 1984).

Much of the seagrass biomass enters marine food-chains as detritus, with relatively efficient trophic transfer which buffers against rapid nutrient cycling and oxygen depletion associated with eutrophic conditions (Twilley et al. 1986; Harrison 1989). Seagrass populations can also stimulate microbial cycling of nutrients and organic matter in sediments (Caffrey and Kemp 1990; Risgaard-Petersen et al. 1998; Blaabjerg et al. 1998) and significantly reduce nutrient concentrations in overlying water (Bulthuis et al. 1984). The physical structure of seagrasses attenuates wave and tidal energy (e.g., Koch and Gust in press), resulting in enhanced deposition (and burial) of suspended particulates (e.g., Harlin et al. 1982), and reduction of water column turbidity (e.g., Ward et al. 1984).

Recent research has also focused on a wide range of other processes related to seagrass ecology and biology. Numerous investigations have examined seagrass photosynthetic responses and adaptations to light availability (e.g., Olesen and Sand-Jensen 1993). This work is particularly important because seagrass depth distribution tends to be limited by light (Duarte 1991) and because decreased water clarity is associated with many anthropogenic disturbances (Dennison et al. 1993). In fact, it has been observed that the minimum light levels required for growth and survival of seagrasses are relatively high compared to those for both algae and terrestrial plants (e.g., Goldsborough and Kemp 1988; Dennison et al. 1993). The underlying reasons for these elevated light requirements remain uncertain. Two postulated explanations are related to (1) light-dependent oxidation of toxic sulfide concentrations in the root zone (Kuhn 1992; Goodman et al. 1995); (2) inherently low photosynthetic efficiency under the characteristically limited inorganic carbon supplies in marine environments (Zimmerman et al. 1997). The phyto-toxic effects of sulfide (Carlson et al. 1994) and the ability of seagrasses to mitigate this effect via oxygen release from roots (e.g., Caffrey and Kemp 1991) are both reasonably well established. Over seasonal cycles, seagrasses adjust to variations in light, temperature, and nutrient conditions, in part, by modifying allocation and storage of non-structural carbohydrates in different plant parts (e.g., Lee and Dunstan 1996), and these patterns may vary in relation to environmental stresses (Burke et al. 1996). Seagrass populations are able to maintain their nutrient economy by reclamation of nutrients in internal cycles (Borum et al. 1989; Pedersen and Borum 1993) and by stimulation of and linkage with external cycles (e.g., Hemminga et al. 1991; Risgaard-Petersen et al. 1998). Indeed, interactions between seagrass growth and nutrient cycling processes are complex (Duarte 1995), and perhaps related to patchiness and fine scale spatial patterns in plant distribution (e.g., Duarte and Sand-Jensen 1990).

A relatively recent phenomenon which has captured the interest of seagrass scientists and managers is the global trend of regional declines in seagrass abundance. The geographic scope of this trend is staggering, and most of these declines appear to be related to human-induced disturbances, many of which are related to reductions in light available for plant photosynthesis (e.g., Short and Wyllie-Echeverria 1996). Major epicenters for seagrass losses are adjacent to areas of dense human habitation, including Europe (den Hartog and Polderman 1975), Australia (Walker and McComb 1992), and North America (e.g., Orth and Moore 1983a). At local scales, seagrass losses have been attributed to dredging for maintenance of navigational channels (Onuf 1994), harvest of shellfish (Ruffin 1998), discharges of silt (Bach et al. 1998) and turbidity plumes (Heck 1976), and scouring associated with motorboat propellers (Zieman 1976) and boat moorings (Walker et al. 1989). Although significant temporal changes in seagrass growth may be related to hydrologic changes associated with natural climatologic cycles (Marba and Duarte 1995), human manipulations of regional hydrology may also be (at least partially) responsible for recent massive reductions in seagrass abundance (Fourqurean and Robblee 1999).

One of the most ubiquitous factors contributing to demise of seagrass populations throughout the world is excessive nutrient enrichment from adjacent watersheds. Many of the best documented cases of seagrass losses resulting from eutrophication have been in temperate regions, including Chesapeake Bay (Kemp et al. 1983); Roskilde Fjord, Denmark (Borum 1985); and Waquoit Bay, MA (Short and Burdick 1996). However, detrimental effects of nutrient enrichment on seagrasses have also been demonstrated for tropical and subtropical regions, including western Australia

(Cambridge and McComb 1984), Bermuda's north shore (McGlathery 1995), and southwest Florida (Tomasko et al. 1996). In many of these situations, seagrass declines are related to overgrowth by epiphytic diatoms and filamentous macroalgae. Although epiphytic algae are an important element of healthy seagrass beds, providing the base for key food-chains (e.g., Klumpp et al. 1992), excessive algal accumulation causes significant attenuation of light available for growth of the host plant (e.g., Twilley et al. 1985). Grazing by herbivorous invertebrates can regulate epiphyte growth in oligotrophic habitats (e.g., Alcovera et al. 1997) and even with nutrient enrichment under certain circumstances (e.g., Neckles et al. 1993). This problem of coastal eutrophication is an insidious one because it is so closely tied to economic development and expansion of human populations (e.g., Peierlis et al. 1991). Dealing with this and other anthropogenic stresses which threaten seagrasses worldwide may be particularly difficult in tropical regions where seagrasses are adapted to relatively clear water and constant environmental conditions.

Managing seagrass habitats for conservation of natural resources thus presents a significant challenge (Fonseca et al. 1998). A keystone of any seagrass management effort is systematic monitoring. Surveys of seagrass distribution, abundance, and diversity need to be conducted in relation to routine measurements of water and sediment quality. In many coastal regions, major sources of anthropogenic stress for seagrasses have been identified and quantified. Changes in these stressors need to be related to changes in seagrass distribution. To the extent that it is possible to establish minimum environmental conditions needed for seagrass survival, these can be used as monitoring targets (e.g., Dennison et al. 1993). However, because of the complex dynamic interactions between environmental factors and seagrass populations, use of such static guidelines may be limited. Numerical models have been developed as research tools to explore seagrass ecosystem dynamics (e.g., Wetzel and Neckles 1986; Fong and Harwell 1994; Madden and Kemp 1996). Approaches for modeling of seagrass ecosystems are still in early stages of development; however, there is a pressing need to produce the next generation of models that can incorporate essential scientific information into a platform suitable for use in resource management. A strategic interplay between monitoring, modeling, and scientific research is needed to meet the challenge of conserving and restoring seagrasses in coastal waters stressed by human activities. Techniques are being developed for facilitating restoration of seagrass populations by means of artificial transplantation (e.g., Orth and Moore 1983b); however, successful restoration first requires establishing suitable environmental conditions. Nowhere are these problems of managing and restoring seagrasses more complex and acute than in tropical and subtropical regions, with their intense pressures of expanding human population.

It is in this context of intellectual excitement and focus on conservation of dwindling seagrass resources that a workshop was convened in the fall of 1998 to consider ecology and management of tropical and subtropical seagrasses. The written proceedings of this workshop are contained in the present book, which is organized into four major sections: ecology and physiology; monitoring and trends; restoration; management. There are many exciting research topics which are considered in this book and which mark many of the major thrusts directing seagrass research for the coming decade. One of the most daunting tasks ahead is to bridge the intellectual and cultural gaps between scientists and managers so that we can bring our collective knowledge to bear on solving the problem of conserving and restoring seagrasses in coastal regions around the world.

REFERENCES

Alcovera, T., C. Duarte, and J. Romero. 1997. The influence of herbivores on *Posidonia oceanica* epiphytes. *Aquat. Bot.* 56:93–104.

Bach, S. S., J. Borum, M. Fortes, and C. Duarte. 1998. Species composition and plant performance of mixed seagrass beds along a siltation gradient at Cape Bolinao, The Philippines. *Mar. Ecol. Prog. Ser.* 174:247–256.

Bell, J. D. and M. Westoby. 1986. Abundance of marcrofauna in dense seagrass is due to habitat preference, not predation. *Oecologia* 68:205–209.

Blaabjerg, V., K. Mouritsen, and K. Finster. 1998. Diel cycles of sulphate reduction rates in sediments of a *Zostera marina* bed (Denmark). *Aquat. Microb. Ecol.* 15:97–102.

Borum, J. 1985. Development of epiphytic communities on eelgrass (*Zostera marina*) along a nutrient gradient in a Danish estuary. *Mar. Biol.* 87:211–218.

Borum, J., L. Murray, and W. M. Kemp. 1989. Aspects of nitrogen acquisition and conservation in eelgrass plants. *Aquat. Bot.* 35:289–300.

Bulthuis, D., G. Brand, and M. Mobley. 1984. Suspended sediments and nutrients in water ebbing from seagrass-covered and denuded tidal mudflats in a southern Australian embayment. *Aquat. Bot.* 20:257–266.

Burke, M., W. Dennison, and K. Moore. 1996. Non-structural carbohydrate reserves of eelgrass *Zostera marina*. *Mar. Ecol. Prog. Ser.* 137:195–201.

Caffrey, J. M. and W. M. Kemp. 1990. Nitrogen cycling in sediments with estuarine populations of *Potamogeton perfoliatus* L. and *Zostera marina* L. *Mar. Ecol. Prog. Ser.* 66:147–160.

Caffrey, J. M. and W. M. Kemp. 1991. Seasonal and spatial patterns of oxygen production, respiration and root-rhizome release in *Potamogeton perfoliatus* L. and *Zostera marina* L. *Aquat. Bot.* 40:109–128.

Cambridge, M. L. and A. J. McComb. 1984. The loss of seagrasses in Cockburn Sound, Western Australia. I. The time course and magnitude of seagrass decline in relation to industrial development. *Aquat. Bot.* 20:229–243.

Carlson, P. R., L. Yarbro, and T. Barber. 1994. Relationship of sediment sulfide to mortality of *Thalassia testudinum* in Florida Bay. *Bull. Mar. Sci.* 54:733–746.

den Hartog, C. and P. J. G. Polderman. 1975. Changes in the seagrass populations of the Dutch Waddenzee. *Aquat. Bot.* 1:141–147.

Dennison, W. C., R. J. Orth, K. A. Moore, J. C. Stevenson, V. Carter, S. Kollar, P. W. Bergstrom, and R. A. Batiuk. 1993. Assessing water quality with submersed aquatic vegetation: habitat requirements as barometers of Chesapeake Bay health. *Bioscience* 43:86–94.

Duarte, C. M. 1991. Seagrass depth limits. *Aquat. Bot.* 40:363–377.

Duarte, C. M. 1995. Submerged aquatic vegetation in relation to different nutrient regimes. *Ophelia* 41:87–112.

Duarte, C. M. and K. Sand-Jensen. 1990. Seagrass colonization: patch formation and patch growth in *Cymodocea nodosa*. *Mar. Ecol. Prog. Ser.* 65:193–200.

Fong, P. and M. Harwell. 1994. Modeling seagrass communities in tropical and subtropical bays and estuaries: a mathematical model synthesis of current hypotheses. *Bull. Mar. Sci.* 54:575–781.

Fonseca, M. S., W. J. Kenworthy, and G. W. Thayer. 1998. Guidelines for the conservation and restoration of seagrasses in the United States and adjacent waters. NOAA Coastal Ocean Program Decision Analysis Series No. 12, Silver Spring, MD, 222 pp.

Fourqurean, J. W. and M. B. Robblee. 1999. Florida Bay: a history of recent ecological changes. *Estuaries* 22(2B):345–357.

Goldsborough, W. G. and W. M. Kemp. 1988. Light response and adaptation for the submersed macrophyte, *Potamogeton perfoliatus:* implications for survival in turbid tidal waters. *Ecology* 69:1775–1786.

Goodman, J. L., K. A. Moore, and W. C. Dennison. 1995. Photosynthetic responses of eelgrass (*Zostera marina* L.) to light and sediment sulfide in a shallow barrier island lagoon. *Aquat. Bot.* 50:37–47.

Harlin, M., B. Thorne-Miller, and J. Boothroyd. 1982. Seagrass-sediment dynamics of a flood-tidal delta in Rhode Island (USA). *Aquat. Bot.* 14:127–138.

Harrison, P. G. 1989. Detrital processing in seagrass systems: a review of factors affecting decay rates, remineralization and detritivory. *Aquat. Bot.* 23:263–288.

Heck, K. L. 1976. Community structure and the effects of pollution in sea-grass meadows and adjacent habitats. *Mar. Biol.* 35:345–357.

Heck, K. L., K. Able, C. Roman, and M. Fahay. 1995. Composition, abundance, biomass and production of macrofauna in a New England estuary: comparison among eelgrass meadows and other nursery habitats. *Estuaries* 18:379–389.

Hemminga, M. A., P. G. Harrison, V. Vanlent. 1991. The balance of nutrient losses and gains in seagrass meadows. *Mar. Ecol. Prog. Ser.* 71:85–96

Kemp, W. M. 1983. Seagrass communities as a coastal resource. *Mar. Technol. Soc. J.* 17:3–5.

Kemp, W. M., W. R. Boynton, R. R. Twilley, J. C. Stevenson, and J. C. Means. 1983. The decline of submerged vascular plants in upper Chesapeake Bay: summary of results concerning possible causes. *Mar. Technol. Soc. J.* 17:78–89.

Klumpp, D. W., J. Salita-Espinosa, and M. D. Fortes. 1992. The role of epiphytic periphyton and macroinvertebrate grazers in the trophic flux of a tropical seagrass community. *Aquat. Bot.* 43:327–349.

Koch, E. M. and G. Gust. In Press. Water flow in tide and wave dominated beds of the seagrass *Thalassia testudinum. Mar. Ecol. Prog. Ser.*

Kuhn, W. A. 1992. Interacting effects of light and sediment sulfide on eelgrass (*Zostera marina* L.) growth. Master's thesis, University of Maryland, College Park, MD.

Lee, K.-S. and K. H. Dunstan. 1996. Production and carbon reserve dynamics of the seagrass *Thalassia testudinum* in Corpus Christi Bay, Texas, USA. *Mar. Ecol. Prog. Ser.* 143:201–210.

Lubbers, L., W. R. Boynton, and W. M. Kemp. 1990. Variations in structure of estuarine fish communities in relation to abundance of submersed vascular plants. *Mar. Ecol. Prog. Ser.* 65:1–14.

Madden, C. J. and W. M. Kemp. 1996. Ecosystem model of an estuarine submersed plant community: calibration and simulation of eutrophication responses. *Estuaries* 19(2B):457–474.

Marba, N. and C. Duarte. 1995. Coupling of seagrass (*Cymodocea nodosa*) patch dynamics to subaqueous dune migration. *J. Ecol.* 83:381.

McGlathery, K. J. 1995. Nutrient and grazing influences on a subtropical seagrass community. *Mar. Ecol. Prog. Ser.* 122:239–252.

Neckles, H. A., R. L. Wetzel, and R. J. Orth. 1993. Relative effects of nutrient enrichment and grazing on epiphyte-macrophyte (*Zostera marina* L.) dynamics. *Oecologia* 93:285–295.

Olesen, B. and K. Sand-Jensen. 1993. Seasonal acclimatization of eelgrass *Zostera marina* growth to light. *Mar. Ecol. Prog. Ser.* 94:91–99.

Onuf, C. P. 1994. Seagrass, dredging and light in Laguna Madre, Texas, USA. *Estuar. Coast. Shelf Sci.* 39:75–91.

Orth, R. J. and K. A. Moore. 1983a. Chesapeake Bay: an unprecedented decline in submerged aquatic vegetation. *Science* 222:51–53.

Orth, R. J. and K. A. Moore. 1983b. Submerged vascular plants: techniques for analyzing their distribution an abundance. *Mar. Technol. Soc. J.* 17:38–52.

Orth, R. J. and J. van Montfrans. 1990. Utilization of marsh and seagrass habitat by early stages of *Callinectes sapidus:* a latitudinal perspective. *Bull. Mar. Sci.* 46:126–144.

Pedersen, M. F. and J. Borum. 1993. An annual nitrogen budget for a seagrass *Zostera marina* population. *Mar. Ecol. Prog. Ser.* 101:169–177.

Peierlis, B. N., N. Caraco, M. Pace, and J. Cole. 1991. Human influence on river nitrogen. *Nature* 350:386–387.

Petersen, C. H. 1986. Enhancement of *Mercenaria mercenaria* densities in seagrass beds: is pattern fixed during settlement season or altered by subsequent differential survival? *Limnol. Oceanogr.* 31:200–205.

Reusch, T. B., A. R. Chapman, and J. P. Groger. 1994. Blue mussels *Mytilus edulis* do not interfere with eelgrass *Zostera marina* but fertilize shoot growth through biodeposition. *Mar. Ecol. Prog. Ser.* 108:265–282.

Risgaard-Petersen, N., T. Dalsgaard, S. Rysgaard, P. B. Christensen, J. Borum, K. McGlathery, and L. P. Nielsen. 1998. Nitrogen balance of a temperate eelgrass *Zostera marina* bed. *Mar. Ecol. Prog. Ser.* 174:281–291.

Ruffin, K. K. 1998. The persistence of anthropogenic turbidity plumes in a shallow water estuary. *Estuar. Coast. Shelf Sci.* 47:579–592.

Short, F. T. and D. M. Burdick. 1996. Quantifying eelgrass habitat loss in relation to housing development and nitrogen loading in Waquoit Bay, Massachusetts. *Estuaries* 19:730–739.

Short, F. T. and S. Wyllie-Echeverria. 1996. Natural and human-induced disturbance of seagrasses. *Environ. Conserv.* 23:17–27.

Short, F. T., B. Ibelings, and C. den Hartog. 1988. Comparison of a current eelgrass disease to the wasting disease in the 1930's. *Aquat. Bot.* 30:295–304.

Stoner, A. W. 1983. Distribution of fishes in seagrass meadows: Role of macrophyte biomass on species composition. *Fish. Bull.* 81:837–846.

Thayer, G. W., K. A. Bjorndal, J. C. Ogden, S. L. Williams, and J. C. Zieman. 1984. Role of larger herbivores in seagrass communities. *Estuaries* 7:351–376.

Tomasko, D. A., C. J. Dawes, and M. O. Hall. 1996. The effects of anthropogenic nutrient enrichment on turtle grass (*Thalassia testudinum*) in Sarasota Bay, Florida. *Estuaries* 19:448–456.

Twilley, R. R., W. M. Kemp, K. W. Staver, J. C. Stevenson, W. R. Boynton. 1985. Nutrient enrichment of estuarine submersed vascular plant communities. 1. Algal growth and effects on production of plants and associated communities. *Mar. Ecol. Prog. Ser.* 23:179–191.

Twilley, R. R., G. Ejdung, P. Romare, and W. M. Kemp. 1986. A comparative study of decomposition, oxygen consumption, and nutrient release for selected aquatic plants occurring in an estuarine environment. *Oikos* 47:190–198.

Walker, D. I. and A. J. McComb. 1992. Seagrass degradation in Australian coastal waters. *Mar. Pollut. Bull.* 25:191–195.

Walker, D. I., R. J. Lukatelich, G. Bastyan, and A. J. McComb. 1989. Effect of boat moorings on seagrass beds near Perth, Western Australia. *Aquat. Bot.* 36:69–77.

Ward, L. G., W. M. Kemp, and W. R. Boynton. 1984. The influence of waves and seagrass communities on suspended sediment dynamics in an estuarine embayment. *Mar. Geol.* 59:85–103.

Wetzel, R. L. and H. Neckles. 1986. A model of *Zostera marina* L. photosynthesis and growth: simulated effects of selected physical-chemical variables and biological interactions. *Aquat. Bot.* 26:307–323.

Zieman, J. C. 1976. The ecological effects of physical damage from motor boats on turtle grass beds in southern Florida. *Aquat. Bot.* 2:127–139.

Zieman, J. C., R. L. Iverson, and J. C. Ogden. 1984. Herbivory effects on *Thalassia testudinum* leaf growth and nitrogen content. *Mar. Ecol. Prog. Ser.* 15:151–158.

Zimmerman, R. C., D. Kohrs, D. Stellar and R. Alberte. 1997. Impacts of CO_2 enrichment on productivity and light requirements of eelgrass. *Plant Physiol.* 115:599–602.

Section I

Ecology and Physiology

2 Establishing Light Requirements for the Seagrass *Thalassia testudinum*: An Example from Tampa Bay, Florida

L. Kellie Dixon

Abstract A continuous record of photosynthetically active radiation (PAR) was collected for one year at the maximum depth limits of *Thalassia testudinum* meadows in Tampa Bay, Florida. Attenuation by epiphytic material, growth, and condition measurements were performed periodically at both the maximum depths and at shallower (non-light-limited) locations. Water clarity (Secchi depth) during the project year was comparable to the previous 5 years, and grassbeds were presumed to be growing at a steady state with respect to light. The deepest station in the study exhibited morphological adaptations to lowered light levels: increased blade lengths, decreased blade production per year, higher ratios of above- to below-ground biomass, and a decline in shoot density in contrast to expected seasonal patterns. Annual water column PAR totals were at or above 4860 moles m^{-2} yr^{-1} for sites with no evidence of stress; 3730 moles m^{-2} yr^{-1} at the light-stressed site. Annual means of epiphytic attenuation were an additional 33.1% of PAR remaining at depth. Annual percentages of immediately subsurface irradiance at canopy height ranged between 20.1 and 23.4%. Incorporating conservatively high estimates of bottom reflectance resulted in estimates of nearly 20.6% as an annual percentage. Total PAR received was a more sensitive predictor of shading stress than annual average attenuation; timing of PAR deficits also appears crucial. Percentages of water column PAR can be used as transferrable light requirements only if epiphytic growth (and attenuation) is comparable. Seasonal patterns of light and growth emphasize that the two parameters are temporally decoupled.

INTRODUCTION

Losses of seagrasses have been documented worldwide in association with anthropogenic eutrophication. Declines, as grassbeds retreat to shallow depths, have been associated with decreases in available light due to increased phytoplankton, epiphytic, or macroalgal growth (Kemp et al. 1988; Cambridge and McComb 1984; Silberstein et al. 1986; Tomasko et al. 1996). By 1982, Tampa Bay (west central Florida) had experienced losses of seagrasses estimated between 46 and 72% of historical levels (Haddad 1989; Lewis et al. 1991). Calculated nitrogen inputs approximately doubled between 1940 and 1991 (Zarbock et al. 1994) and chlorophyll concentrations in this nitrogen-limited estuary were positively correlated with nitrogen loads (Johansson 1991). Maximum depths of remaining grassbeds (Janicki et al. 1995) within the Bay exhibited a spatial gradient that generally corresponded with gradients of salinity (EPCHC 1989) and nutrient loadings (Zarbock

0-8493-2045-3/99/$0.00+$.50
© 2000 by CRC Press LLC

et al. 1994). Since the late 1970s, reduction of nutrient loads to an upper portion of the Bay has coincided with declines in chlorophyll concentrations (Johansson 1991) and macroalgal abundance. It has also coincided with a return of seagrasses to the area (Avery 1996). In Tampa Bay chlorophyll was demonstrated to be an important water column attenuator (Miller and McPherson 1993) from which attenuation coefficients could be predicted (Janicki and Wade 1996).

Seagrass restoration efforts continue in Tampa Bay, and target goals have been set at 1950s era areas. At that time, seagrass depth limits were 2.0 to 2.5 m in most of the Bay and 1.0 to 1.5 m in the upper Bay (Janicki et al. 1995). In the 1990s, maximum depths are relatively unchanged for the lower Bay, but are shallower (between 0.5 and 2.0 m) for the upper Bay segments. To achieve target acreages, the present day maximum depths of seagrasses must be extended by improving water clarity in the upper Bay segments. Since water clarity data are not available for the 1950s, seagrass light requirements determined in the present day can be coupled with desired maximum depths to compute water clarity conditions. These conditions are necessary to achieve regrowth to 1950s level depths (and coverages). As water clarity can be predicted from chlorophyll concentrations, and as chlorophyll concentrations can be predicted from nitrogen loads, resource-based management efforts entail setting nutrient load reduction goals for nitrogen in order to foster the restoration of seagrasses.

This study was designed to determine light requirements of the seagrass *Thalassia testudinum* by an empirical approach. The light climate at the deep edges of *T. testudinum* meadows at four locations in Tampa Bay was measured over a calendar year. These data were coupled with measurements of above-ground growth, morphology, and light reduction by epiphytic material at both light-limited and non-light-limited stations. Biotic parameters were used as a framework to evaluate photosynthetically active radiation (PAR) levels, with particular emphasis on the stability and condition of the deep edge over the course of the project. Comparisons between deep and shallow station parameters were used for inference on the effects of light limitation. Providing the project year was "typical" of recent water clarity, the observed light climate would be used to set management goals for water clarity in Tampa Bay.

MATERIALS AND METHODS

Study Sites

Bay bathymetry and patterns of decline in seagrass acreage have resulted in seagrasses concentrated in the lower portion of Tampa Bay (Janicki et al. 1995) where the influence of coastal waters is greater and relative anthropogenic nutrient enrichment (domestic and industrial waste discharges and non-point source loadings) is less than in upper Bay segments (Zarbock et al. 1994). As a result, the four study sites were located in the lower Bay in regions with stable *T. testudinum* coverage, as determined by geographical information system (GIS) change analysis (SWFWMD-SWIM 1994) and visual inspection of 1:24,000 scale aerial photography. There were no bathymetric controls, i.e., offshore sandbars or channels, and grasses appeared light-limited based on the gradual reduction of shoot density as depth increased.

Site designations and locations (Figure 2.1) were Site 1 (27°38.3′ 82°41.4′), Site 2 (27°41.8′ 82°38.5′), Site 3 (27°37.5′ 82°35.0′), and Site 4 (27°31.8′ 82°40.1′). Previous work near Site 1, Mullet Key (Hall et al. 1991), identified the deepest beds in this region as light-limited. Site 4 was located near the mouth of the Manatee River, a major freshwater source to Tampa Bay which seasonally discharges highly colored, humic acid-laden waters. The remaining sites (Site 2 near Point Pinellas and Site 3 near Port Manatee) were located farther up the Bay, along gradients of decreasing salinity, and increasing nutrients, chlorophyll, color, and water clarity (EPCHC 1989). In addition to the deep station, a non-light-limited station was established at each site in water depths 0.9 m shallower than the deep edge.

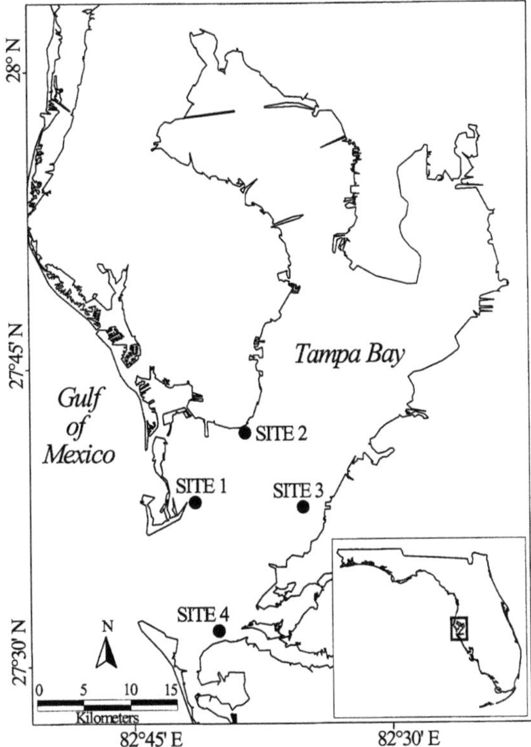

FIGURE 2.1 Station locations for the continuous measurements of photosynthetically active radiation at the maximum depth limits of *Thalassia testudinum* seagrass beds in Tampa Bay, Florida.

Instrumentation and Tides

Two scalar quantum irradiance sensors (4π; LI-COR 193SA) and a data logger were installed immediately outside the deep edge of the grassbed at each site and recorded 15 minute average irradiance. The lower sensor was located at canopy height (0.2 m above the bottom); the upper sensor was located 0.5 m above the lower sensor. Sensors were covered with clear plastic food wrap to permit easy removal of fouling organisms. Instrument arrays (including temperature loggers) were operated from late November 1993 through early December 1994. Depths and vertical PAR profiles were recorded on each site visit with PAR sensors (LI-COR 193SA [4π] and 190SA [air, 2π]). The profiles often included additional upward and downward facing sensors (LI-COR 192SA [2π]) to determine irradiance reflectance. Salinity (Jaeger 1973) was determined weekly. Tidal variations in water depth were estimated from data recorded within the Bay (CDM 1994; Tampa PORTS 1995), adjusting time and height offsets as necessary (U.S. Department of Commerce 1994), and solving for tidal harmonics constituents (Boon and Kiley 1978; Hayward 1993; Shureman 1971) at each site.

Biotic Measurements

Above-ground growth rates at each station were determined on six occasions, using blade marking techniques (n = 15 per station) after Tomasko and Dawes (1989). On collection, the marked shoot was scraped of all epiphytic material, and old and new tissues dried at 103–105°C. Calculations included determination of above-ground biomass; leaf-, shoot-, and area-based growth rates; leaf areas; turnover times; and plastochrone interval (Brouns 1985). Measurements were transformed

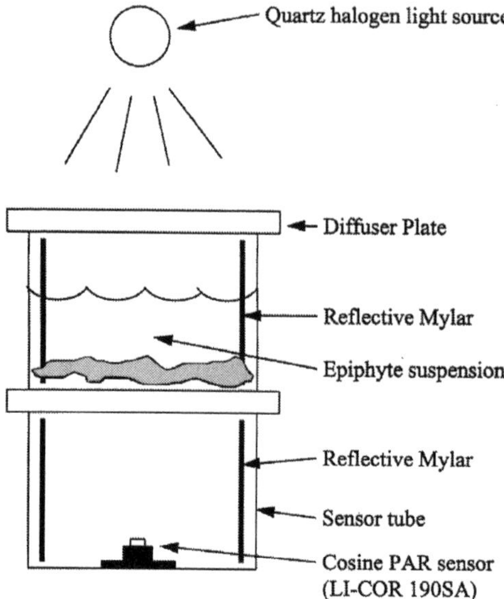

FIGURE 2.2 Apparatus for determination of attenuation due to epiphytic materials (not to scale).

to areal units (per m^2) based on quadrat counts (25 cm × 25 cm, n ~ = 40 per station) of short shoot densities. Attached rhizome also was collected with the marked shoots and a 1–3 cm section analyzed for carbohydrates by the colorimetric trichloroacetic acid-phenol-sulfuric acid method of Dubois et al. (1956) using glycogen standards (Dawes 1981).

Cores (12.5 cm × 12.5 cm, 20 cm depth, n = 10 per station) were collected on four occasions. Biomass and tissue apportionment were calculated as above-ground (blades) to below-ground (short shoots, roots, and rhizomes) ratios of tissue weights. Biomass measurements were transformed to areal units using shoot density in cores and quadrats. Cores for sediment samples (3.8 cm diameter, 15 cm depth, n = 5 per station) were collected initially for the determination of grain size distribution by whole-phi increments (Folk 1974) and percent organic content by ignition to 550° ± 50°C (APHA 1989; 2540E).

Epiphytic Attenuation

Attenuation of PAR due to epiphytic materials (defined as attached epiphytes, epifauna, and associated sediment or detritus) was determined on shoots processed for growth determinations. Necrotic portions were removed and blades of an entire shoot were scraped with a razor blade to remove all epiphytic material. The blade area scraped was determined. Epiphytic attenuation was determined by the method of Dixon and Kirkpatrick (1995), following Sand-Jensen and Sondergaard (1981), with the apparatus illustrated in Figure 2.2. The suspension of scraped material was placed into a transparent (acrylic) dish. Walls of the transparent dish were lined with reflective mylar, and the dish illuminated with a quartz-halogen light source (Dyna-Lume 240–350) and a white acrylic diffusing slide. A cosine PAR sensor (LI-COR 190SA) was placed 12 cm below the sample. The sensor tube was also lined with reflective mylar. For each sample, a blank reading of the irradiance passing through deionized water (E_{Blank})was followed by multiple readings of the irradiance penetrating the epiphyte suspension (E_{Sample}). The inverse logarithmic relationship between the epiphyte suspension density and transmitted irradiance was used to correct for inequalities between dish area and blade area scraped.

$$y = \frac{\ln(E_{Sample}) - \ln(E_{Blank})}{(\text{Dish Area}^{-1})} \cdot \frac{1}{\text{Blade Area}} + \ln(E_{Blank})$$

$$\text{Epiphyte Attenuation } (\%) = \frac{E_{BLANK} - e^{y}}{E_{Blank}} \cdot 100\%$$

Epiphytic attenuations represent area-averaged conditions, i.e., as if the entire epiphyte burden of the productive blades was evenly distributed across all blade surfaces. Epiphytic communities have been demonstrated to have a broad absorption spectrum (Neckles 1993). As a result, PAR attenuation of an epiphytic suspension is an appropriate approximation to calculate PAR available to seagrass blades. This is used in the absence of more rigorous spectral attenuation information for water column and epiphytes, and photosynthetic action spectra for *T. testudinum*.

Reconstructive Aging

Reconstructive aging analysis was performed (Gallegos et al. 1992, 1993; Durako 1994). Leaf scars of short shoots with intact rhizomes in biomass cores were enumerated. Plastochrone intervals were used to normalize leaf scars on all shoots to the final October sampling, and frequency analysis was performed on all shoots from a station. Scar classes were used to identify annual cohorts (Duarte and Sand-Jensen 1990). The frequency of the resulting age of shoots by year classes was computed. This frequency was used to identify gross shoot recruitment and net annual shoot production.

PAR Data Reduction

Sensors were cleaned every 5 days and PAR data were corrected for attenuation produced by fouling communities on the sensors. Degree of fouling was measured directly every 2 weeks and converted to a fouling rate, or percentage of PAR reduction per day of deployment. Linear interpolation was used to adjust individual data for fouling based on deployment time since cleaning. Scalar water column attenuation coefficients (K_0) were calculated from corrected sensor data. Absolute levels of light received at the canopy height of the deep edge were computed data between 0500 and 2100 hr.

The PAR remaining at the bottom of the water column (% PAR_w), as a percentage of the immediately subsurface scalar irradiance, was computed from each attenuation coefficient (K_0) and mean station water level (z).

$$\%PAR_w = e^{-K_0 \cdot z} \cdot 100\%$$

Site- and time-specific interpolated water depths obtained from tidal data were also used to compute percentages of light remaining at the bottom of the water column under simulated tidal conditions.

The % PAR_w represents water column attenuation only and does not include reductions of incident solar irradiance due to air–water surface reflection. For comparison with other authors, reflectance at the air–water interface was estimated. Wind, wave, and solar elevation will alter the percentage of reflection, but, as these properties are not subject to management actions, the percentage reduction by the water column alone was the desired goal of the study.

The effects of epiphytic attenuation were also incorporated. Bottom sensor data and percentages of PAR remaining in the water column at depth (%PAR_w) were further reduced by the amount of

PAR reduction caused by epiphytes to estimate the PAR actually available to *T. testudinum* blades ($\%PAR_p$).

$$\%\ PAR_p = \%\ PAR_w \cdot (1 - \text{Epiphyte Attenuation}\% \cdot 100^{-1})$$

To maintain comparability with other Bay monitoring programs, monthly, seasonal, and annual averages of K_0, $\%PAR_w$, and $\%PAR_p$ are based on data only between 1000 and 1400 hours.

Bottom Reflectance

Bottom reflectance was of concern in this project due to the use of 4π sensors deployed near the bottom of the water column. In the event of substantial bottom reflectance, an attenuation coefficient determined near the bottom of the water column will be lower than if determined for the entire water column, thus resulting in an artificially high $\%PAR_w$. Bottom reflectance decreases with a dark bottom type, vegetation, and when the light field reaching the bottom is diffuse. Bottom reflectance must be distinguished from upwardly scattered light (upwelling light). Upwelling light is produced by the interaction of downwelling light and water molecules or, to a lesser extent, particulates. It is present at all depths in the water column and is not an exclusive function of bottom reflectance. Upwelling light is attenuated similarly to downwelling light and typically consists of between 4 to 10% of downwelling light (Kirk 1994). Estimates of bottom reflectance were obtained from irradiance reflectance (ratio of upwelling:downwelling PAR) computed at all measurement depths from upward and downward facing cosine (2π) sensors and from comparison of the surface irradiance reflectance with that measured at bottom.

Statistical Testing

Statistical analysis of the data was performed using Sigmastat Statistical Analysis system Version 1.0, Northwestern Analytical NWASTATPAK, Version 4.1 (1986), and SYSTAT Version 7.0 for Windows. Analyses included *t* test and one-way analysis of variance (ANOVA). With non-normally distributed data, Mann–Whitney, Friedman's two-way ANOVA, Kruskal–Wallis, and Wilcoxon signed rank tests were used. Pairwise comparison employed Student Newmann–Keuls as Dunn's test. Significance levels were set at $p < = 0.05$.

RESULTS

Monitoring Representativeness

The Environmental Protection Commission of Hillsborough County (EPCHC) provided monthly water quality data from 1974 to 1994 from stations nearest the study sites (within 2 km) to evaluate the representativeness of the monitoring year. There were no significant differences in Secchi depths by year for 1989–1994. Annual mean Secchi depths ranged from 2.1 to 2.6 m.

Physical Parameters

Mean water depths at the deep edges were 2.10 m, 2.37 m, 2.15 m, and 1.98 m for Site 1, 2, 3, and 4, respectively. Tidal range (MLLW to MHHW) of the region is near 0.7 m. Annual average salinity at the study sites ranged between 28.7 and 30.6 ppt, with seasonal fluctuations generally between 22 and 33 ppt. During the late summer and fall wet season, freshwater discharges produced abrupt and short-term (< 1 week) depressions in salinity at Site 4 and Site 3, to minima of 21 and 17 ppt, respectively. Temperatures were comparable between stations with project means between 23.7°C and 25.0°C. During January 1994, minima (between 10.8°C and 12.7°C) were substantially lower than long-term averages of 16°C for this month (EPCHC, unpublished data). Maximum values were similar between stations, ranging from 31.5°C to 33.0°C in July and August, and were warmer than long-term averages of near 30°C (EPCHC, unpublished data). Temperatures during November–December 1994, were 4°C to 5°C warmer than both historical averages (EPCHC, unpublished data) and December 1993. Sediments at the sites were all classified as fine sands with mean and median sizes at the deep stations ranging between 0.140 and 0.200 mm, percent silt–clay fractions generally less than 1.5% by weight, and percent organics typically less than 0.75%. Site 2 sediments were significantly finer and higher in percent organics, but the range in values between sites was not great.

Biotic Parameters

Seasonal biotic data for deep stations are summarized in Table 2.1. Deep station short shoot densities varied between 36 and 109 shoots m^{-2}, with a minimal seasonal signal in comparison to non-light-limited stations. Two sites (Site 1 and Site 4) displayed no significant net changes in shoot density between the beginning and end of the study. Density at Site 3 nearly doubled (+29 shoots m^{-1}) while counts at Site 2 were significantly reduced (–19 shoots m^{-1}) by nearly one third. The significant changes detected were less than the annual variation.

Above-ground biomass was typically less than 25 g dwt m^{-1} for the deep edge stations with slight seasonal variations. Biomass at the deep edge of Site 1 was unchanged between initial and final samplings, was significantly higher at Site 2 and Site 3 ($p < 0.001$), and was lower at Site 4. Short-term salinity reductions at Site 4 resulted in exfoliation and a decrease in biomass, leaf area per shoot, and leaf area index by the final October sampling. Increases in biomass at Site 3 are consistent with expected seasonal patterns, observed temperatures, and increased shoot density. Increase in biomass at Site 2 is counter to the reduction in shoot density noted there. Site 2 displayed significantly higher above-ground biomass among the four deep stations, both overall and during all but one sampling. For all stations, above-ground biomass was highest during April or June, with minima in December and February. Seasonal changes were more pronounced at the shallow, non-light-limited stations. Above-ground biomass, shoot density, blade width, leaf area per shoot, and leaf area index significantly decreased with increased depth.

For three of four sites (1, 3, and 4), above-ground to below-ground ratios of biomass at both depths were generally near 0.2 (Figure 2.3). The seasonal exception was the April sampling when higher above-ground biomass increased ratios to near 0.5. At the maximum depth limits of Site 2, however, there were significantly higher ratios for all samplings (mean ratio, 0.8) and particularly during May (mean ratio, 1.5). Examination of raw weights indicated that rhizome tissues at Site 2 were approximately half and significantly lower than the other deep stations during all samplings. Blade weights at Site 2 were significantly greater than the other stations (particularly during the August and October samplings). Other than at Site 2, ratios of above- to below-ground biomass displayed no significant trends with depth.

Carbohydrate levels were significantly higher in rhizomes from deep stations than from non-light-limited stations at all sites throughout the year. Typically, deep station rhizomes contained more than 300 mg g^{-1} carbohydrate while shallow station rhizomes contained less. Significant seasonal variation was limited to deep stations as a whole with lower carbohydrate concentrations during June. Among the individual stations, however, only the deep edges of Sites 1 and 2 and the non-light-limited station at Site 2 displayed significant seasonal variation. Although insufficient to produce significant seasonal variations, the shallow stations at Sites 1, 2, and 4 exhibited decreased mean carbohydrate concentrations during the June sampling.

During the winter months, mean blade lengths averaged near 10 cm, increasing to near 20 cm by April or June. At Site 2, however, blade length increased through August and remained high through October. Lengths at the end of the study were significantly greater than during the initial

TABLE 2.1
Biotic parameters (mean/standard deviation) measured at the maximum depth limits of *Thalassia testudinum* beds in Tampa Bay, Florida. All stations pooled by sampling.

Sampling	December-93	February-94	April-94	June-94	August-94	October-94
Epiphytic attenuation (%)	58.8/19.7	43.5/16.5	39.0/14.8	23.0/7.2	21.1/14.2	16.5/16.7
AFDW epiphyte: AGBM	0.29/0.39	0.15/0.08	0.12/0.07	0.05/0.05	0.04/0.08	0.02/0.04
AFDW epiphyte: DW Epiphyte	0.23/0.11	0.22/0.15	0.29/0.13	0.33/0.11	0.30/0.12	0.37/0.18
Shoot Density (shoot m^{-1})	55.6/13.2	62.5/4.8	68.2/9.4	76.8/14.4	82.1/18.9	52.3/8.7
AGBM (gdw m^{-2})	4.54/1.78	6.81/2.82	22.2/10.6	18.4/6.8	15.7/10.8	6.9/4.0
AGBM:BGBM	0.39/0.31	—	0.75/0.65	—	0.30/0.24	0.20/0.19
Leaf area per shoot (cm^2 $shoot^{-1}$)	14.8/5.4	18.7/7.4	60.8/27.4	53.4/19.4	38.8/29.4	28.9/21.8
Leaf area index (m^2 m^{-2})	0.080/0.030	0.119/0.051	0.413/0.194	0.403/0.154	0.325/0.262	0.144/0.098
Areal production (gdw m^{-2} day^{-1})	0.009/0.011	0.103/0.077	0.702/0.417	0.541/0.273	0.283/0.285	0.110/0.089
Production (mgdw $shoot^{-1}$ day^{-1})	0.18/0.26	1.63/1.16	10.26/5.50	7.17/3.39	3.3/3.1	2.1/1.8
Leaf relative growth (mgdw g^{-1} day^{-1})	2.1/2.5	14.2/8.5	31.1/11.1	29.2/9.9	16.4/10.8	14.2/7.8
Turnover (% day^{-1})	0.21/0.25	1.42/0.86	3.10/1.11	2.92/0.99	1.62/1.09	1.42/0.78
Rhizome carbohydrates (mg g^{-1})	336/87	339/87	335/102	305/91	360/106	363/83
Mean blade length (cm)	7.9/2.6	8.4/2.9	17.7/5.6	19.4/6.0	18.4/9.3	15.1/7.4
Mean daily PAR (mol m^{-2} day^{-1})	13.1/0.4	16.6/2.7	17.8/6.4	12.7/1.6	6.3/1.6	6.1/1.2
Mean hours > I_k	6.5/0.3	7.5/0.4	7.8/2.3	8.0/1.0	3.8/1.0	4.2/1.1
Mean hours > I_c	8.8/0.4	9.9/0.3	10.8/0.8	13.0/0.3	8.0/0.9	8.3/0.3

Note: AFDW = Ash-free dry weight, DW = Dry weight, AGBM = Above-ground biomass

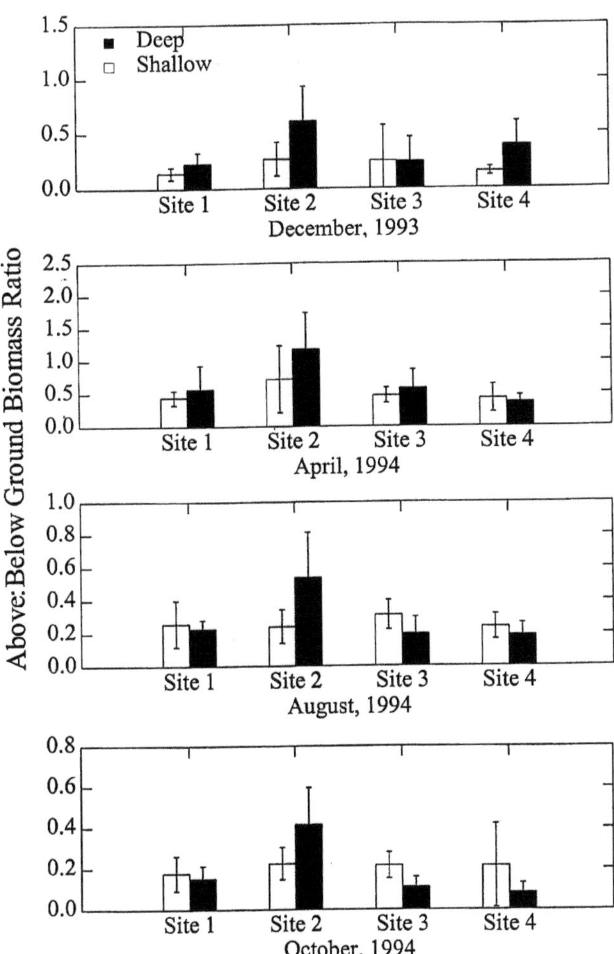

FIGURE 2.3 Above- to below-ground biomass ratios (mean, sd) by site for deep and shallow (non-light-limited) locations, Tampa Bay, Florida.

sampling at all sites. Mean blade length of the deep stations overall was significantly shorter than the shallow stations. At Site 2, however, blades from the depth-limited site were longer than the shallow site, although the differences were not always significant (Figure 2.4). Of the deep stations, Site 2 consistently exhibited significantly longer blades than other stations, but differences were most pronounced in August and October. Although blade weights per shoot were similarly higher at Site 2 during these months, blade density, as gdw cm^{-2} of blade area, was significantly lower than the remaining stations.

Similar spatial and temporal patterns were observed for leaf area per shoot. At the deep stations of Site 2, the maximum leaf area (near 75 cm^2) was delayed until August, compared to April or June maxima at other depth-limited stations. Leaf area index was similarly delayed at the deep stations of Site 2. Among deep stations, Site 2 had the greatest leaf area per shoot from June through October and was higher than other shallow stations during both August and October.

Growth Rates

Leaf relative growth rate (LRGR) demonstrated no consistent depth-related patterns, but reached seasonal maxima (25 to 35 mgdw gdw^{-1} d^{-1}) at depth limits during April and June. Above-ground

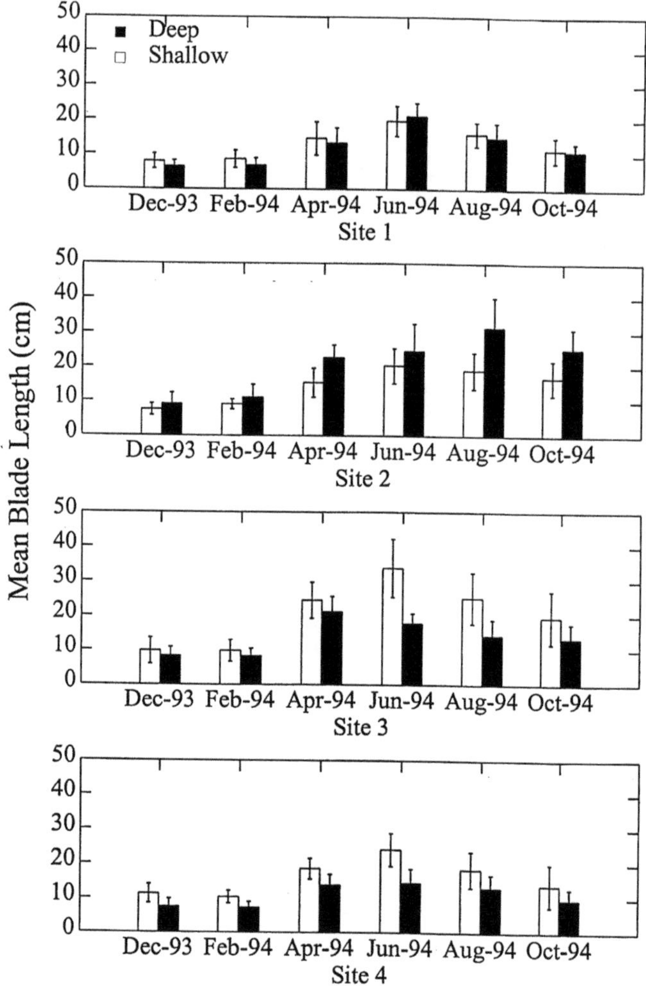

FIGURE 2.4 *Thalassia testudinum* blade lengths (cm; mean, sd) from December 1993 through October 1994 for deep and shallow (non-light-limited) locations, Tampa Bay, Florida.

growth per shoot was significantly higher at shallow stations during all but seasons of peak growth (April, June). At depth limits, significantly higher LRGR and growth per shoot were evident for Site 2 in August and October, and were also manifest in the longer mean blade lengths and higher number of blades per shoot at this site. Site 2 was also noteworthy in the low numbers of blades per year (6.5 blades yr^{-1}) produced in comparison to other deep stations (8.1 to 9.7 blades yr^{-1}) as indicated both by plastochrone intervals and by the average inter-cohort distance determined from reconstructive aging analyses. For all deep stations, however, no significant station-to-station differences in shoot distribution by year classes were observed. Areal growth generally peaked in April (0.5 to 1 gdw m^{-2} d^{-1} at depth limits). Areal growth integrates shoot densities and shoot relative growth rates; therefore, a real growth was always higher at the shallower stations (3 to 6 gdw m^{-1} d^{-1} in April).

Epiphytic Attenuation

Annual average epiphytic attenuation at the deep edges ranged between 31.6 and 36.0% with no significant differences between sites overall; however, for a given sampling, there were generally

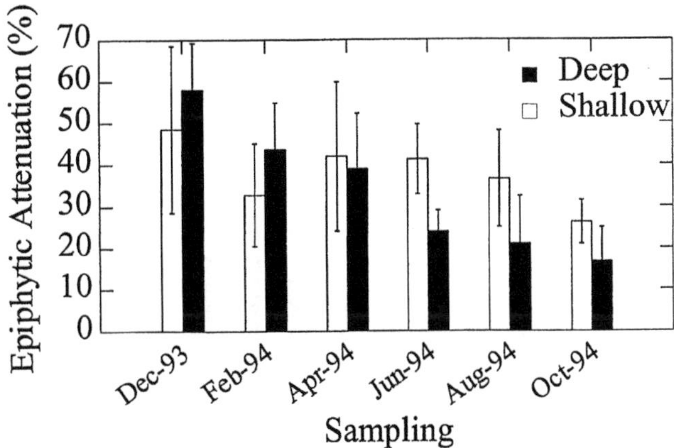

FIGURE 2.5 Seasonal variation in epiphytic attenuation (%; mean, sd) at deep and shallow (non-light-limited) stations.

significant differences by station. Seasonally, values of epiphytic attenuation at individual sites ranged between 67.3 and 6.7%, and on individual shoots ranged as high as 97% during the course of the study. Epiphytic growths were heavy in December and January, when epiphyte loads at the deep stations were as much as three times higher than *T. testudinum* above-ground biomass. Epiphytic loads decreased during the spring and summer growing season and remained low throughout the October sampling, typically consisting of 50% of *T. testudinum* above-ground biomass. Where differences between the deep edge and non-light-limited shallower station existed at a given site, epiphytic attenuation was greater at the shallower location, particularly during the June and August samplings (Figure 2.5).

The PAR attenuation due to epiphytes exhibited an exponential relationship with epiphytic loads with the relationship of attenuation to ash-free (AF) epiphytic biomass (Figure 2.6) somewhat better than for dry weight epiphytic loads. Ash-free weights of epiphytes were, on average, 30% of the total epiphytic dry loads.

Attenuation Coefficients

Maximum mean monthly water clarity (lowest K_0) occurred in winter and early spring (December through May; Figure 2.7). By June, water clarity declined due to seasonal increases in chlorophyll

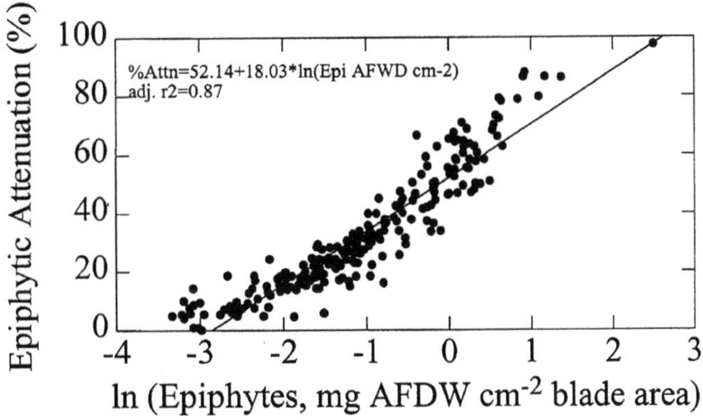

FIGURE 2.6 Epiphytic attenuation (%) as a function of epiphytic loading (ln [mg ash-free dry weight cm^{-2} blade area]).

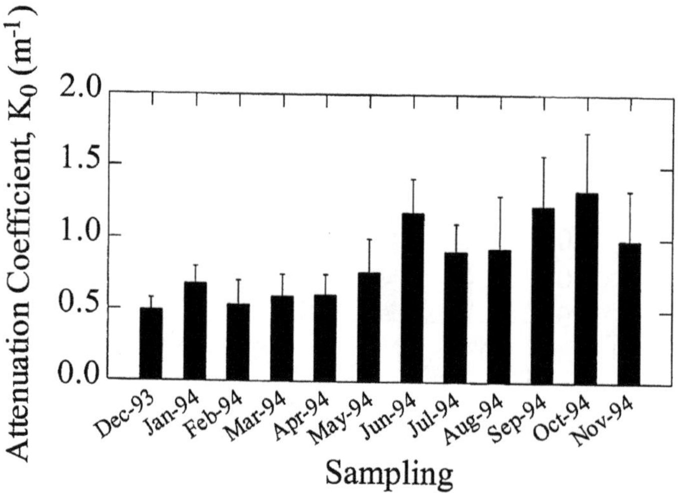

FIGURE 2.7 Seasonal variation in monthly mean attenuation coefficients, mean and sd of four stations.

while September and October clarity reductions were due to increasing amounts of fresh water and color during the wet season. The seasonal signature was less evident at Site 2 with higher K_0 during June and lower K_0 for October and November than at the remaining sites.

Annual mean attenuation coefficients, computed as the mean of monthly means to reduce bias from missing data, were 0.84 m^{-1}, 0.80 m^{-1}, 0.84 m^{-1}, and 0.94 m^{-1} for Sites 1, 2, 3 and 4, respectively. Based on the biological parameters, a 6-month growing season of April through September was also defined, and mean attenuation coefficients recalculated as above. Values of K_0 were somewhat larger during the growing season, at 0.90 m^{-1}, 0.93 m^{-1}, 0.84 m^{-1}, and 1.06 m^{-1}, for Sites 1, 2, 3, and 4, respectively. Inter-station differences (examined as the difference of K_0 at a single station from the mean of all four stations for each 15-min period) were generally significant during all quarterly periods and for the project as a whole.

Percentages of PAR

Under constant water depths, annual percentage of PAR reaching the maximum depth limits (%PAR_w) ranged between 20.1% (Site 2) and 23.4% (Site 4) with a mean among stations of 22.1% (Table 2.2). During the growing season of April–September, the deep edges received an average of between 16.4 (Site 2) and 20.2% (Site 3) of subsurface scalar PAR. The effect of tidally varying water depths fluctuated with site and month, dependent on whether daylight high or low tides predominated, and is expected to precess in accordance with tidal epochs. Inclusion of the tidal variation, as estimated for the project year, resulted in a difference of between –0.5 to +0.7% in annual PAR percentages computed. The %PAR_w values were reduced further by the percentage attenuation due to epiphytic load to obtain %PAR_p. The annual average %PAR_p amounted to between 13.0 (Site 2) and 14.2% (Site 3) of subsurface irradiance, and between 11.9 (Site 1) and 14.6% (Site 3) during the growing season.

Bottom Reflectance

Irradiance reflectance at surface depths (0.2 m) averaged 5.6%, well within ranges reported by other investigations (summarized in Kirk 1994). The difference between surface irradiance reflectance and irradiance reflectance at the depths closest to the bottom averaged 10.1% and was taken as an approximation of the bottom reflectance, or the amount of PAR reaching the bottom which is redirected upward into the water column. The 25th and 75th percentiles of reflectance values

TABLE 2.2
Percentage of PAR remaining in the water column at maximum depth limits of *Thalassia testudinum* between 1000 to 1400 hours (%PAR$_w$); monthly means and standard deviations.

Month	Site 1	Site 2	Site 3	Site 4
December-93	36.8, 19.1	37.3, 9.1	28.8, 13.5	33.3, 14.6
January-94	24.8, 11.8	21.6, 11.3	No Data	34.4, 10.5
February-94	29.9, 11.3	23.7, 10.8	46.4, 11.5	46.0, 20.6
March-94	27.1, 12.8	25.5, 10.2	40.9, 14.9	42.7, 13.5
April-94	33.8, 11.1	18.8, 11.6	31.6, 8.8	36.3, 9.3
May-94	28.2, 10.1	19.0, 11.5	22.5, 9.8	31.1, 8.4
June-94	11.4, 5.7	10.2, 5.6	14.7, 5.9	14.0, 9.4
July-94	12.5, 4.3	18.4, 4.6	18.5, 6.1	13.3, 6.4
August-94	16.4, 7.1	18.8, 7.3	27.1, 14.3	8.4, 5.0
September-94	9.6, 2.8	13.4, 4.9	7.0, 2.5	5.4, 4.1
October-94	11.0, 5.1	11.2, 4.5	6.4, 5.7	4.7, 3.6
November-94	19.3, 8.6	22.9, 7.3	9.2, 6.1	11.2, 5.6
Annual	21.7	20.1	23.0	23.4
April–September	18.6	16.4	20.2	18.1

were 8.8 and 12.7%. There were no significant differences among the deep stations and no station had any single bottom reflectance estimate in excess of 15.9%. Bottom reflectance values at the vegetated (darker substrate) shallow sites averaged less than 2%.

An approximate model of scalar attenuation in the presence of bottom reflectance was constructed in which downwelling (2π) light was attenuated with depth, irradiance reflectance (backscatter) was assumed constant at 5.6% of downwelling light and bottom reflectance was assumed to be a fixed percentage (10.1%) of the downwelling light reaching the bottom (which was subsequently attenuated in an upward direction using the same coefficient used for downwelling light). The sum of downwelling, upwelling (backscattered), and bottom-reflected light was used to approximate a scalar profile of irradiance such that the effects of bottom reflectance could be assessed. Applying the model to observed %PAR$_w$ values resulted in a decrease in the mean annual %PAR$_w$ from 22.1 to 20.6% with a corrected range among stations of 18.6 to 21.8%. Assumptions of the maximum reflectance decreased calculated %PAR$_w$ by less than 2%.

Annual Light Regimes

Monthly averages of total PAR per day were computed both before and after (Figure 2.8) epiphytic attenuation. Site 4 received much higher daily PAR totals available to *T. testudinum* than other stations during April and May, a function of both water clarity and reduced epiphytic load. Monthly means of daily total PAR at Site 4 for July through September, however, were the lowest of any of the sites. By late fall at Site 4, the low epiphyte load offset the high water column attenuation, such that average moles m^{-2} d^{-1} received at Site 4 was very similar to other stations. The depth-limited Site 2 station received notably lower levels of PAR than any other site during the 5 months of February through July. Monthly averages of daily totals at Site 2 never exceeded 10 moles m^{-2} d^{-1} during any month at this site. Annual averages of total PAR were 6.43 moles m^{-2} d^{-1} for Site 2, and 8.64, 8.12, and 8.45 moles m^{-2} d^{-1} for Site 1, Site 3, and Site 4, respectively. Annual totals of water column total PAR were 4860, 3730, 4960, and 4920 moles m^{-2} yr^{-1} for Site 1, Site 2, Site

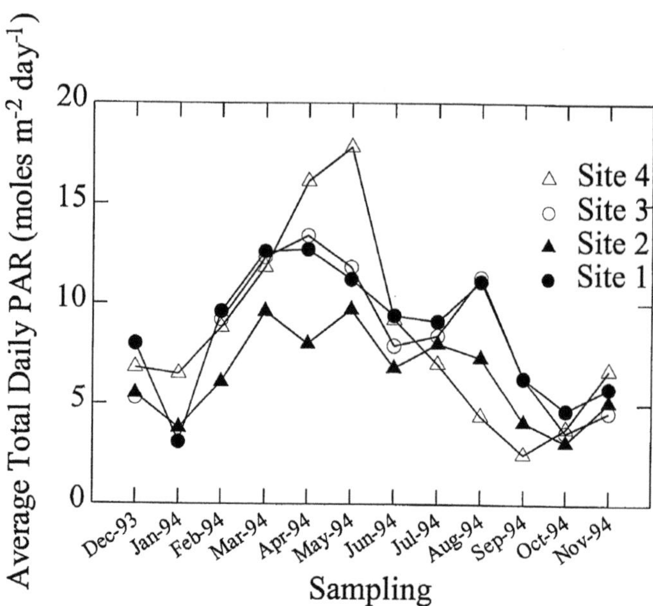

FIGURE 2.8 Monthly means of daily total plant-available photosynthetically active radiation received at the maximum depth limits of *Thalassia testudinum* beds in Tampa Bay, Florida.

3, and Site 4, respectively. Plant available total PAR totaled 3150, 2350, 3130, and 3080 moles m^{-2} yr^{-1} for the same four sites, respectively.

DISCUSSION

Maximum depths of *T. testudinum* may be expected to reach a "steady state" under stable water clarity conditions and in the absence of other stressors. Measured light climates at maximum depth limits will be a robust minimum light requirement if grassbeds exhibit no net increase or decrease. Annual light levels received then represent an ecological compensation point, or the PAR necessary to support a stable grassbed in which growth is balanced with normal mortality, grazing, and repetitive minor stresses. Water clarity of the prior several years must be examined since low light stress may take some time to be evident in *T. testudinum* (Hall et al. 1991). As water clarity during the project year was comparable to the prior 5 years, PAR measurements at stable *T. testudinum* beds as presented here are likely to represent maintenance level light requirements.

SHOOT DENSITY AND BIOMASS

T. testudinum shoot densities (and biomass) at maximum depths were lower than shoot densities reported from similar depths along the west central Florida coast (Dawes and Tomasko 1988; Tomasko and Dawes 1990; Hall et al. 1991). This is attributed to station selection of this project in which the very edge of a light-limited bed was sampled. Approximately 1 m shoreward from the marked deep edges, shoot densities were 2 to 6 times higher (unpublished data). Similar to many other investigations, shoot densities (and biomass) were much higher at the shallower stations, a pattern attributed to changes in water depth and therefore light (Buesa 1974; 1975; Kemp et al. 1988; Hall et al. 1991). Seasonal variation in shoot density was also more evident at the shallow stations.

In the analysis of *T. testudinum* bed stability at maximum depth limits over the course of the project, shoot density at Site 2 suffered a significant net decline between the December and the October sampling. Seasonal patterns in shoot density (Iverson and Bittaker 1986; Durako and

Moffler 1985; Tomasko and Dawes 1990; Hall et al. 1991) predict higher densities in October than December. Other stations in the study either displayed the expected increase in shoot density or were unchanged.

From April through October, the highest above-ground biomass of the deep stations occurred at Site 2. Despite a decline in shoot density at this station, above-ground biomass also increased between the initial and final samplings. Higher blade weights per area are the product of the longer blade lengths and shoot relative growth noted at the station, although LRGR remained generally comparable among the deep stations. Reductions in biomass noted at Site 4 between the beginning and end of the project were attributed to salinity stress.

Above- to Below-Ground Biomass

The above- to below-ground biomass ratios calculated in this study were consistently lower than those reported from nearby by Dawes and Tomasko (1988), despite the fact that a longer corer was used in the earlier work. Project values near 0.2, however, are very comparable to the ratio of 0.17 derived from tissue proportions given in Fourqurean and Zieman (1991).

Above- to below-ground ratios of biomass emphasize that the deep Site 2 station differed from the remaining deep stations. While Burkholder et al. (1959) reported an increase in above- to below-ground ratios (0.14–0.33) with finer sediments (~0.700 to 0.044 mm grain size), the range of sediments between the Tampa Bay sites was much smaller and insufficient to explain the difference in biomass ratios. Dunton (1994) reports an increase in above- to below-ground ratios under low light stress for the seagrass *Halodule wrightii*. Buesa (1975) and Kemp et al. (1988) have also linked higher ratios to reductions in light intensity. Dawes and Tomasko (1988) reported higher ratios at the deep edge of *T. testudinum* meadows at Anclote Key, although the pattern differed at the mouth of Tampa Bay leading to speculation that the latter site was not light-limited. In the present study, a consistent trend in above- to below-ground ratios across the depths monitored was not observed, and the conditions at the deep Site 2 station may indicate responses to low light stress.

Increased blade length and blade area per shoot noted at Site 2 are viewed as an early morphological change to decreases in PAR. The blade density, as ratios of blade area to tissue weights, also signify that the increased blade area at Site 2-D had less structural support. Hall et al. (1991) found a short-term increase in leaf area in *T. testudinum* that had been shaded to reduce available light. However, the response was not maintained and, by 9 months post-shading, leaf area per shoot was lower in shaded plots than in controls at both deep and shallow stations. Goldsborough and Kemp (1988) also found increased blade lengths to be one of the typical morphological responses produced by shading. Dennison and Alberte (1986) and Bulthuis (1983) found similar results for other species.

Growth

LRGRs of 27 to 35 mg $gdwt^{-1}$ d^{-1} during June were quite comparable to the rates determined by Dawes and Tomasko (1988). Similar to Dawes and Tomasko (1988) and Hall et al. (1991), no consistent pattern of growth with depth was detected during this project, while Tomasko and Dawes (1990) showed higher LRGR at deeper stations compared to shallow ones. During the present study, however, blade elongation, presumably to optimize light capture was not limited to deeper grasses since LRGRs were comparable between light-limited and non-light-limited (shallow) stations. Areal growth rates determined for the shallow stations in this study were comparable with those reported by Hall et al. (1991). Rates at deep stations, however, were substantially lower than those established by others (Hall et al. 1991; Dawes and Tomasko 1988) for similar depths and were attributed to the much lower shoot density recorded in this study.

Blade production per year (of the significantly longer blades) was lower at Site 2 than other deep stations, and the leaf scar results indicated that low blade production had been going on for some years. The lower below-ground biomass (and higher above- to below-ground ratios) seen at Site 2 during all samplings also indicate that plant resources have been diverted from typical allocation patterns for some time. It is unknown at this time whether the stress experienced by Site 2 can be tolerated indefinitely or will result in further declines in the shoot density of the deep edge.

Carbohydrates

Rhizomes from deep stations had higher carbohydrate concentrations than shallow stations, similar to observations by Tomasko and Dawes (1990), indicating that plants at the deep edge were not sacrificing rhizome carbohydrate reserves to maintain above-ground growth. Seasonal patterns seen by Dawes et al. (1979) and Dawes and Lawrence (1980), however, were not observed in the present study, nor were the winter and spring levels of carbohydrates as low as observed by Dawes and Lawrence (1980). The seasonal pattern observed in this study was a decline in June carbohydrate concentrations at the deep stations. Growth conditions of the preceding year may well affect wintertime carbohydrate levels. If a "deficit" is present, energy storage can be replenished during periods of high productivity, appearing as a seasonal cycle which peaks during the growing seasons. Alternatively, if carbohydrate levels are above some threshold, then little investment may be made in storing nutrients, but rather used to fuel increased growth.

At the deep, apparently light-stressed Site 2 station, the reduced below-ground biomass had carbohydrate concentrations comparable to the remaining deep stations during all samplings. Changes in mean carbohydrate concentrations, although not seasonally significant, were observed in three of the shallow stations during June, during which a reduction in carbohydrate concentration was coupled with a decline in shoot density, areal biomass, and leaf area index. The changes can be attributed to thermal stress. Carbohydrate decline was not attributed to the coincident decreases in water clarity, as reductions in carbohydrates were not observed in all deep station rhizomes (even at stations receiving lesser amounts of PAR). Deep stations also had higher concentrations of rhizome carbohydrate than did shallow stations.

Epiphytes

Increases in epiphytes reliably reflect increased nutrient loads (Tomasko and Lapointe 1991; Dunton 1990; Neckles et al. 1994; Neverauskas 1987; Silberstein et al. 1986; Lapointe et al. 1994), at times even more so than water column chlorophyll levels (Borum 1985). Based on the similarity of epiphytic attenuation between sites in Tampa Bay, the total nutrient loads relative to receiving water volumes and flushing time appear comparable for all four locations. Annual average epiphytic attenuation ranged between 32.0 and 36.2% at the deep edge and varied significantly by station within a given sampling.

The range of ratios of epiphyte to macrophyte biomass observed in Tampa Bay, 0.3 to 3.0, was comparable to that observed by Kemp et al. (1988), in some cases higher than that observed by Lapointe et al. (1994) or Tomasko and Lapointe (1991), and within the range of the more saline stations of Tomasko and Hall (1999). The reduction in PAR is an exponential relationship with epiphytic loads for numerous species (Bulthuis and Woelkerling 1983; Sand-Jensen and Borum 1983; Neckles 1993; Kemp et al. 1988) and is demonstrated here for *T. testudinum* in Tampa Bay. Relationships of ash-free epiphytic biomass with PAR attenuation are somewhat better than for dry weights, despite the fact that ash-free weights are only approximately 20 to 40% of the total epiphytic load.

The epiphytic load on *T. testudinum* blades exhibits a consistent seasonal pattern of increased epiphytic biomass during winter periods of low growth (Zieman and Zieman 1989; Humm 1974; Leverone 1991; Lapointe 1992; Dixon and Kirkpatrick 1995). When blades are growing more rapidly, colonization rates of epiphytic species are apparently insufficient to keep pace, and proportions of epiphytic biomass decreases (Zieman and Zieman 1989). This mechanism is supported by the relatively constant rates of fouling on the PAR sensors during the project year (unpublished data). Other factors such as nutrient supply, grazing removal rates, PAR, temperature, and currents may also play a role in regulating epiphytic growth (Neckles 1993). Seasonal patterns of epiphytic attenuation during this study were slightly unusual in that epiphytic growths were initially heavy in December, decreased during summer, and remained low through the October sampling. This is attributed to the fact that above-ground growth was still continuing, that the average age of leaves on shoots was younger than during the winter and early spring, and that water temperatures were 4–5°C above normal in the fall.

When epiphytic attenuation varied between deep and shallow locations at a site, the shallow stations were more likely to have a heavier epiphytic covering, particularly during summer and fall. This pattern is in contrast to other authors (Leverone 1991; Kemp et al. 1988), who have determined a positive correlation of epiphytic load with depth. Tomasko and Lapointe (1991), however, observed higher epiphytic loads with higher ambient light conditions under nutrient enriched conditions.

REFLECTANCE

The mean correction for reflectance (1.5% of subsurface irradiance) is conservatively high and represents an upper boundary of potential bias. The modeled estimation does not incorporate the angular component of both upward and downward propagating light. If the angular components were included, they would result in a larger increase in the downwelling component than in the backscattered or reflected, and would thus further reduce the differences between the two attenuation coefficients and calculated PAR percentages. Accordingly, $\%PAR_w$ and $\%PAR_p$ data are discussed below without correction for reflectance. Total PAR values are unaffected.

$\%PAR_w$

Stations in this study annually received 20.1–23.4% of immediately subsurface irradiance with the condition of the grassbed maintained if annual $\%PAR_w$ was 21.7% or greater. Percentages during the growing season were typically somewhat less. Seagrass PAR requirements, however, are often reported as percent of surface, in-air, irradiance rather than as the percentage of immediately subsurface light as reported in this work. Based on calculated solar elevation between 1000 and 1400 hours at the study latitude (from 10 to 60° from vertical), estimates of reflectance at the air-water interface range between 5 and 7% under a variety of wind speeds (Kirk 1994). Using very generalized approximation of 6% reflectance at the air-water interface, the stations received between 18.9 and 22.0% of surface irradiance (SI). Percentages of SI are consistent with those noted for other species: greater than 5% (Dunton and Tomasko 1991), 10–20% (Fourqurean and Zieman 1991), 10–15% (Kenworthy et al. 1991), 15–25% (Onuf 1991), and 23–37% (Kenworthy 1993). For *T. testudinum*, Czerny and Dunton (1995) noted declines in leaf elongation within 1 month at 10%SI and within 9 months at 14%SI, concluding that *T. testudinum* would not be maintained at these levels of irradiance. As one of the sites in this project appeared to exhibit shading responses, annual PAR requirements for *T. testudinum* appear to be refined even further: 18.9%SI at maximum depth limits produced a shading response, while 20.4%SI and greater maintained a stable deep edge. It is important to note that water column light requirements such as these do not incorporate the effects of epiphytes and are not appropriate water clarity targets for other areas unless epiphytic attenuations are comparable.

$\%PAR_p$

Epiphytic cover has been found to reduce PAR available for photosynthesis (Day et al. 1989; Kemp et al. 1988; Sand-Jensen and Sondergaard 1981) through attenuation and, by so doing, results in reduced productivity (Kemp et al. 1988; Tomasko 1993). For west central Florida alone, annual attenuation by epiphytes on *T. testudinum* has been documented to range between 33 and 56% (Dixon and Kirkpatrick 1995) of PAR at depth. Determination of light requirements should, therefore, include the impacts of epiphytic attenuation.

The $\%PAR_p$ received by the light-stressed station was 13.0% and roughly equivalent to 12.2%SI; stations with no shading response received between 13.8 and 14.2%PAR_p, or 13.0 to 13.3%SI. Between the Tampa Bay work and that in Texas (Czerny and Dunton 1995), latitudes and, therefore, calculated daily insolation should be identical, barring potential climatic differences in cloud cover. The fact that *T. testudinum* in Texas exhibited marked declines under 14%SI while Tampa Bay grasses were maintained at 13%SI (available to the plant) may be attributed to potential differences in solar insolation, differing seasonal variations in water clarity, uncertainties in estimation of %SI, or differing epiphytic attenuation between the two locales. Epiphytic attenuation in Tampa Bay averaged 33.1% of the PAR remaining at depth.

Variance in %PAR among stations (as percent relative standard deviation of annual means) was reduced when the effects of epiphytes were included. The reduction in variance is noted whether data are limited to the non-light-stressed stations or whether all four are included. Comparison of light requirements between Tampa and Sarasota Bays (Dixon and Kirkpatrick 1995) illustrate the reduction as well. The effect will be more apparent when data are collected across a wide range of nutrient and epiphytic regimes.

Total PAR

Annual totals of PAR appeared to be a more sensitive indicator of light stress in this project than %PAR. While $\%PAR_w$ or $\%PAR_p$ had a relatively small variance across all sites, annual total PAR at the light-stressed Site 2 was 75% of the remaining stations, both before (3730 moles m^{-2} yr^{-1}) and after accounting for epiphytic attenuation (2350 moles m^{-2} yr^{-1}). Sites maintaining biotic condition over the study received 4910 moles m^{-2} yr^{-1} and 3120 moles m^{-2} yr^{-1} before and after epiphytic attenuation, respectively. Annual totals were consistent with other PAR data on *T. testudinum* (Czerny and Dunton 1995) in which plants receiving 1819 moles m^{-2} yr^{-1} or less declined over a 9-month study while plants receiving 6500 moles m^{-2} yr^{-1} were maintained.

Timing of reduced levels of PAR also appeared to influence outcomes. While Site 2 had consistently lower light levels than all other deep sites, it also had the highest number of months when the monthly mean total PAR values in the water column were below 10 moles m^{-2} d^{-1}. After epiphytic attenuation, no mean monthly value exceeded 10 moles m^{-2} d^{-1} at Site 2. In 2 consecutive fall months at Site 2, *T. testudinum* received less than 5 moles m^{-2} d^{-1} on average. Long-term reductions in PAR to 5 moles m^{-2} d^{-1} or below have been linked to reduction in biomass and decline or disappearance of *Halodule* and *T. testudinum* in Texas (Dunton 1994; Czerny and Dunton 1995). While *T. testudinum* at Site 4 received less than 5 moles m^{-2} d^{-1} on average during 3 consecutive fall months, the growing season at this site had previously recorded 3 months in which plant-available PAR averaged between 12 and 18 moles m^{-2} d^{-1} and stability of the deep edge was maintained.

At all sites, maximum monthly means of daily PAR were typically observed in April and June and generally coincided with periods of maximum shoot and LRGR. In the case of both shallow and deep Site 2 stations, however, growth occurred even though there was a minimal corresponding increase in available PAR. Additionally observed growth was comparable to Site 1 where PAR levels received were much greater. Examination of individual shoot growth rates together with $\%PAR_p$ calculated from shoot-specific epiphytic attenuation revealed no significant relationship of growth with %PAR. As others have observed (Czerny and Dunton 1995; Hall et al. 1991), *T. testudinum* growth can be decoupled from ambient light, and growth alone is not a reliable indicator of low light stress. Thermal cues (Tomasko and Hall 1999) and endogenous seasonal rhythms may also supersede light availability.

SUMMARY

In conclusion, changes in shoot density and biomass at the deep edges of three of the sampling sites were consistent with the normal seasonal patterns of biomass. In contrast, blade length, leaf area index, and shoot relative growth were maintained at Site 2 well beyond the peak times observed at other stations, despite seasonal declines in available PAR and despite similarity of available PAR among stations during the latter months of the project. Reduced shoot density values from the beginning to end of the project, increased blade length, reduced below-ground biomass, and reduced blade production rates at Site 2 are more indicative of a response to lowered PAR and a slight "retreat" of the deep edge at this site. The level of change, while significant, is very slight. The long-term effects of the lower PAR received at this station are unknown.

With the knowledge that PAR at one of the deep stations may be at unacceptable levels, or at the least that morphological adaptations to shading were triggered, the light climate may be examined knowing that the data encompasses a threshold value. Annual totals of 4860 moles m^{-2} yr^{-1} in the water column maintained grassbeds at existing depths while shading responses were evident at 3730 moles m^{-2} yr^{-1}.

The mechanism accepted for seagrass decline under increased nutrient levels, however, is one of decreased available light and a net failure to maintain a positive energy balance of production against respiration, grazing, and other stresses or energy deficits. To compute the balance, or to identify the light requirements necessary for balance, one must consider the contribution of all materials which reduce plant-available PAR. The appearance of epiphytic cover at the study sites did not indicate highly eutrophic conditions or excessive overgrowth, yet it resulted in the annual reduction of over one third of the PAR in the water column. Higher epiphytic attenuation values have been observed elsewhere. It is readily apparent that neglecting epiphytic attenuation will overestimate *T. testudinum* requirements. Conversely, the use of the water column light requirements determined in this study would not support seagrasses in estuaries with higher epiphytic loads. Under this study, seagrass beds were maintained if annual PAR available to *T. testudinum* was near 3120 moles m^{-2} yr^{-1}, while shading responses were observed at the site which received 2350 moles m^{-2} yr^{-1} of plant-available PAR. It is imperative that the plant requirement value be adjusted upward for local epiphytic attenuation before water column management targets are identified.

Additionally, total annual PAR values appeared much more sensitive for evaluating light stress. While the site with shading response did receive lower %PAR or %SI values, the difference between this site and the remaining areas consisted of a few percentage points and may be very difficult to identify under typical monitoring programs with biweekly or weekly samplings. In contrast, total PAR values at the site with shading were 75% of the sites which maintained condition and highlighted the uniqueness of the biotic response at this particular site. Given the plastic response of *T. testudinum* to light stress and the extent to which it can endure stresses, it may be that light requirements of *T. testudinum* communities will not be more rigidly defined in natural systems. To identify sites with potential light stress, however, total PAR is a more useful parameter than %SI or %PAR.

Timing as well as totals of PAR received also may be critical to maintaining the condition of light-limited *T. testudinum* beds. While all sites received less than 10 moles m^{-2} d^{-1} during 7 or more months of the year, the *T. testudinum* at the light-stressed site in this study never exceeded a monthly mean total PAR of 10 moles m^{-2} d^{-1}. At all but the light-stressed site, PAR levels were above 10 moles m^{-2} d^{-1} during March, April, and May. All sites endured months in which plant-available PAR averaged less than 5 moles m^{-2} d^{-1}. Both the light-stressed site and another recorded 3 consecutive months in which monthly mean PAR was less than 5 moles m^{-2} d^{-1}; both the light-stressed site and another received approximately 3 moles m^{-2} d^{-1} during January of 1993. The key to maintaining bed condition appears to have been the receipt of adequate (in this study, greater than 11 moles m^{-2} d^{-1}) PAR at the plant during March, April, and May. Future experimental shading studies should consider the application of varying light reduction levels during different seasonal periods, as well as account for epiphytic attenuation.

ACKNOWLEDGMENTS

Funding for this project was provided by the Surface Water Improvement and Management Division of the Southwest Florida Water Management District. In addition, there were many who gave of their time and expertise. Special thanks go to J.R. Leverone without whose efforts this project would not have been completed. I would also like to thank D. Dell'Armo, E.D. Estevez, M.O. Hall, R.C. Kurz, K.A. Moore, A. Nissanka, C.P. Onuf, R.J. Orth, J.S. Perry, D.A. Tomasko, R.L. Wetzel, and the Tampa Bay Seagrass Focus Group members, as well as the Marine Research Institute of the Florida Department of Environmental Protection, the Sarasota Bay National Estuary Program, the Tampa Bay National Estuary Program, and the University of South Florida.

REFERENCES

American Public Health Association. 1989. *Standard Methods for the Examination of Water and Wastewater.* 17th Ed., Washington, D.C.

Avery, W. M. 1996. Distribution and abundance of macroalgae and seagrass in Hillsborough Bay, Florida, from 1986 to 1995. In *Proceedings, Tampa Bay Area Scientific Information Symposium 3,* S. F. Treat (ed.). Tampa Bay Regional Planning Council, Clearwater, FL.

Boon, J. D., III and K. P. Kiley. 1978. Harmonic analysis and tidal prediction by the method of least squares. Special Report No. 186, Virginia Institute of Marine Science, Gloucester Point, VA.

Borum, J. 1985. Development of epiphytic communities on eelgrass (*Zostera marina*) along a nutrient gradient in a Danish Estuary. *Marine Biology* 87:211–218.

Brouns, J. J. W. M. 1985. The plastochrone interval method for the study of the productivity of seagrasses: possibilities and limitations. *Aquatic Botany* 21:71–88.

Buesa, R. J. 1974. Population and biological data on turtle grass (*Thalassia testudinum* Konig, 1805) on the northwestern Cuban shelf. *Aquaculture* 4:207–226.

Buesa, R. J. 1975. Population biomass and metabolic rates of marine angiosperms on the northwestern Cuban shelf. *Aquatic Botany* 1:11–23.

Bulthuis, D. A. 1983. Effects of *in situ* light reduction on density and growth of the seagrass *Heterozostera tasmanica* (Martens ex Aschers.) den Hartog in Western Port, Victoria, Australia. *Journal of Experimental Marine Biology and Ecology* 67:91–103.

Bulthuis, D. A. and W. J. Woelkerling. 1983. Biomass accumulation and shading effects of epiphytes on leaves of the seagrass *Heterozostera tasmanica*, in Victoria, Australia. *Aquatic Botany* 16:137–148.

Burkholder, P. R., L. M. Burkholder, and J. A. Rivero. 1959. Some chemical constituents of turtle grass, *Thalassia testudinum. Bulletin of the Torrey Botanical Club* 86(2):88–93.

Cambridge, M. L. and A. J. McComb. 1984. The loss of seagrass in Cockburn Sound, Western Australia. I. The time course and magnitude if seagrass decline in relation to industrial developments. *Aquatic Botany* 20:229–243.

Camp Dresser & McKee, Inc. 1994. Tampa Bay tributary loading project. Report to the Tampa Bay National Estuary Program, TBNEP Technical Report No. 03-94, St. Petersburg, FL.

Czerny, A. B. and K. H. Dunton. 1995. The effects of *in situ* light reduction on the growth of two subtropical seagrasses, *Thalassia testudinum* and *Halodule wrightii. Estuaries* 18:418–427.

Dawes, C. J. (ed.). 1981. *Marine Botany.* John Wiley & Sons, New York.

Dawes, C. J. and D. L. Lawrence. 1980. Seasonal changes in the proximate composition of the seagrasses. *Aquatic Botany* 8:371–380.

Dawes, C. J. and D. A. Tomasko. 1988. Depth distribution of *Thalassia testudinum* in two meadows on the west coast of Florida: a difference in effect of light availability. Pubblicazioni della Stazione zoologica di Napoli I: *Marine Ecology* 9(2):123–130.

Dawes, C. J., K. Bird, M. Durako, R. Goddard, W. Hoffman, and R. McIntosh. 1979. Chemical fluctuations due to seasonal and cropping effects on an algal-seagrass community. *Aquatic Botany* 6:79–86.

Day, J. W., Jr., C. A. S. Hall, W. M. Kemp, and A. Yáñez-Arancibia. 1989. *Estuarine Ecology*. John Wiley & Sons, New York.

Dennison, W. C. and R. S. Alberte. 1986. Photoadaptation and growth of *Zostera marina* L. (eelgrass) transplants along a depth gradient. *Journal of Experimental Marine Biology and Ecology* 98:265–282.

Dixon, L. K. and G. J. Kirkpatrick. 1995. Light attenuation with respect to seagrasses in Sarasota Bay, Florida. Mote Marine Laboratory Technical Report No. 406. Submitted to the Sarasota Bay National Estuary Program, Sarasota, FL.

Duarte, C. M. and K. Sand-Jensen. 1990. Seagrass colonization: biomass development and shoot demography in *Cymodocea nodosa* patches. *Marine Ecology Progress Series* 67:97–103.

Duarte. C. M., N. Marbá, N. Agawin, J. Cebrián, S. Enríquez, M. D. Fortes, M. E. Gallegos, M. Merino, B. Olesen, K. Sand-Jensen, J. Uri, and J. Vermaat. 1994. Reconstruction of seagrass dynamics: age determinations and associated tools for the seagrass ecologist. *Marine Ecology Progress Series* 107(1):195.

Dubois, M., K. A. Gilles, J. K. Hamilton, P. A. Rebers, and F. Smith. 1956. Colorimetric method for determination of sugars and related substances. *Analytical Chemistry* 28(3):350–356.

Dunton, K. H. 1990. Production ecology of *Ruppia maritima* L.s.l. and *Halodule wrightii* Aschers. in two subtropical estuaries. *Journal of Experimental Marine Biology and Ecology* 143:147–164.

Dunton, K. H. 1994. Seasonal growth and biomass of the subtropical seagrass *Halodule wrightii* in relation to continuous measurements of underwater irradiance. *Marine Biology* 120:479–489.

Dunton, K. H. and D. A. Tomasko. 1991. Seasonal variation in the photosynthetic performance of *Halodule wrightii* measured *in situ* in Laguna Madre, Texas. In *The Light Requirements of Seagrasses: Proceedings of a Workshop to Examine the Capability of Water Quality Criteria, Standards, and Monitoring Programs to Protect Seagrasses,* W. J. Kenworthy and D. E. Haunert (eds.). NOAA Technical Memorandum, NMFS-SEFC-287.

Dunton, K. H. and D. A. Tomasko. 1994. *In situ* photosynthesis in the seagrass *Halodule wrightii* in a hypersaline subtropical lagoon. *Marine Ecology Progress Series* 107:281–293.

Durako, M. J. 1994. Seagrass die-off in Florida Bay (USA): changes in shoot demographic characteristics and population dynamics in *Thalassia testudinum. Marine Ecology Progress Series* 110(1):59.

Durako. M. J. and M. D. Moffler. 1985. Spatial influences on temporal variations in leaf growth and chemical composition of *Thalassia testudinum* Banks ex Konig in Tampa Bay, Florida. *Gulf Research Reports* 8(1)43–49.

Environmental Protection Commission, Hillsborough County. 1989. *Surface Water Quality, Hillsborough County, Florida, 1988–1989,* R. Boler (ed.). Tampa, FL.

Environmental Protection Commission, Hillsborough County. Unpublished water quality monitoring data. Tom Cardinale. Tampa, FL.

Folk, R. L. 1974. *Petrology of Sedimentary Rocks.* Hemphill Publishing Company, Austin, TX.

Fourqurean, J. W. and J. C. Zieman. 1991. Photosynthesis, respiration and whole plant carbon budget of the seagrass *Thalassia testudinum. Marine Ecology Progress Series* 69:161–170.

Gallegos, M. E., M. Merino, N. Marbá, and C. M. Duarte. 1992. Flowering of *Thalassia testudinum* Banks ex Konig in the Mexican Caribbean: age dependence and interannual variability. *Aquatic Botany* 43(3):249.

Gallegos, M. E., M. Merino, N. Marbá, and C. M. Duarte. 1993. Biomass and dynamics of *Thalassia testudinum* in the Mexican Caribbean: elucidating rhizome growth. *Marine Ecology Progress Series* 95(1):185–192.

Goldsborough, W. J. and W. M. Kemp. 1988. Light responses of a submersed macrophyte: implications for survival in turbid tidal waters. *Ecology* 69(6):1775–1786.

Haddad, K. 1989. Habitat trends and fisheries in Tampa and Sarasota Bays. In *Tampa and Sarasota Bays: Issues, Resources, Status, and Management,* E. D. Estevez (ed.). Estuary-of-the-Month Seminar Series, No. 11, National Oceanic and Atmospheric Administration, Washington, D.C.

Hall, M. O., D. A. Tomasko and F. X. Courtney. 1991. Responses of *Thalassia testudinum* to *in situ* light reduction. In *The Light Requirements of Seagrasses: Proceedings of a Workshop to Examine the Capability of Water Quality Criteria, Standards, and Monitoring Programs to Protect Seagrasses,* W. J. Kenworthy and D. E. Haunert (eds.). NOAA Technical Memorandum, NMFS-SEFC-287.

Hayward, D. H. 1993. Tides: software for the derivation of tidal constants and prediction of tidal amplitude. Mote Marine Laboratory, Sarasota, FL.

Humm, H. J. 1974. Benthic algae. Anclote Environmental Project Annual Report, G. F. Mayer and V. Maynard (eds.). Department of Marine Science, University of South Florida, St. Petersburg, FL, pp. 455–478.

Iverson, R. L. and H. F. Bittaker. 1986. Seagrass distributions in the Eastern Gulf of Mexico. *Estuarine Coastal Shelf Science* 22:577–602.

Jaeger, J. E. 1973. The determination of salinity from conductivity, temperature and pressure measurements. In *Proceedings, Second S/T/D Conference and Workshop,* San Diego, CA.

Janicki, A. and D. Wade. 1996. Estimating critical nitrogen loads for the Tampa Bay Estuary: an empirically based approach to setting management targets. Coastal Environmental, Inc. Report to the Tampa Bay National Estuary Program, TBNEP Technical Report No. 03-95, St. Petersburg, FL.

Janicki, A., D. Wade, and D. Robison. 1995. Habitat protection and restoration targets for Tampa Bay. Coastal Environmental, Inc. Report to the Tampa Bay National Estuary Program, TBNEP Technical Report No. 07-93, St. Petersburg, FL.

Johansson, J. O. R. 1991. Long-term trends of nitrogen loading, water quality and biological indicators in Hillsborough Bay, Florida. In *Proceedings, Tampa Bay Area Scientific Information Symposium 2,* S. F. Treat and P. A. Clark (eds.). Tampa Bay Regional Planning Council, Clearwater, FL.

Kemp, W. M., W. R. Boynton, L. Murray, C. J. Madden, R. L. Wetzel, and F. Vera Herrera. 1988. Light relations for the seagrass *Thalassia testudinum*, and its epiphytic algae in a tropical estuarine environment. In *Ecology of Coastal Ecosystems in the Southern Gulf of Mexico: The Terminos Lagoon Region,* A. Yáñez-Arancibia and J. W. Day, Jr. (eds.). Instituto de Ciencias del Mar y Limnologia, Universidad Nacional Autonoma de Mexico, Mexico, D.F., p. 193.

Kenworthy, W. J. 1993. Defining the ecological light compensation point of seagrasses in the Indian River Lagoon. In *Proceedings and Conclusions of Workshops on Submerged Aquatic Vegetation Initiative and Photosynthetically Active Radiation,* L. J. Morris and D. A. Tomasko (eds.). Indian River Lagoon National Estuary Program, Palatka, FL.

Kenworthy, W. J., M. S. Fonseca, and S. J. DiPiero. 1991. Defining the ecological light compensation point for seagrasses *Halodule wrightii* and *Syringodium filiforme* from long-term submarine light regime monitoring in the southern Indian River. In *The Light Requirements of Seagrasses: Proceedings of a Workshop to Examine the Capability of Water Quality Criteria, Standards, and Monitoring Programs to Protect Seagrasses,* W. J. Kenworthy and D. E. Haunert (eds.). NOAA Technical Memorandum, NMFS-SEFC-287.

Kirk, J. T. O. 1994. *Light and Photosynthesis in Aquatic Ecosystems.* 2nd Ed., Cambridge University Press, Cambridge, U.K.

Lapointe, B. E. 1992. Final Report: eutrophication and the trophic structuring of marine plant communities in the Florida Keys. Submitted to the Florida Department of Environmental Regulation and Monroe County.

Lapointe, B. E., D. A. Tomasko, and W. R. Matzie. 1994. Eutrophication and trophic state classification of seagrass communities in the Florida Keys. *Bulletin of Marine Science* 54(3): 696–717.

Leverone, J. R. 1991. Seagrass epiphytes. In *Anclote Monitoring Studies,* J. K. Culter and L. K. Dixon (eds.). Mote Marine Laboratory Technical Report. Submitted to Florida Power Corporation, St. Petersburg, FL, Chap. 8.

Lewis, R. R., III, K. D. Haddad, and J. O. R. Johansson. 1991. Recent areal expansion of seagrass meadows in Tampa Bay, Florida: real bay improvement or drought induced? In *Proceedings, Tampa Bay Area Scientific Information Symposium 2,* S. F. Treat and P. A. Clark (eds.). Tampa Bay Regional Planning Council, Clearwater, FL.

Miller, R. and B. McPherson. 1993. Modeling the effect of solar elevation angle and cloud cover on the vertical attenuation and the fraction of incident photosynthetically active radiation entering the water of Tampa Bay, FL. In *Proceedings and Conclusions of Workshops on Submerged Aquatic Vegetation Initiative and Photosynthetically Active Radiation,* L. J. Morris and D. A. Tomasko (eds.). Indian River Lagoon National Estuary Program, Palatka, FL.

Neckles, H. A. 1993. The role of epiphytes in seagrass production and survival: microcosm studies and simulation modeling. In *Proceedings and Conclusions of Workshops on Submerged Aquatic Vegetation Initiative and Photosynthetically Active Radiation,* L. J. Morris and D. A. Tomasko (eds.). Indian River Lagoon National Estuary Program, Palatka, FL.

Neckles, H. A., E. T. Koepfleer, L. W. Haas, R. L. Wetzel, and R. J. Orth. 1994. Dynamics of epiphytic photoautotrophs and heterotrophs in *Zostera marina* (eelgrass) microcosms: responses to nutrient enrichment and grazing. *Estuaries* 17(3):597–605.

Neverauskas, V. P. 1987. Accumulation of periphyton biomass on artificial substrates deployed near a sewage sludge outfall in south Australia. *Estuarine Coastal Shelf Science* 25:509–517.

Onuf, C. P. 1991. Light requirements of *Halodule wrightii, Syringodium filiforme,* and *Halophila engelmannii* in a heterogeneous and variable environment inferred from long-term monitoring. In *The Light Requirements of Seagrasses: Proceedings of a Workshop to Examine the Capability of Water Quality Criteria, Standards, and Monitoring Programs to Protect Seagrasses,* W. J. Kenworthy and D. E. Haunert (eds.). NOAA Technical Memorandum, NMFS-SEFC-287.

Sand-Jensen, K. and M. Sondergaard. 1981. Phytoplankton and epiphyte development and their shading effect on submerged macrophytes in lakes of different nutrient status. *Internationale Revue der gesamten Hydrobiologie* 66(4):529–552.

Sand-Jensen, K. and J. Borum. 1983. Regulation of growth of eelgrass (*Zostera marina* L.) in Danish coastal waters. *Marine Technology Society Journal* 17(2):15–21.

Shureman, P. 1971. Manual of harmonic analysis and prediction of tides. Special Publication No. 98, U.S. Department of Commerce, Coast and Geodetic Survey, Washington, D.C.

Silberstein, K., A. Chiffings, and A. McComb. 1986. The loss of seagrass in Cockburn Sound, western Australia. III. The effect of epiphytes on productivity of *Posodonia australis* Hook F. *Aquatic Botany* 24: 355–371.

Southwest Florida Water Management District. 1994. Map Series. GIS change analysis of seagrass coverage in Tampa Bay, Florida, 1988–1992. Tampa, FL.

Tampa Physical Oceanographic Real-Time System (PORTS). 1995. University of South Florida, Department of Marine Science, St. Petersburg, FL.

Tomasko, D. A. 1993. Assessment of seagrass habitats and water quality in Sarasota Bay. In *Proceedings and Conclusions of Workshops on Submerged Aquatic Vegetation Initiative and Photosynthetically Active Radiation,* L. J. Morris and D. A. Tomasko (eds.). Indian River Lagoon National Estuary Program, Palatka, FL.

Tomasko, D. A. and C. J. Dawes. 1989. Evidence for the physiological integration between shaded and unshaded short shoots of *Thalassia testudinum. Marine Ecology Progress Series* 54:299–305.

Tomasko, D. A. and C. J. Dawes. 1990. Influences of season and depth on the clonal biology of the seagrass *Thalassia testudinum. Marine Biology* 105:345–351.

Tomasko, D. A. and B. E. Lapointe. 1991. Productivity and biomass of *Thalassia testudinum* as related to water column nutrient availability and epiphyte levels: field observations and experimental studies. *Marine Ecology Progress Series* 75:9–17.

Tomasko, D. A. and M. O. Hall. 1999. Productivity and biomass of the seagrass *Thalassia testudinum* along a gradient of freshwater influence in Charlotte Harbor, Florida (USA). *Estuaries* 22:592–602.

Tomasko, D. S., M. O. Hall, and C. J. Dawes. 1996. The effects of anthropogenic nutrient enrichment on turtle grass (*Thalassia testudinum*) in Sarasota Bay, Florida. *Estuaries* 19:448–456.

U.S. Department of Commerce. 1994. Tide Tables 1994: high and low water predictions for east coast of North and South America, including Greenland. National Oceanic and Atmospheric Administration, National Ocean Service, Riverdale, MD.

Zarbock, H., A. Janicki, D. Wade, D. Heimbuch, and H. Wilson. 1994. Estimates of total nitrogen, total phosphorus, and total suspended solids loadings to Tampa Bay, Florida. Coastal Environmental, Inc. report to the Tampa Bay National Estuary Program, TBNEP Technical Report No. 04-94, St. Petersburg, FL.

Zieman, J. C. and R. T. Zieman. 1989. The ecology of the seagrass meadows of the west coast of Florida: a community profile. *U.S. Fish and Wildlife Service Biological Report* 85(7.25).

3 Somatic, Respiratory, and Photosynthetic Responses of the Seagrass *Halodule wrightii* to Light Reduction in Tampa Bay, Florida Including a Whole Plant Carbon Budget

Merrie Beth Neely

Abstract Treatment plots (1.5 m^2) within a shallow, monospecific seagrass bed of *Halodule wrightii* (Aschers.) located off Mullet Key (Fort DeSoto Park), Pinellas County, Florida, were subjected to *in situ* light reductions of 43, 60, and 86% from September 1994 to early March 1995 and compared with control plots for changes in morphology and photosynthesis vs. irradiance (PE) response. Winter die-back caused a 50% reduction in biomass and shoot density in control plots between September and February/March. A further reduction in total and non-photosynthetic biomass of 50% compared to control occurred as a result of treatment effect regardless of the amount of light reduction. However, reduction in photosynthetic biomass was related to the amount of light reduction ($r^2 = 0.89$).

There was an inverse relationship between light reduction and shoot number ($r^2 = 0.98$). Leaves on shoots within the 43% light reduction plots were longer than leaves on shoots in either control or other light reduction treatment plots after the second month of the experiment. This is probably a morphological response to light reduction, as reported for other species.

Respiration and PE responses (alpha, P_{MAX}, I_C, and I_K) and leaf chlorophyll content in *H. wrightii* were variable and did not exhibit trends related to light reduction. The I_{CPLANT} values ranged from 30 to 100 $\mu Em^{-2}s^{-1}$, and averaged 39 to 62 $\mu Em^{-2}s^{-1}$. This corresponds to approximately 4.5–7% of surface irradiance (SI) in September and February/March, respectively, in Tampa Bay.

Carbon budget calculations, using the H_{SAT} model and accounting for water column attenuation and shade in the treatments, indicate net areal productivity by *H. wrightii* in control plots was 0.25 g C $m^{-2}d^{-1}$, and in 43% light reduction plots was 0.025 g C $m^{-2}d^{-1}$, in September. Other light reduction treatments were at a carbon deficit in September, and all treatments and controls were at a carbon deficit in February/March.

These deficits explain the loss of biomass as a result of winter die-back and with a > 60% reduction in light. The discrepancy between the morphological changes that occurred in *H. wrightii* as a result of a 43% light reduction and a calculated positive carbon budget suggest that other sources of light attenuation (epiphytes, surface scattering, etc.) must be accounted for, or it may reflect carbon budget precision or assumptions.

0-8493-2045-3/99/$0.00+$.50
© 2000 by CRC Press LLC

INTRODUCTION

An important consequence of anthropogenic nutrient inputs to coastal waters and estuaries is increased light attenuation. Light attenuation has been shown to affect the morphology (Carlson and Acker 1985; Dennison and Alberte 1986; Gordon et al. 1994), abundance (Czerny and Dunton 1995), and productivity of seagrasses (Fourqurean and Zieman 1991a; Czerny and Dunton 1995). Shoot density, blade length and width, chlorophyll content, and percent tissue biomass also vary with changes in light attenuation in many seagrasses (Neverauskus 1988; Williams and Dennison 1990; West 1990).

To stem the global loss of seagrasses and establish conditions conducive to seagrass recolonization, a multitude of studies have been conducted on the light requirements of seagrasses (Kenworthy and Haunert 1991; Morris and Tomasko 1993). In tropical waters, these studies have focused primarily on *T. testudinum* (Banks ex Konig.) because it is the climax species and because it is usually the dominant species (Fourqurean and Zieman 1991b). *H. wrightii*, a seagrass common in Tampa Bay, has been shown to exhibit higher light requirements (Czerny and Dunton 1995) and is generally found at shallower depths than *T. testudinum* (Duarte 1991). Because of this characteristic, it is the primary species found in stressed-fringed-perennial, ephemeral, and newly-colonizing-perennial seagrass meadows, but it is also usually a component of all types of Tampa Bay seagrass meadows (Lewis et al. 1985). Because it is a colonial species and common in Tampa Bay, deleterious effects to *H. wrightii* meadows caused by light attenuation would substantially impact the Tampa Bay ecosystem.

Relatively few studies describe the morphological effects of light limitation in *H. wrightii* (Dunton 1990; Dunton 1994; Czerny and Dunton 1995; Tomasko and Dunton 1995). There has been no published investigation of photosynthetic or biomass changes in *H. wrightii* as a result of less than a 50% decrease in light penetration *in situ*.

The photosynthetic capacity of *H. wrightii* varies with season (Kenworthy and Haunert 1991). This may be a function of photoperiod, ambient light levels, water temperature, or a combination of these factors. Increased chlorophyll concentrations in response to reduced light availability have been demonstrated in other species (Dennison and Alberte 1982; Dennison and Alberte 1986; Dennison 1987; Dawes and Tomasko 1988; Goldsborough and Kemp 1988) and are consistent with photoacclimation to low light. Several studies have also reported significant differences among photosynthesis vs. irradiance (PE) responses in other species under a variety of conditions. Although variable results have been found in the PE response when used as an indicator of light stress in seagrasses in past studies, it is useful in estimating productivity (Dunton and Tomasko 1991; Fourqurean and Zieman 1991a).

The requirement for a net positive balance between photosynthesis and respiration, i.e., the plant carbon budget, is basic to understanding light requirements for seagrasses (Fourqurean and Zieman 1991a). From measured photosynthetic and respiration rates, a minimum light requirement can be calculated for a species based on a theoretical carbon budget (Fourqurean and Zieman 1991a). The minimum light requirement for seagrass survival is the lowest irradiance at which carbon produced through photosynthesis equals carbon lost via whole plant respiration, i.e., the light compensation point, I_{CPLANT}. Below this threshold, a plant functions at a carbon deficit, using more carbon for respiration than it generates via photosynthesis, and eventually dies. Above this level, a plant can use the extra carbon gained from photosynthesis for maintenance, somatic growth, and sexual reproduction. I_{CPLANT} values are instantaneous rates and are not equivalent to a daily minimum light requirement for seagrasses (a figure representing the duration of time I_{CPLANT} irradiances are achieved *in situ*). The H_{SAT} model incorporates the I_{CPLANT} as the minimum light requirement (H_{COMP} level) for use in calculating the daily production.

Minimum light requirements for maintenance (I_{CPLANT}) have been investigated for *T. testudinum, H. wrightii*, and *Syringodium filiforme* in Florida Bay (Fourqurean and Zieman 1991b); *T. testudinum* in Tampa Bay (Dixon and Leverone 1995); and *H. wrightii* in Laguna Madre, TX (Dunton and Tomasko 1991). Based on mean maintenance values for *H. wrightii,* the minimum light requirements were 65 $\mu Em^{-2}s^{-1}$ in Florida Bay (Fourqurean and Zieman 1991b) and 73 $\mu Em^{-2}s^{-1}$ (reported as μmol photons $m^{-2}s^{-1}$) in Laguna Madre (Dunton and Tomasko 1991). Variations in the relative proportion of tissue types, i.e., photosynthetic vs. non-photosynthetic tissue per plant, will affect the minimum light necessary for a species (Fourqurean and Zieman 1991a,b; Czerny and Dunton 1995; Dixon and Leverone 1995; Tomasko and Dunton 1995). This is controlled by environmental variables such as nutrient availability and sediment type (Fourqurean and Zieman 1991a,b) and may also be altered in response to low light stress. Therefore, the carbon budget, photosynthesis, and respiration rates, and the minimum light requirements for plants from Florida Bay or Laguna Madre may not be applicable to Tampa Bay populations.

Generally 10–20% penetration of the average daily SI is sufficient for *H. wrightii* to maintain a net positive carbon budget for Florida Bay (Fourqurean and Zieman 1991a) and the southern Indian River (Kenworthy et al. 1991). A broader range of 7–23% SI was required for *H. wrightii* in Laguna Madre, TX (Tomasko and Dunton 1991; Dunton 1994; Czerny and Dunton 1995). These light levels generally exceed the 10% SI necessary for *T. testudinum* maintenance previously reported by Iverson and Bittaker (1986) and what was generally accepted by resource managers and researchers as adequate for other benthic marine macrophytes prior to 1990 (Kenworthy et al. 1991). None of the light requirement experiments previously mentioned accounted for attenuation by epiphytes or irradiance reflectance of photosynthetically active radiation (PAR). In addition, some calculations did not account for heterotrophic tissue respiration rates which may equal, or exceed, leaf respiration (Dennison 1987; Dunton and Tomasko 1991; Fourqurean and Zieman 1991a,b). The comprehensive research of Dixon and Leverone (1995) indicated *T. testudinum* in Tampa Bay required at least 22.5% of immediately subsurface irradiance for growth and maintenance and suggested *H. wrightii* likely required a higher irradiance due to its shallower zonation.

Establishing a minimum light requirement for *H. wrightii* in Tampa Bay waters would clearly benefit resource managers since it is the colonial seagrass species and has been shown to exhibit higher light requirements than *T. testudinum* in Laguna Madre and Florida Bay. The impacts of light reduction on the morphology and PE responses of *H. wrightii* would provide insight into potential consequences of increased light attenuation on seagrasses. This study was undertaken to provide that information through the following specific objectives: (1) to determine the effects of shading on *H. wrightii* morphology as indicated by shoot density, leaf length, leaf number and biomass of tissue types; (2) to evaluate the effects of shading on PE responses of *H. wrightii*; and (3) to estimate a range of critical light necessary for survival and growth of *H. wrightii* in Tampa Bay based on a calculated whole plant carbon budget.

METHODS

MORPHOLOGY

Experimental plots for this study were established off Mullet Key (Fort DeSoto Park), Pinellas County, Florida. The experimental plots were located along the west shoreline (Figure 3.1) within a shallow cove approximately 600 m across by 1800 m long, bounded on three sides by land, and open to Tampa Bay on the north.

The site chosen for the experimental plots contained a monoculture of dense *H. wrightii.* All plots exhibited equivalent depths (approximately 0.5 m below mean sea level (MSL), ± 0.1 m) and remained submerged at mean low tide. However, aerial exposure occurred during four extremely low tides in winter.

The study-plot frames were constructed of 1.5 m segments of 2.5 cm PVC pipe forming a square and imbedded approximately 0.5 m above the sediment level. Neutral density screening was used to attain three levels of light reduction (30, 50, and 80%, ± 5%). Duplicate plots for all three light reduction treatments and control were used, totaling eight plots. Light transmittance at each

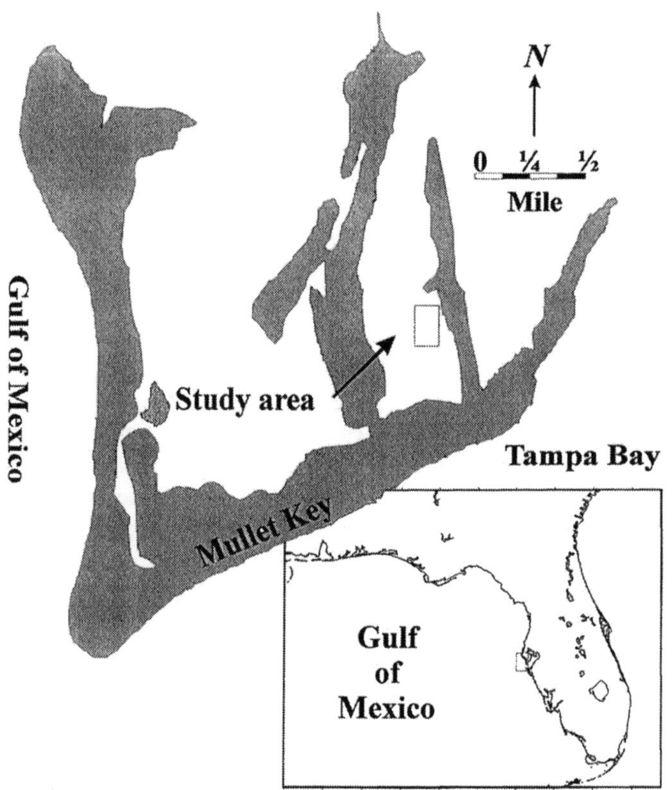

FIGURE 3.1 Map highlighting the study area.

of these levels was therefore 70, 50, and 20% of ambient light at depth, respectively. Due to fouling of the screens by epiphytes, the following ranges of light reduction are more applicable: the 30% light-reduction screen was found to attenuate ambient light at depth by 28–50%, averaging 43%; the 50% screens attenuated ambient light at depth by 46–74%, averaging 60%; the 80% screens attenuated ambient light at depth by 78–92%, averaging 86%. The treatments are referred to hereafter by the unfouled light reduction level. Treatments were indiscriminantly assigned to the plots in replicates. All screens were cleaned (defouled) with a scrub brush at least weekly as needed to maintain the experimental light transmittance levels. Irradiance measurements were made at depth in each of the plots before and after cleaning using LI-COR Model LI185B 2π quantum sensors (190SB in-air and 192SB subsurface).

Shoot density estimates and leaf length measurements were taken monthly beginning September 16, 1994. Water temperature, salinity, time, tide, screen fouling, plot damage, and epiphyte coverage, etc. were also recorded for each sampling event for the duration of the 28-week study.

Immediately after the plots were installed, and monthly thereafter, the rhizomes along the perimeter of each experimental plot were severed by a long knife to prevent translocation of materials from individuals outside the study area. *Ruppia maritima*, which eventually invaded the two control plots and one plot with 30% light reduction, was carefully removed from the plots when observed.

A 1 m^2 area at the center of each plot was set aside for the collection of leaf tissue and measurements of initial (T_o), final (T_f) shoot density. A permanent 0.1 m^2 (10 cm × 10 cm) quadrat was used during the monthly estimates to facilitate repeated measurements of the same area within each plot and because of time and effort constraints. This same permanent quadrat was used to measure blade length. Shoots were never removed from the permanent quadrat for other sampling.

The number of shoots removed within the remainder of each plot, however, were accounted for in the shoot density estimates.

The percent contribution of photosynthetic (above-ground) and heterotrophic (below-ground) tissue to total biomass was used in calculating the whole plant carbon budget. At the start of the experiment, tissue apportionment was determined from four core samples (0.014 $m^2 \times$ approximately 0.15m deep) taken adjacent to each plot with a post hole digger to minimize disturbance within the plot. At the conclusion of the experiment, biomass measurements were taken from four core samples within each plot using a 0.017 m^2 metal sediment corer. Samples were stored at –20°C until analyzed. After thawing, tissue was separated into photosynthetic (leaves and green portion of the upper short shoot) and non-photosynthetic tissue (non-green portion of the short shoot, roots, and rhizomes), dried in an oven (60°C) until the next sampling event (approximately 6 or 7 d), and weighed on a Mettler AE 163 automatic electrobalance to the nearest milligram. Leaf number per shoot was measured on four indiscriminantly harvested shoots per treatment, per sampling event.

The Kolmogorov–Smirnov test was used to determine normality, and the Bartlett's Chi-Square and/or the Cochran's C-test were used to determine sample homogeneity of variances. Nonparametric statistics (StatgraphicsPLUS software) were employed to test for significant differencs, primarily the Kruskal–Wallis (K-W) analysis and the Mann–Whitney U test. No biomass, leaf length, or shoot-number data required transformation to fit a normal distribution. No PE response or chlorophyll-content data required transformation when separated by sampling date. The leaf number data did not fit a normal distribution regardless of transformation. All tests were set at 95% confidence levels.

PE Response

Samples for this phase were taken weekly for the first 6 weeks of the study and every other week thereafter. Two subsamples, containing two or more shoots per subsample, were taken from each plot per sampling event. The sample shoots were severed below the surface and manually transferred underwater to a plastic Ziploc™ bag filled with unfiltered seawater from the study area. The bags were stored at ambient temperature in the dark in an insulated chest and immediately transported to the laboratory. Analyses were performed within 6 hours of collection.

The length, width, and number of blades were recorded for each shoot. Four or six 2 cm-long blade segments cut from the mid-blade region of two young shoots per treatment were used in this analysis. Where possible, sections were taken from the portion of the blade without epiphytes. PE responses were measured using a Hansatech DW/1 oxygen electrode system coupled with a LI-COR 2π quantum sensor, and a Kodak projector lamp with neutral-density filters over 13 light levels ($\approx$10 to >1000 $\mu E\ m^{-2}s^{-1}$ PAR) following the procedure outlined by Durako et al. (1993).

PE analyses were accomplished in the order collected. All PE curves were normalized to both chlorophyll content of the leaf segments used in the PE run and dry weight (using duplicate leaf segments dried at 60°C and weighed to the nearest µg).

Leaf segments were rinsed in deionized water and frozen at –20°C immediately following each PE analysis. Within 2 weeks of the PE analysis, the chlorophyll was extracted by freezing the leaf segments with liquid nitrogen and pulverizing them with a mortar and pestle. It was extracted in 100% methanol in a refrigerator at 4°C overnight. Since pigment concentrations in the initial extraction were always above measurable limits, all samples were diluted 1:10 with 100% methanol for analysis on a Turner Designs Model 10 fluorometer. The fluorometer was calibrated using pure chlorophyll *a* from spinach following the procedure outlined by Holm Hansen and Reimann (1978).

The PE characteristics α, P_{MAX}, I_K, I_C and respiration were determined for each PE curve. The initial slope (α) was calculated from a linear regression using the first three to six photosynthetic rates and the dark respiration rate, whichever yielded the highest r^2 value. Light saturated photosynthetic rate (P_{MAX}) was calculated using the average of all photosynthetic rates above 400 $\mu Em^{-2}s^{-1}$ PAR, the irradiance at which most PE curves exhibited maximum oxygen production. The irradiance

TABLE 3.1
Coefficients used in the carbon budget calculations.

Coefficients	September				February			
	CT	30	50	80	CT	30	50	80
H_{SAT}[a]	5.2	2.5	1.2	0	8.4	8.7	2.8	0
H_{COMP}[a]	11.2	10.8	10.5	9.2	11.7	10.4	11.3	10.0
H_{HIGH}[a]	1.1	1.7	1.4	0.0	0.6	0.4	1.6	0.0
H_{MID}[a]	1.3	1.8	1.9	0.0	0.5	0.8	2.1	0.3
H_{LOW}[a]	1.5	1.7	2.6	3.0	0.5	0.4	2.0	4.9
$H_{COMPDIFF}$[a]	2.0	3.0	3.4	6.2	1.7	0.1	2.8	4.8
P_{MAX}[b]	3.25E-04	3.25E-04	3.25E-04	3.25E-04	5.24E-04	3.92E-04	5.44E-04	4.84E-0
PH_{HIGH}[b]	2.56E-04	2.56E-04	2.56E-04	2.56E-04	4.07E-04	3.16E-04	4.18E-04	—
PH_{MID}[b]	1.90E-04	1.90E-04	1.90E-04	1.90E-04	2.90E-04	2.40E-04	2.91E-04	2.56E-0
PH_{LOW}[b]	1.23E-04	1.23E-04	1.23E-04	1.23E-04	1.73E-04	1.63E-04	1.64E-04	1.41E-0
$PH_{COMPDIFF}$[b]	3.10E-05	3.10E-05	3.10E-05	3.10E-05	1.50E-05	3.00E-05	1.00E-05	1.00E-0
$Resp_{LEAF}$[b]	5.52E-05	5.52E-05	5.52E-05	5.52E-05	5.64E-05	8.73E-05	3.76E-05	2.73E-0
$Resp_{ROOT/RHIZOME}$[b,c]	1.50E-05	1.50E-05	1.50E-05	1.50E-05	3.00E-05	3.00E-05	3.00E-05	3.00E-0
Leaf Biomass[d]	37.20	37.20	37.20	37.20	3.59	1.20	1.14	0.49
Root+Rhizome Biomass[d]	82.14	82.14	82.14	82.14	57.93	35.64	32.19	31.60
PQ[e,f]	1.2	1.2	1.2	1.2	1.2	1.2	1.2	1.2
RQ[f,g]	1.0	1.0	1.0	1.0	1.0	1.0	1.0	1.0

[a] units are h/day
[b] units are mol O_2/g dwt/h
[c] from Tomasko and Dunton (1991)
[d] units are g dwt/m^2
[e] units are mol O_2/mol C
[f] from Fourqurean and Zieman (1991a)
[g] units are mol C/mol O_2

at light saturation (I_K) was determined from the intersection of α and P_{MAX}. The leaf light-compensation point (I_C) was determined by dividing the measured respiration rate by α. Another method of determining this value, using the initial slope and setting the y intercept at 0, was investigated and is referred to as I_{CCALC}. The equation for I_{CCALC} is: $X = (0 - Y \text{ intercept value})/\alpha$, where the regression-determined y-intercept value will be negative. A comparison of I_C and I_{CCALC} data using the Wilcoxon Signed Ranks test revealed significant differences in the values obtained for this PE characteristic. Both methods of calculation were used in the statistical analysis and are presented here.

The carbon budget was calculated following Fourqurean and Zieman (1991a), using a photosynthetic quotient (PQ) of 1.2 and respiratory quotient (RQ) of 1. The PE responses measured were normalized to biomass. I assumed a depth of 0.5 m of water over the seagrass bed and used the average daily SI (using an integrated diurnal curve for 2 weeks before and after the initial and final sampling dates) corrected for the water column light attenuation at depth. SI data was measured on a LI-COR light meter (air sensor only). The carbon budget calculations included the H_{SAT} (hours at or above P_{MAX}) and H_{COMP} (hours at or above I_{CPLANT}) values presented in Table 3.1. These values were calculated for Tampa Bay using the average SI transmitted within treatment plots and the extinction coefficients (k) of 1.2 for summer (September) and 0.8 for winter (February/March) from field measurements. Light attenuation by epiphytes for *H. wrightii* averaged 44–56% of PAR in nearby Sarasota Bay and 32–36% for *T. testudinum* in Lower Tampa Bay (Dixon and Leverone 1995). In addition, Dixon and Leverone (1995) approximated irradiance reflectance in Tampa Bay at between 7 and 15%. In the absence of specific measurements in this study, and because epiphytes were removed from the leaf tissue prior to the PE runs, epiphyte attenuation and irradiance reflectance were not accounted for in the carbon budget estimations.

Root and rhizome respiration rates were not measured in this study. However, for calculation purposes, the respiration rate of the nonphotosynthetic tissue (root and rhizome) was assumed to be 15 μmol O_2 g dry wt^{-1} h^{-1} in September and 30 μmol O_2 g dry wt^{-1} h^{-1} in February/March based on values measured at similar temperatures for *H. wrightii* in Laguna Madre, TX (Dunton and Tomasko 1991).

Areal net photosynthesis (Pnet) was determined by subtracting the areal above- and below-ground respiration ($Resp_{PLANT}$) from the sum of areal gross photosynthesis (Total Pgross). Net photosynthesis between H_{SAT} and H_{COMP} was determined by integrating the gross photosynthesis beneath the daily irradiance curve at five intervals: H_{SAT}, H_{HIGH} , H_{MID}, H_{LOW}, and $H_{COMPDIFF}$. H_{HIGH}, H_{MID}, H_{LOW}, and $H_{COMPDIFF}$ include only those morning and evening hours between H_{COMP} and H_{SAT}. Productivity was based upon the measured PE responses normalized to the tissue biomass and areal density for each treatment.

The coefficients are presented in Table 3.1 and the equations are as follows:

Total Pgross = Pgross at H_{SAT} + Pgross at H_{HIGH} + Pgross at H_{MID} + Pgross at H_{LOW} + Pgross at $H_{COMPDIFF}$

Pgross at H_{SAT} = [(Avg. mol O^2 h^{-1} at P_{MAX} × 1/PQ) + (|mol O^2 h^{-1} in $Resp_{LEAF}$| × RQ] × 12 g C/mol C × leaf biomass in g m^{-2} × h at H_{SAT}/day

Pgross at H_{HIGH} = [(Avg. mol O^2 h^{-1} at P_{HHIGH} × 1/PQ) + (|mol O^2 h^{-1} in $Resp_{LEAF}$| × RQ] × 12 g C/mol C × leaf biomass in g m^{-2}× h at H_{HIGH}/day

Pgross at H_{MID} = [(Avg. mol O^2 h^{-1} at P_{HMID} × PQ) + (|mol O^2 h^{-1} in $Resp_{LEAF}$| × RQ] × 12 g C/mol C × leaf biomass in g m^{-2} × h at H_{MID}/day

Pgross at H_{LOW} = [(Avg. mol O^2 h^{-1} at P_{HLOW} × PQ) + (|mol O^2 h^{-1} in $Resp_{LEAF}$| × RQ] × 12 g C/mol C × leaf biomass in g m^{-2} × h at H_{LOW}/day

Pgross at $H_{COMPDIFF}$ = [(Avg. mol O^2 h^{-1} at $P_{HCOMPDIFF}$ × PQ) + (|mol O^2 h^{-1} in $Resp_{LEAF}$| × RQ] × 12 g C/mol C × leaf biomass in g m^{-2} × h at $H_{COMPDIFF}$/day

$Resp_{LEAF}$ = (|mol O^2 h^{-1} in respiration| × RQ) × 12 g C/mol C × green biomass in g m^{-2} × 24 h/day

$Resp_{ROOT/RHIZOME}$ = (|mol O^2 h^{-1} in respiration| × RQ) × 12 g C/mol C × nongreen biomass in g m^{-2} × 24 h/day

$Resp_{PLANT}$ = $Resp_{LEAF}$ + $Resp_{ROOT/RHIZOME}$

Pnet = (Total Pgross – |$Resp_{PLANT}$|)

RESULTS

MORPHOLOGY

Average biomass and shoot numbers in the control plots decreased significantly from the beginning to the end of the experiment due to winter leaf kill (Figures 3.2 and 3.3). Treatment plots had

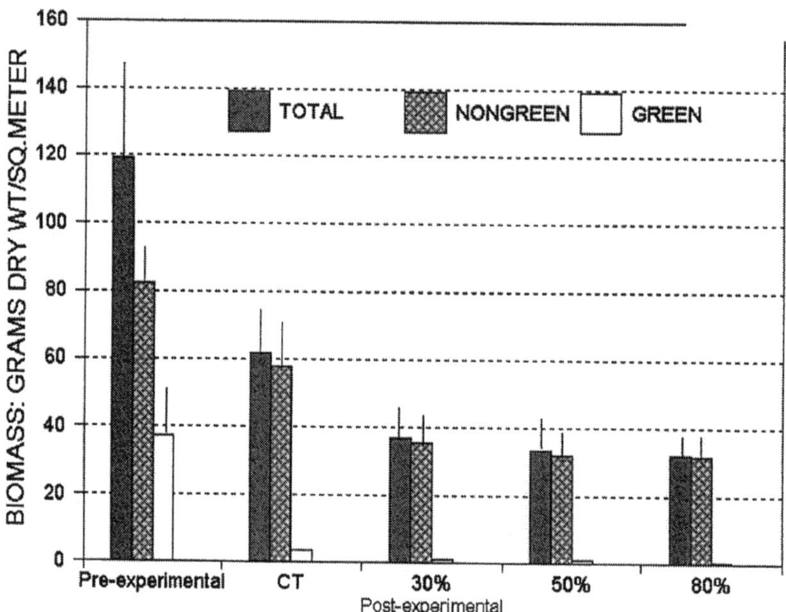

FIGURE 3.2 The mean biomass (g dry wt/m²) from pre- and post-experimental cores for treatments (30, 50, and 80% light reduction) and control plots. Values are taken from four core samples adjacent to (pre-experimental) or within (post-experimental) each plot. Error bars represent one standard error of the mean (pre-experimental n = 64, post-experimental n = 8 per treatment or control).

approximately half the biomass as controls (Figure 3.2) post-experimentally, indicating a treatment effect as well. Green biomass was approximately two thirds lower than biomass in controls, and 80% light-reduction treatments had just under half as much green tissue as other treatments and one seventh that of controls (Figure 3.2). A relationship ($r^2 = 0.89$) between light reduction and green biomass loss was also found.

Pre-investigation (September) total biomass for *H. wrightii* in this study (50.56–264.25 g dry wt m^{-2}) is within the range of values reported from Florida's east coast (10–300 g dry wt m^{-2}) (Virnstein 1982; Pulich 1985). The post-investigation (February/March) total biomass values for controls of 32.19–88.60 g dry wt m^{-2} also fall within the expected range, although they are lower than the September figures, and reflect the expected winter die-back of *H. wrightii* in Tampa Bay (Phillips 1960).

Average leaf biomass in this study was 3.59 (February/March) to 37.20 (September) g dry wt m^{-2} and root/rhizome biomass ranged from 57.93 (February/March) to 82.14 (September) g dry wt m^{-2}. This is within the range of the previously reported values for the east coast of Florida by Virnstein (1982) (5–54 and 150 g dry wt m^{-2} for leaves, and 10–200 and 290 g dry wt m^{-2} for roots and rhizomes) and similar to that reported for Tampa Bay (38–50 and 60–140 g dry wt m^{-2}, for leaf and root/rhizome biomass, respectively) (Lewis and Phillips 1980). The seasonal difference in biomass indicates that photosynthetic biomass in *H. wrightii* from this portion of Tampa Bay comprises half the total biomass in late summer and less than one tenth the total biomass in winter.

Shoot number also showed a pre-/post-experimental effect with a reduction from an average summer density of 3360 shoots m^{-2} to an average winter density of 1846 shoots m^{-2} (control plots). A treatment effect was also found with post-investigation average shoot densities of approximately 66, 50, and 37% that of controls in the 30, 50, and 80% light-reduction plots, respectively (Figure 3.3). An inverse relationship ($r^2 = 0.98$) between percent light reduction and shoot number was found.

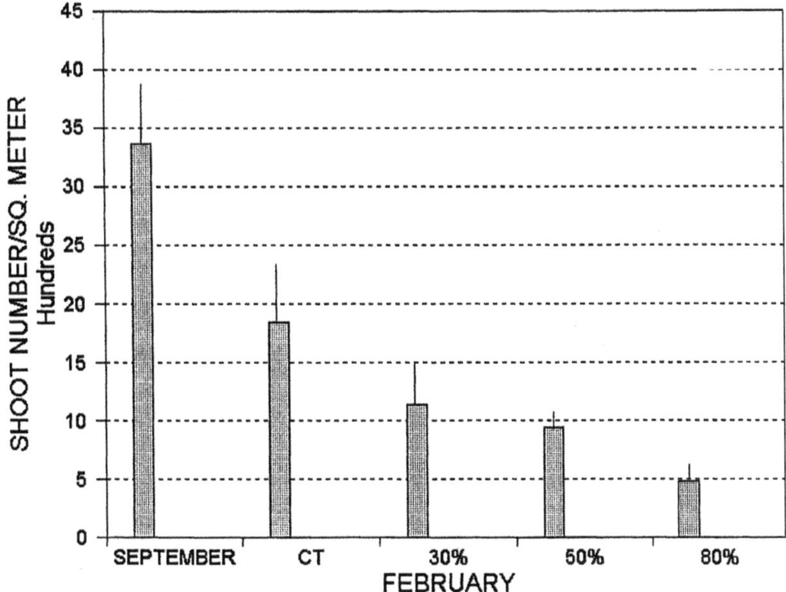

FIGURE 3.3 The mean shoot number (shoot number/m^2) from pre- and post-experimental cores for treatments (30, 50, and 80% light reduction) and control plot biomass cores. Values are taken from four core samples adjacent to (pre-experimental) or within (post-experimental) each plot and adjusted for shoot loss from other sampling. Error bars represent one standard error of the mean (pre-experimental n = 64, post-experimental n = 8 per treatment or control).

When tissue-type biomass is normalized to the shoot number for each treatment, an estimate of the weight of each shoot can be made (Table 3.2). After 6 months of shading, all treatments have green-biomass per shoot similar to the control, indicating a loss of shoots with light reduction rather than reallocation of tissue type. Therefore, *H. wrightii* does not generate fewer, larger shoots in association with light reduction. The 30 and 50% treatments have similar nongreen- and total-biomass per shoot to the controls, but the nongreen- (and thereby the total-) biomass per shoot for the 80% light-reduction treatment is nearly twice that of the control and the other treatments. There were fewer shoots, but similar below-ground biomass at high light reduction. This result warrants further investigation and may indicate a minimum amount of below-ground reserves that *H. wrightii*

TABLE 3.2
The biomass to shoot number ratio for September (pre-experimental) and February/March (post-experimental) cores in treatment (30, 50, and 80% light-reduction) and control plots.

	September	February/March			
Ratio	Control	Control	30% Reduction	50% Reduction	80% Reduction
Total-biomass	0.0355	0.0333	0.0324	0.0354	0.0663
Green-biomass	0.0111	0.0019	0.0011	0.0012	0.0010
Non-green	0.0244	0.0314	0.0313	0.0342	0.0653

Note: The values are taken from four core samples adjacent to (pre-experimental) or within (post-experimental) each plot.

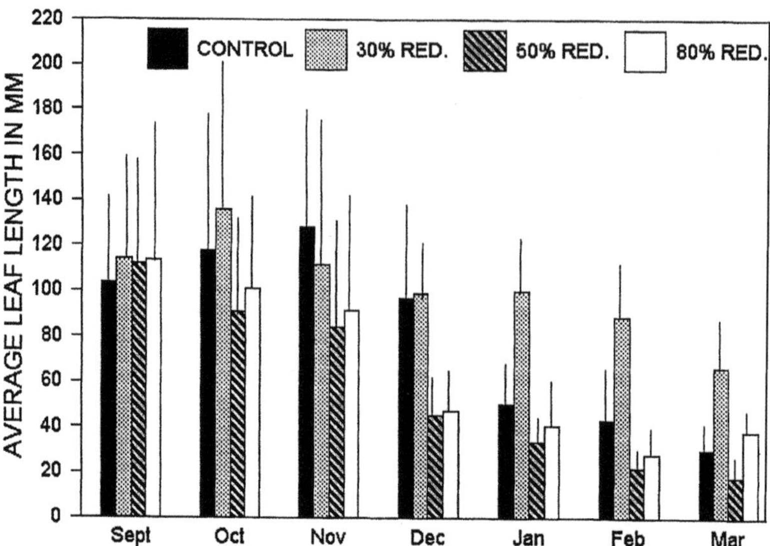

FIGURE 3.4 The mean monthly leaf length from permanent 0.1m² quadrats located within plots for treatments (30, 50, and 80% light reduction) and control plots. Error bars represent one standard error of the mean.

must maintain in order to rebound after winter or during periods of low light, or a maximum biomass the remaining blades could support.

The mean monthly leaf lengths in the permanent quadrat within each plot for the duration of the experiment are presented in Figure 3.4. Leaf length was shorter among plots with 50 and 80% light reduction compared to control plots. However, longer leaf lengths were found in the 30% light-reduction plots than in either the control or other treatment plots for all but one monthly sampling event (November). This difference was especially noticeable from January through March. Based on the Kruskal–Wallis analysis, no significant differences in leaf lengths existed between treatments in any of the permanent quadrats at the start of the investigation (September measurements). Significant differences between treatments were evident from October through the end of the project. Controls were significantly different from all treatments in October and November. In December through February, leaf lengths in the control plots differed from leaf lengths in other treatments, and leaf lengths in the 30% light-reduction treatment were significantly different from leaf lengths in the 50 and 80% treatments. At the conclusion of the experiment, leaf lengths in all control and treatment plots differed signficantly from each other, and only a few shoots remained in some permanent quadrats.

It should be noted that the small number of measurements within the permanent 0.1m² quadrats late in the study may mask the clear trend of shorter leaves and fewer shoots with greater than 50% light reduction observed in the field. Shoot number within the permanent quadrats generally decreased among all treatments and controls over the course of the experiment, and fewer shoots were found in treatment quadrats than in control quadrats by the end of the investigation (Figure 3.5).

PE Response

PE responses for all treatments were separated by sampling event (15 events) and normalized to both chlorophyll content and dry weight biomass. Results were highly variable for both data sets. Although at least one comparison for every PE response was found to be significantly different among treatments, the PE responses generally did not vary in association with light reduction. Where significant differences were found, the differences occurred between controls and treatment plots and/or 30% light-reduction treatment plots vs. 50 and 80% light-reduction treatment plots.

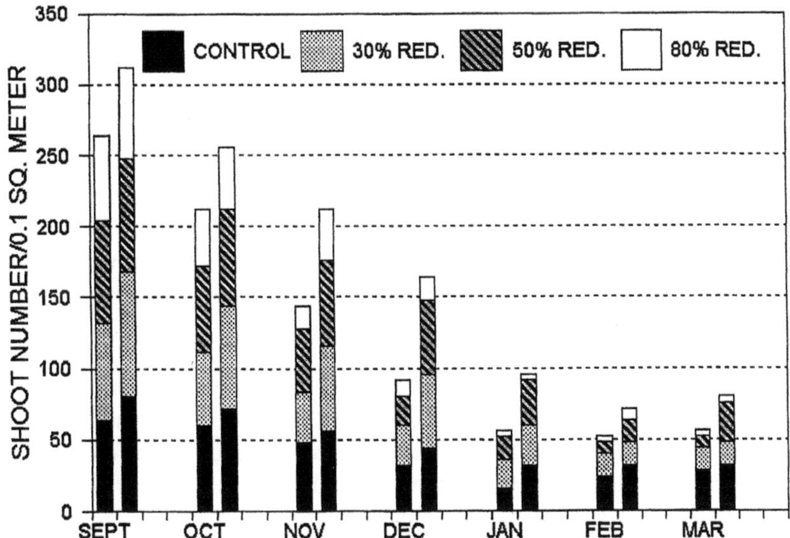

FIGURE 3.5 The actual shoot number from $0.1 m^2$ permanent quadrats located within plots for treatments (30, 50, and 80% light reduction) and control plots. Duplicate light-reduction plots were used, therefore, two bars are used to represent the shoot number from each fixed quadrat monthly.

PE response values were comparable to those observed by Dunton and Tomasko (1991) for *H. wrightii* in Laguna Madre, except for winter P_{MAX} values about half those found in Texas.

Chlorophyll content was also variable at each light level from September through early December, then decreased to a minimum in late January, followed by an increase to levels comparable to September and October (Figure 3.6). Again, there was no pattern in significant differences between treatments and controls as a result of light reduction.

I_C and I_{CCALC} ranged from 15 to 70 $\mu Em^{-2}s^{-1}$ and 5 to 50 $\mu Em^{-2}s^{-1}$, respectively. I_C, based upon the actual measured respiration rate of the leaf section, was usually higher and would indicate a higher light requirement for *H. wrightii* than the theoretical I_{CCALC}.

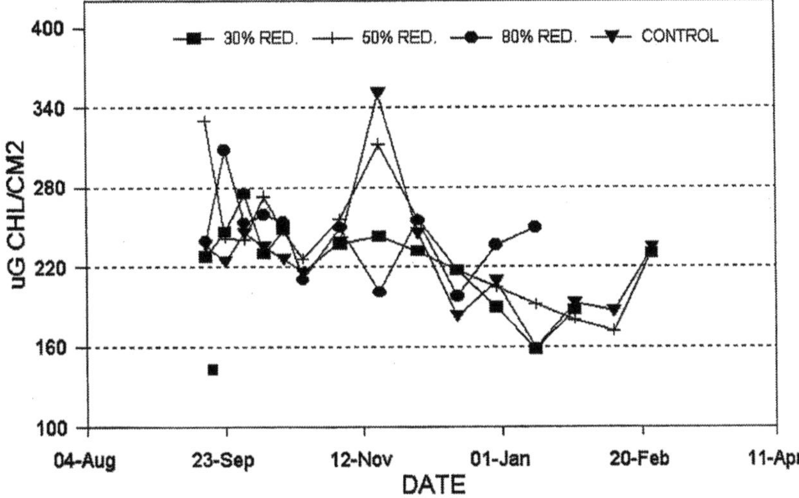

FIGURE 3.6 The chlorophyll *a* content from treatment (30, 50, and 80% light reduction) and control plots by sampling date. Each point is the average of 3 or 4 replicates.

Goldsborough and Kemp (1988) found significant differences in PE characteristics of shaded plots of *Potamogeton perfoliatus* in as little as 3–17 d. In examining the PE responses for the initial sampling dates in this experiment (September 16–October 22), there was still no change in PE response as a result of light reduction.

A general depression of all PE responses except I_K was noted from December through the middle of February, which coincided with periods of low temperature. This would indicate a suppression of photosynthesis due to low temperature, which may have masked any treatment effect.

The areal net productivity of unshaded *H. wrightii* was calculated to be 0.25 g C $m^{-2}d^{-1}$ in September. In February/March, the control plots exhibited a carbon deficit with an average carbon loss of –0.36 g C $m^{-2}d^{-1}$. This carbon deficit in winter would explain the seasonal loss of biomass even in control plots (Figure 3.2).

The September areal net productivity in the 30% light-reduction treatment plot was 0.025 g C $m^{-2}d^{-1}$. The plants in the 50 and 80% light-reduction treatment plots were functioning at a carbon deficit even in September with carbon budgets of –0.12 and –0.51 g C $m^{-2}d^{-1}$, respectively. The February/March carbon budgets for the treatments were not significantly different from the controls, indicating that treatment plots were functioning at a carbon deficit as well (–0.28 g C $m^{-2}d^{-1}$ for 30% light-reduction plots, –0.25 g C $m^{-2}d^{-1}$ for 50% light-reduction plots, and –0.27 g C $m^{-2}d^{-1}$ for 80% light-reduction plots). The slightly lower carbon loss value for the plants in the 50% light-reduction treatment is a result of slightly higher average P_{MAX} values.

Average areal productivity rates reported for *H. wrightii* from North Carolina ranged from 0.5–2 g C $m^{-2}d^{-1}$ (Zieman and Zieman after Dillon 1971 (1989)) using C^{14} uptake measurements. Areal production along Florida's east coast, using leaf clipping and *in situ* photography, was estimated at 1.3 g C $m^{-2}d^{-1}$ (Virnstein 1982). These studies did not account for the below-ground tissue respiratory demand known to meet or exceed the leaf respiration rate during some portions of the year (Dennison 1987; Dunton and Tomasko 1991; Fourqurean and Zieman 1991a,b), or below-ground growth, respectively (Virnstein 1982). In addition, the biomass reported in these studies was at least twice that observed in the present work.

Kemp et al. (1986) found that both C^{14} and oxygen evolution measurements were acceptable techniques for predicting productivity. This was substantiated by Tomasko and Dunton (1995), who found similar predictions of areal productivity using C^{14} and oxygen evolution *in situ* within chambers, but found that leaf clipping measurement underestimated production. The plant responds adversely to leaf clipping in two ways: (1) by mobilizing below ground reserves, thus altering the constituents of new tissue, and (2) by stunting new tissue growth.

DISCUSSION

MORPHOLOGY

The observed longer leaf lengths for 30% light reduction may be a morphological response to lower light as reported for other species (Dennison and Alberte 1982; Dennison and Alberte 1986; Hall et al. 1991). Some findings regarding *T. testudinum* suggest that blade width is evidence of sexual dimorphism (Durako and Moffler 1985), but more recent research indicates photoacclimation is also a cause (Carlson and Acker 1985; Hall et al. 1991). In contrast, blade width did not vary in *Posidonia sinuosa* (Gordon et al. 1994) when exposed to reduced light, and Dunton (1994) and Czerny and Dunton (1995) found blade width did not vary with light attenuation in *H. wrightii* off Texas.

If the observed increased blade length is a morphological response to light reduction, then a critical range might exist wherein leaf length would increase under light reduction up to a point beyond which this adaptation is not efficient for the plant. This would explain leaf elongation in the 30% light-reduction plots and not in the 50 and 80% plots.

Dunton (1994) monitored ambient underwater PAR and leaf elongation rates (among other parameters) of *H. wrightii* in several areas of Laguna Madre and found a clear circannual pattern in *H. wrightii* leaf elongation related to shoot production. Leaf elongation was unrelated to underwater PAR levels. He speculated that an external factor such as daylength or temperature might actually control leaf elongation in *H. wrightii*. Further investigation of leaf elongation by *H. wrightii* in response to light reduction of up to 50% ambient light at depth is warranted.

The leaf number per shoot by treatment for each sampling event was compared using the K-W analysis. Two of the fifteen sampling events showed significant differences among treatments. No trend in significant differences was found, indicating that *H. wrightii* was not generating more leaves per shoot as a result of light reduction.

The experimental treatments showed that light reduction caused loss of both photosynthetic and heterotrophic tissue (up to a point), but that loss of the former was most severe and most important to the primary production potential for the plant and community structure. Based upon my results, a 66% loss of photosynthetic tissue (despite longer leaves) and 33% reduction in shoot density would occur if light attenuation were increased by just 30% over the levels observed during this study during the late summer, fall, and winter.

PE Response

PE response trends appear to be a result of seasonality, not treatment effect. There was no effect of 30, 50, and 80% light reduction on PE responses or chlorophyll content in *H. wrightii* leaf tissue. Thus, it seems that physiological photoacclimation was not a strategy used by *H. wrightii* plants for adjusting to *in situ* light reduction.

This finding is consistent with another study of PE responses for *Zostera marina* (Dennison and Alberte 1982), which indicated morphological responses, but no significant differences in PE response with *in situ* shading. Other studies have reported variability in PE responses in other seagrass species where statistical calculations could not detect significant differences. There was no difference in PE responses between shallow and deep water individuals of *T. testudinum* in Tampa Bay (Dixon and Leverone 1995). Transplanted *Z. marina* did exhibit changes in PE responses as a result of increased or decreased light attenuation, but not to a significant degree (Dennison and Alberte 1986). P_{MAX} values were lower or higher under decreasing light due to depth (Dawes and Tomasko 1988) and also whether P_{MAX} was normalized to chlorophyll or dry-weight biomass (Dennison and Alberte 1982). Alpha decreased with light reduction due to depth (Tomasko 1993) and increased with light reduction due to shade (Goldsborough and Kemp 1988).

This variability in measurements from the literature may be a function of differences in methodology, including using whole shoots vs. shoot sections and different acclimation regimes prior to analysis. Some of the difference in PE responses may also be due to high variability within seagrass beds, including sampling different plants, different aged leaf blades (Mazella and Alberte 1986; Durako and Kuss 1994), or different portions of the leaf blade (Mazella and Alberte 1986).

The highest net productivity rate observed (0.252 g C $m^{-2}d^{-1}$) was far less than previously reported for the east coast of Florida and North Carolina populations. The calculations in the present work did account for estimated below-ground respiratory demands and utilized the H_{SAT} model, which was found to be the best estimator of primary production rates of seagrass (Tomasko and Dunton 1995). Williams and McRoy (1982) measured net productivity in *H. wrightii* (from Laguna Madre, Cuba, and Florida Bay) using C^{14} on both above- and below-ground tissue, obtaining results comparable to those in my study (when broken down to an hourly rate). However, Williams and McRoy (1982) did not measure productivity on a daily basis; there was variability among measurements, and the estimates are only for spring and summer.

Tomasko and Dunton (1995) found production rates of 0.04-0.49 mol C $m^{-2}d^{-1}$ (0.48-5.88 g C $m^{-2}d^{-1}$) in *H. wrightii* for winter and spring, respectively, using the H_{SAT} model and accounting for below-ground respiration. It is curious that Tomasko and Dunton (1995) did not observe a carbon

TABLE 3.3
The mean surface irradiance and ambient irradiance at depth in treatment plots (correcting for water column attenuation and light-reduction treatment) for September 1994 and February/March 1995.

Treatment	September (μE $m^{-2}s^{-1}$)	February/March (μE $m^{-2}s^{-1}$)
Mean Surface Irradiance	775	850
Control	400	530
30% Light Reduction	250	300
50% Light Reduction	200	250
80% Light Reduction	80	70

deficit during the winter despite differences in seasonal biomass. This disparity, again, may be explained by the higher biomass (200–500 g dry wt m^{-2}) in Texas, which was more than four times the biomass found in my study plots. The February/March carbon budget in my study was calculated from measurements made on *H. wrightii* at low temperatures (low metabolic rate), which may bias the results toward an unusually low daily average. In addition, Tomasko and Dunton (1995) used spring productivity in their calculations because it exceeded summer productivity. A study period of a year or more may reconcile these differences between the carbon budget estimate here and values from other studies.

The negative carbon budget for the control plots in February/March would account for the significant seasonal loss of biomass. In addition, the negative carbon budget in the 30, 50, and 80% light-reduction treatments throughout the study would account for the significant biomass loss as a result of treatment effect.

Average I_{CPLANT} values ranged from 30 to 100 μE $m^{-2}s^{-1}$, and averaged 39 to 62 μE $m^{-2}s^{-1}$. This translates to a minimum instantaneous light requirement of 4.5–7% of September and February/March SI for *H. wrightii* in Tampa Bay (Table 3.3). In waters with an extinction coefficient of 1.2, this corresponds to approximately 2.2–2.5 m of water depth.

Given these approximate irradiances (which account only for the water column extinction coefficient and the individual shade treatment), the leaves of *H. wrightii* in even the 80% light reduction plots should have received irradiances at, or above, the I_{CPLANT} (H_{COMP} level in the model) for more than 10 hours per day. However, the biomass portion of my study indicates that insufficient light was present to maintain a positive carbon budget for the plant.

Other sources of attenuation not accounted for in my study would decrease the calculated irradiance received by the leaves within the treatments. Using the average value of 47% attenuation of PAR at depth by epiphytes on *H. wrightii* for Sarasota Bay (Dixon and Leverone 1995) and a 10% surface reflectance correction term, the average irradiance within the 50 and 80% treatments is equivalent to, or less than, the range of I_{CPLANT} values for February/March. However, the 30% treatments would still have received sufficient light for maintenance using these correction factors.

Correcting for attenuation by epiphytes and 10% surface reflectance in the carbon budget calculations, the September carbon budget for the 30% treatment is at a deficit. However, the September carbon budget for the control plot also reflects a slight carbon deficit. This discrepancy may indicate an error in the carbon budget assumptions, e.g., using the below-ground tissue respiration rates reported by Dunton and Tomasko (1991).

The results of this experiment indicate that *H. wrightii* in this region of Tampa Bay responds to light reduction of 30, 50, or 80% during the late summer, fall, and winter via leaf lengthening and biomass loss, rather than through modifying PE responses or increasing chlorophyll content. Altered leaf length and shoot number were the most characteristic responses to light reduction and quite easily discernible to the casual observer. The consequential reduction in seagrass primary productivity, brought about by a loss of photosynthetic tissue without apparent increase in photosynthetic efficiency or capacity, is clearly a deleterious effect. This results in a proportional loss to the primary productivity of the seagrass bed and ecosystem.

REFERENCES

Carlson, P. R., Jr. and J. G. Acker. 1985. Effects of *in situ* shading on *T. testudinum*: preliminary experiments. In *Proceedings of the 12th Annual Conference on Wetlands Restoration and Creation*, F. J. Webb, Jr. (ed.). Hillsborough Community College Environmental Studies Center, Tampa, FL, pp. 64–73.

Czerny, A. B. and K. H. Dunton. 1995. The effects of *in situ* light reduction on the growth of two subtropical seagrasses, *Thalassia testudinum* and *Halodule wrightii*. *Estuaries* 18(2):418–427.

Dawes, C. J. and D. A. Tomasko. 1988. Depth distribution of *Thalassia testudinum* in two meadows on the west coast of Florida: a difference in effect of light availability. *Pubblicazioni della Stazione zoologica di Napoli I: Marine Ecology* 9:123–130.

Dennison, W. C. 1987. Effects of light on seagrass photosynthesis, growth, and depth distribution. *Aquatic Botany* 27:15–26.

Dennison, W. C. and R. S. Alberte. 1982. Photosynthetic responses of *Zostera marina* L. (eelgrass) to *in situ* manipulations of light intensity. *Oecologia* 55:137–144.

Dennison, W. C. and R. S. Alberte. 1986. Photoadaptation and growth of *Zostera marina* L. (eelgrass) transplants along a depth gradient. *Journal of Experimental Marine Biology and Ecology* 98:265–282.

Dixon, L. K. and J. R. Leverone. 1995. Light requirements of *Thalassia testudinum* in Tampa Bay, Florida, final report, Mote Marine Laboratory Technical Report No. 425, Sarasota, FL.

Duarte, C. M. 1991. Seagrass depth limits. *Aquatic Botany* 40:363–377.

Dunton, K. H. 1990. Production ecology of *Ruppia maritima* L. s.l. and *Halodule wrightii* Aschers. in two sub-tropical estuaries. *Journal of Experimental Marine Biology and Ecology* 143:147–164.

Dunton, K. H. 1994. Seasonal growth and biomass of the subtropical seagrass *Halodule wrightii* in relation to continuous measurements of underwater irradiance. *Marine Biology* 120:479–489.

Dunton, K. H. and D. A. Tomasko. 1991. Seasonal variations in the photosynthetic performance of *Halodule wrightii* measured *in situ* in Laguna Madre, Texas. In *The Light Requirements of Seagrasses: Proceedings of a Workshop to Examine the Capability of Water Quality Criteria, Standards, and Monitoring Programs to Protect Seagrasses*, W. J. Kenworthy and D. E. Haunert (eds.). NOAA Technical Memorandum, NMFS-SEFC-287, pp. 71–79.

Durako, M. J. and M. D. Moffler. 1985. Observations on the reproductive ecology of *Thalassia testudinum* (Hydrocharitaceae) II. Leaf width as a secondary sex character. *Aquatic Botany* 21:265–275.

Durako, M. J. and K. M. Kuss. 1994. Effects of *Labyrinthula* infection on the photosynthetic capacity of *Thalassia testudinum*. *Bulletin of Marine Science* 54(3):727–732.

Durako, M., W. J. Kenworthy, S. M. R. Fatemy, H. Valavi, and G. Thayer. 1993. Assessment of the toxicity of Kuwait crude oil on the photosynthesis and respiration of seagrasses of the northern Gulf. *Marine Pollution Bulletin* 27:223–227.

Fourqurean, J. W. and J. C. Zieman. 1991a. Photosynthesis, respiration and whole plant carbon budget of the seagrass *Thalassia testudinum*. *Marine Ecology Progress Series* 69:161–170.

Fourqurean, J. W. and J. C. Zieman. 1991b. Photosynthesis, respiration and whole plant carbon budgets of *Thalassia testudinum, Halodule wrightii,* and *Syringodium filiforme*. In *The Light Requirements of Seagrasses: Proceedings of a Workshop to Examine the Capability of Water Quality Criteria, Standards, and Monitoring Programs to Protect Seagrasses*, W. J. Kenworthy and D. E. Haunert (eds.). NOAA Technical Memorandum, NMFS-SEFC-287.

Goldsborough, W. J. and W. M. Kemp. 1988. Light responses of a submersed macrophyte: implications for survival in turbid waters. *Ecology* 69(6):1775–1786.

Gordon, D. M., K. A. Grey, S. C. Chase, and C. J. Simpson. 1994. Changes to the structure and productivity of a *Posidonia sinuosa* meadow during and after imposed shading. *Aquatic Botany* 47:265–275.

Hall, M. O., D. A. Tomasko, and F. X. Courtney. 1991. Responses of *Thalassia testudinum* to *in situ* light reduction. In *The Light Requirements of Seagrasses: Proceedings of a Workshop to Examine the Capability of Water Quality Criteria, Standards, and Monitoring Programs to Protect Seagrasses*, W. J. Kenworthy and D. E. Haunert (eds.). NOAA Technical Memorandum, NMFS-SEFC-287, pp. 85–94.

Holm Hansen, O. and B. Reimann. 1978. Chlorophyll a determination: improvements in methodology. *Oikos* 30:438–447.

Iverson, R. L. and H. F. Bittaker. 1986. Seagrass distribution in the eastern Gulf of Mexico. *Estuarine Coastal Shelf Science* 22:577–602.

Kemp, W. M., M. R. Lewis, and T. W. Jones. 1986. Comparison of methods for measuring production by the submersed macrophyte, *Potamogeton perfoliatus* L. *Limnology and Oceanography* 31:1322–1334.

Kenworthy, W. J. and D. E. Haunert (eds.). 1991. *The Light Requirements of Seagrasses: Proceedings of a Workshop to Examine the Capability of Water Quality Criteria, Standards, and Monitoring Programs to Protect Seagrasses.* NOAA Technical Memorandum, NMFS-SEFC-287.

Kenworthy, W. J., M. S. Fonseca, and S. J. DiPiero. 1991. Defining the ecological light compensation point for seagrasses *Halodule wrightii* and *Syringodium filiforme* from long-term submarine light regime monitoring in the southern Indian River. In *The Light Requirements of Seagrasses: Proceedings of a Workshop to Examine the Capability of Water Quality Criteria, Standards, and Monitoring Programs to Protect Seagrasses,* W. J. Kenworthy and D. E. Haunert (eds.). NOAA Technical Memorandum, NMFS-SEFC-287, pp. 106–113.

Lewis, R. R. and R. C. Phillips. 1980. Seagrass mapping project, Hillsborough County, Florida. Tampa Port Authority, Tampa, FL.

Lewis, R. R., M. J. Durako, M. D. Moffler, and R. C. Phillips. 1985. Seagrass meadows of Tampa Bay: a review. In *Proceedings of the Tampa Bay Area Scientific Information Symposium,* S. F. Treat, J. L. Simon, R. R. Lewis, III, and R. L. Whitman, Jr. (eds.). Florida Sea Grant Project No. IR/82-2, Report No. 65, pp. 221–246.

Mazella, L. and R. S. Alberte. 1986. Light adaptation and the role of autotrophic epiphytes in primary production of the temperate seagrass, *Zostera marina* L. *Journal of Experimental Marine Biology and Ecology* 100:165–180.

Morris, L. J. and D. A. Tomasko (eds.). 1993. *Proceedings and Conclusions of Workshops on Submerged Aquatic Vegetation Initiative and Photosynthetically Active Radiation.* Special publication of the St. Johns River Water Management District, SJ93-SP13, Palatka, FL.

Neverauskus, V. P. 1988. Response of a *posidonia* community to prolonged reduction in light. *Aquatic Botany* 31:361–366.

Phillips, R. C. 1960. Environmental effects on leaves of *Diplanthera* du Petit-Thomas. *Bulletin of Marine Science* 10:346–353.

Pulich, W. M., Jr. 1985. Seasonal growth dynamics of *Ruppia maritima* and *Halodule wrightii* Aschers. in southern Texas and evaluation of sediment fertility status. *Aquatic Botany* 23:53–66.

Tomasko, D. A. 1993. Physiological measures of seagrass health. In *Proceedings and Conclusions of Workshops on Submerged Aquatic Vegetation Initiative and Photosynthetically Active Radiation,* L. J. Morris and D. A. Tomasko (eds.). Special publication of the St. Johns River Water Management District, SJ93-SP13, Palatka, FL, pp. 55–60.

Tomasko, D. A. and K. H. Dunton. 1991. Growth and production of *Halodule wrightii* in relation to continuous measurements of underwater light levels in South Texas. In *The Light Requirements of Seagrasses: Proceedings of a Workshop to Examine the Capability of Water Quality Criteria, Standards, and Monitoring Programs to Protect Seagrasses,* W. J. Kenworthy and D. E. Haunert (eds.). NOAA Technical Memorandum, NMFS-SEFC-287, pp. 79–84.

Tomasko, D. A. and K. H. Dunton. 1995. Primary productivity in *Halodule wrightii*: a comparison of techniques based on daily carbon budgets. *Estuaries* 18(18):271–278.

Virnstein, R. W. 1982 Leaf growth rate of the seagrass *Halodule wrightii* photographically measured *in situ.* *Aquatic Botany* 12:209–218.

West, R. J. 1990. Depth-related structural and morphological variations in an Australian *posidonia* seagrass bed. *Aquatic Botany* 6:153–166.

Williams, S. L. and C. P. McRoy. 1982. Seagrass productivity: the effects of light on carbon uptake. *Aquatic Botany* 12:321–344.

Williams, S. L. and W. C. Dennison. 1990. Light availability and diurnal growth of a green macroalga (*Caulerpa cupressoides*) and a seagrass (*Halophila decipiens*). *Marine Biology* 106:437–443.

Zieman, J. C. and R. T. Zieman. 1989. The ecology of the seagrass meadows of the West Coast of Florida: a community profile. U.S. Fish and Wildlife Service Biological Report No. 85(7.25).

4 The Effects of Dock Height on Light Irradiance (PAR) and Seagrass (*Halodule wrightii* and *Syringodium filiforme*) Cover

Jeffrey L. Beal and Brandon S. Schmit

Abstract The impacts of structures over seagrasses are not clearly understood. Construction parameters such as height and orientation regulate the total amount of solar radiation received within the area shadowed by over-water structures. The effects of dock height and piling presence on *Halodule wrightii* and *Syringodium filiforme* were examined for 12 m transects at three sites at Indian River Lagoon, Florida. Seagrass cover, shoot density, and photosynthetically active radiation (PAR) were measured *in situ* over a 12-month period for treatments of pilings without decking, 0.91 m docks (above Mean High Water) and 1.52 m docks. Treatment and control transects showed post-construction decreases in cover and shoot density, partially because of an algal bloom produced by an unusually wet winter. Average percent decreases between 14 weeks pre-construction and 33 weeks post-construction showed predicted trends: cover at controls declined 12.3%; cover at pilings declined 21.3%; cover at 1.51 m docks declined 23.8%; and cover at 0.91 m docks declined 28.4%. Shoot counts nearest structures decreased 56% whereas shoot counts farthest from the structures decreased 40%. Analysis of seagrass change along treatment transects reveals a trend toward decreases nearest the docks and pilings, showing the greatest average treatment impacts closest to the structures. Monthly light data collected at the seagrass bed edge at the three sites predicted minimum seagrass light requirements of 33.3, 28.6, and 26.7% I_0. Daylong PAR surveys showed irradiance values for 0.91 and 1.51 m docks as significantly lower than control and pilings values in peak growing season (July). Current trends and PAR measurements were used to predict the decline and perhaps demise of the seagrasses directly influenced by dock shading.

INTRODUCTION

The importance of seagrasses as primary producers (500–1000 g C m^{-2} yr^{-1}) in coastal ecosystems is well documented (Wood et al. 1969; Zieman and Wetzel 1980). Seagrasses baffle and stabilize sediments, fix nutrients, and provide critical habitat for many organisms, including the Florida manatee (*Trichechus manatus latirostris*). Strong evidence suggests that light availability (i.e., quality and quantity of photosynthetically active radiation (PAR); 400–700 nm) is the major factor governing seagrass distribution and productivity (see Kenworthy 1994 for review). Current research has demonstrated the usefulness of determining minimum light requirements (and maximum depth of growth (Z_{max})) of submerged aquatic vegetation (SAV) for ecological study and management purposes, including conservation and restoration (Dennison 1987; Fonseca et al. 1987; Kenworthy and Haunert 1991; Kenworthy 1994). For example, correlative relationships between incident (surface) light (I_0: given as average annual percent) and Z_{max} can be readily established for given

0-8493-2045-3/99/$0.00+$.50
© 2000 by CRC Press LLC

species and regions (Gallegos and Kenworthy 1996; Kenworthy and Fonseca 1996). Based on ecological requirement studies, optical water quality models and state and federal water transparency standards can be updated based on scientifically defensible criteria. Generally, analyses for seagrass growth criteria have been developed via two approaches: correspondence analysis (Dennison et al. 1993) and optical modeling (Gallegos and Kenworthy 1996).

Urbanization, industrialization, and agriculturization of coastal areas have contributed greatly to the degradation of water quality and subsequent loss of seagrasses via dredge and fill activities, alteration of natural water delivery from upland watersheds, increased sediment and pollutant loads, watercraft operation, and coastal construction, especially over submerged lands (Livingston 1987; Kenworthy et al. 1989; Kenworthy and Haunert 1991; Sargent et al. 1995). The cumulative impacts of the latter causative agent are not fully understood. This study was designed to quantify the potential differential effects of dock elevation (0.91 and 1.51 m above Mean High Water (MHW)) on light attenuation and seagrass (*Halodule wrightii* and *Syringodium filiforme)* abundance (as measured by quadrat percent cover; Neckles 1994; Heidelbaugh and Nelson 1995). Secondarily, this study was designed to test the short-term effects of dock construction and the long-term effects of piling presence on seagrass cover. Standard techniques of ecological compensation depth determinations were used to gain further understanding of seagrass (*H. wrightii* and *S. filiforme*) minimum light requirements (Kruczynski and Flemer 1994) in the humic-stained Indian River Lagoon (Gallegos and Kenworthy 1996).

Study Location

The Indian River Lagoon, along the coast of east central Florida, is the most diverse estuarine system in the continental U.S. *H. wrightii* and *S. filiforme* are the two most abundant seagrasses in the lagoon and occur in mixed beds at intermediate depths. Three study sites with similar features (contiguous seagrass beds, water depths ~0.75 m, adjacent shoreline of mangrove marsh) were chosen along the eastern side of the Indian River Lagoon. Both species were present, with *S. filiforme* predominant. Figure 4.1 shows the three study sites: Bear Point, Middle Point, and Herman's Bay Point. Preliminary sampling was conducted October through December. Haphazard throws of a 1-m^2 quadrat for percent cover in 2-m^2 area determined similarity of seagrass abundance among sites. Pre-construction sampling of transects also revealed no significant difference in percent cover ($p = 0.178$) among sites.

MATERIALS AND METHODS

Experimental Design

In October 1997, four north-south transects (each 12 m long) were established at each of the three sites. A 1-m^2 quadrat (subdivided into a 10 cm × 10 cm grid of squares) was used to quantify seagrass cover in 2-m^2 areas (Heidelbaugh and Nelson 1995) along each transect in a belt fashion. That is, the quadrat was placed at one endpoint and flipped over toward the other endpoint along each transect and both sides of each transect line were sampled. The 12 2-m^2 samples per transect are hereafter referred to as transect points 1–12, south to north. Squares occupied by seagrass were enumerated and recorded at each sample date. Also, shoot counts were recorded in 15 randomly determined squares at transect points 1, 6, and 12 for each sample date. Transects were sampled 14 weeks (January 1998) and 1 week (March 1998) prior to dock construction. After construction in April, transects were sampled at 1, 5, 10, 14, 19, 25, and 33 weeks.

At each site, the four transects were each defined by one treatment: one 2.44 m × 1.22 m dock 1.51 m above MHW, one 2.44 m × 1.22 m dock 0.91 m above MHW, and one partial dock (consisting of four pilings forming a 2.44 m × 1.22 m rectangle), or one control. Docks and partial docks were aligned east-west and bisected the transects just south of center such that transect point 6 was

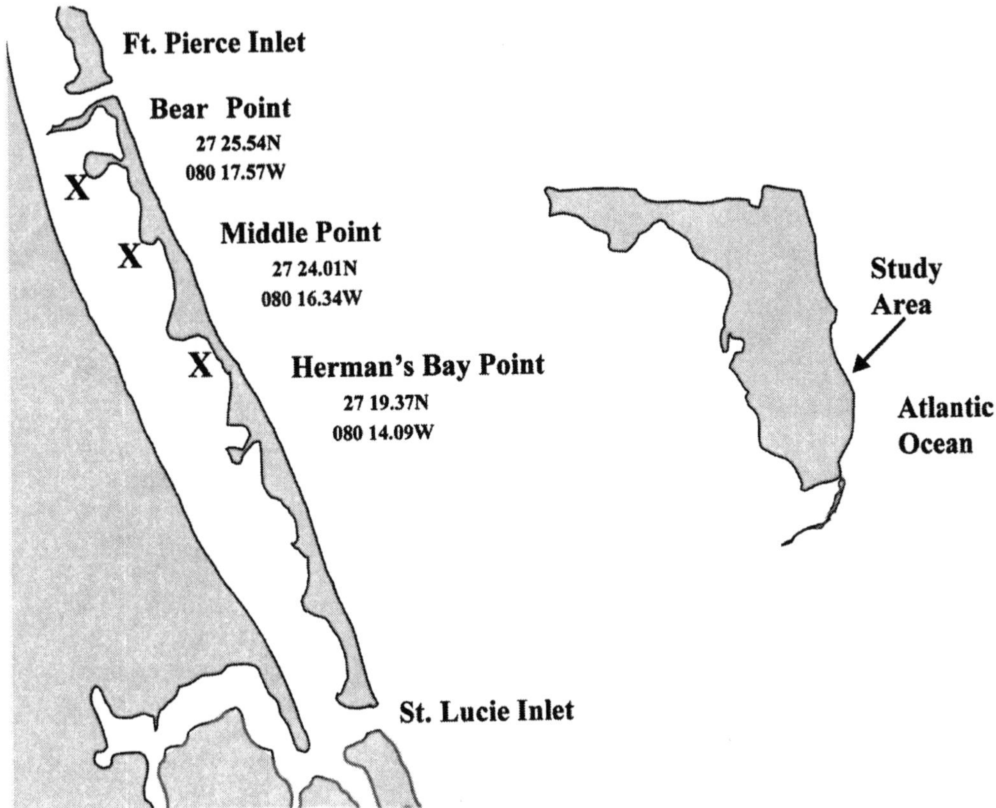

FIGURE 4.1 Map of Indian River Lagoon showing location of study sites in St. Lucie County, Florida.

shaded by decking most of the year. Docks consisted of 20.3 cm decking with 1.27 cm deck spacing. Treatment configurations were established such that no two sites were alike.

Light (PAR) data were collected at the deepest edge of the seagrass bed using LI-COR LI-193SA spherical quantum sensors and a LI-COR datalogger (LI-COR, Inc., Lincoln, NE). Beginning December 1997 at the deepest edge, a submarine light profile (10 cm intervals) of PAR was collected every month (between 10:00 a.m. and 2 p.m.) and used to calculate average annual k (attenuation coefficient found as the slope of the regression, including 10 mo log-transformed light readings on water depth) (*sensu* Kenworthy and Fonseca 1996). Attenuation coefficient, in turn, is used to estimate the minimum light requirement for the two seagrass species at the study site; using the Beer–Lambert equation:

$$I_z = I_o e^{-kz}$$

where I_o is % PAR at 10 cm depth (100%), I_z is % PAR at depth z (average of ten monthly observations for maximum depth) in meters, and k (m^{-1}) is the light attenuation coefficient.

Light (PAR) data were also collected at the dock treatment stations in July and November. Sensors collected PAR data at one site per day on 3 consecutive days. The sampling days showed similar weather (less than 20% chance of rain and winds less than 12 knots). The sensor array consisted of one open air, one control, one partial dock, and two dock (0.91 and 1.51 m) treatments. All sensors except open air were fixed in place at seagrass canopy height (~30 cm above substrate) and in the predominant shadow at dock treatments. Data were collected from sunrise to sunset at 5 s intervals integrated every 30 min. Daylong PAR data were compared to minimum light calculations for each site.

Statistical Analysis

For seagrass cover, a four-way repeated-measures analysis of variance (ANOVA) was used to analyze seagrass abundance with time as the repeated measure, and treatment (0.91 m, 1.51 m, pilings, and control), location, and transect points as the other factors. Shoot count data were analyzed using a four-way repeated-measures ANOVA with time as the repeated measure, and treatment, location, and transect points (1, 6, and 12) as the other factors. For both data sets, a one-way ANOVA was used for specific factors where appropriate. For each factor showing significant ($p < 0.05$) treatment effect, Tukey's multiple comparisons test was performed to determine which means differed significantly ($p < 0.05$). To meet assumptions of ANOVA, data were $\log_{10}(x+1)$ transformed prior to analysis (Sokal and Rohlf 1981). All ANOVAs were analyzed using a balanced design, i.e., equal sample sizes. Data were back transformed for presentation in tables and figures.

PAR data collected at the docks were analyzed for each site using a three-way repeated-measures ANOVA with time as the repeated measure, and treatment (0.91 m, 1.51 m, pilings, and control) and location as the other factors (Winer 1971). When location and time were not significant factors, data were pooled to test the factor of treatment. For each factor showing significant ($p < 0.05$) treatment effect, Tukey's multiple comparisons test was performed to determine which means differed significantly ($p < 0.05$). To meet assumptions of ANOVA, data were $\log_{10}(x + 1)$ transformed prior to analysis (Sokal and Rohlf 1981). All ANOVAs were analyzed using a balanced design, i.e., equal sample sizes. Data were back transformed for presentation in tables and figures.

RESULTS

Seagrass Cover and Shoot Counts

For the present study, the means for 2-m^2 seagrass cover, i.e., number of squares containing seagrass out of a possible 200, prior to dock construction were 196.93, 198.0, and 199.47 for the three sites for the two sampling dates combined. Also, single classification ANOVA revealed that location was not significant for the variate of shoot counts for the two pre-construction sampling dates. Repeated samples of seagrass cover taken between 14 weeks pre-construction and 33 weeks post-construction revealed significant changes excluding the interaction of date and transect points (Table 4.1). All single factors and most interaction terms were significant ($p < 0.001$). The factor of date analyzed alone was significant ($p < 0.001$) and Tukey's multiple comparisons test revealed that most of the variation stemmed from a sharp decline in seagrass coverage on and after the 14 week (July) sample dates (Table 4.2). For all Tukey's tables presented, means whose intervals do not overlap were significantly different. Prior to dock construction and for the first three samplings, date was not a large contributor to variation. The overall decline in seagrass cover (average controls vs. average treatments) is shown in Figure 4.2. Seagrass declined for all controls and treatments over time, but treatment transects revealed the greatest average decrease. The factor of treatment analyzed independently for seagrass cover revealed significance ($p < 0.001$) and the following hierarchy (smallest to greatest decrease): controls > pilings > 0.91 m docks > 1.51 m docks. Compared to 14 weeks pre-construction, the 33 weeks post-construction sample dates showed that controls decreased 12.3% (from 199.2 to 174.8), pilings decreased 21.3% (from 195.9 to 154.3), 1.51 m docks decreased 23.8% (from 199.1 to 151.7), and 0.91 m docks decreased 28.4% (from 198.4 to 142.1).

Transect point seagrass cover analyzed independently was significant ($p < 0.001$) and most of the variability stemmed from the south ends of some transects (Table 4.3). Transect points 1–4 exhibited reduced cover relative to points 5–12 for all sample dates and locations combined. Because location was significant for cover, each location was analyzed separately for the factor of transect

TABLE 4.1
Results of balanced ANOVA of $\log_{10}(x + 1)$ transformed data testing the effects of location, date, treatment, and transect point on seagrass coverage (number of occupied 100 cm^2 squares per 2 m^2).

Source	DF	SS	MS	F Values	P Values
Location	2	1.10051	0.55026	19.82	0.000***
Date	8	6.31776	0.78972	28.44	0.000***
Treatment	3	0.72153	0.24051	8.66	0.000***
Meter	11	1.63138	0.14831	5.34	0.000***
Location · Date	16	2.30398	0.14400	5.19	0.000***
Location · Tr. Point	22	3.99477	0.18158	6.54	0.000***
Location · Treatment	6	3.65864	0.60977	21.96	0.000***
Date · Tr. Point	88	2.34421	0.02664	0.96	0.586
Date · Treatment	24	1.04455	0.04352	1.57	0.040*
Treatment · Tr. Point	33	4.00058	0.12123	4.37	0.000***
Error	1082	30.04226	0.02777		
Total	1295	57.16018			

Note: * $p < 0.05$, *** $p < 0.001$, DF = degrees of freedom, SS = sum of squares, MS = mean squares.

point using ANOVA. Tables 4.4–4.6 show results of Tukey's multiple comparisons tests by location for transect points. Herman's Bay Point and Bear Point transects exhibited decreases concentrated in the center of the transects (except for point 1 at Herman's Bay Point) whereas Middle Point showed most decreases to the south of center at points 2–4. Middle Point transects were on average lower than the other two sites for all transect points whereas Bear Point had the highest average readings. Control transect points analyzed separately for all dates and locations showed only point 1 as lower than all other points.

TABLE 4.2
Tukey's multiple comparisons test for seagrass coverage by date alone in 2-m^2 areas (number of squares with seagrass present out of 200) shown as 95% confidence intervals for mean based on pooled standard deviation.

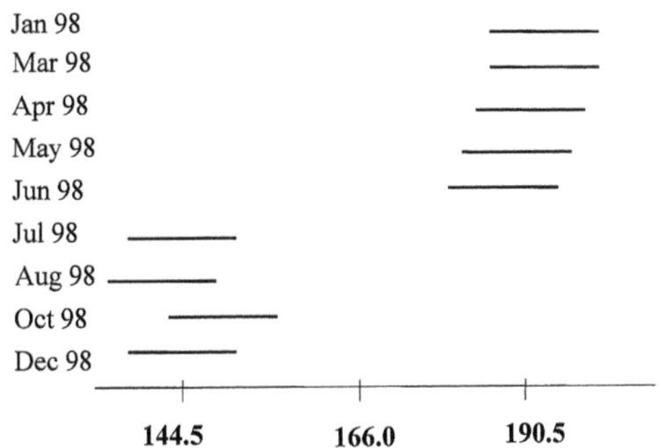

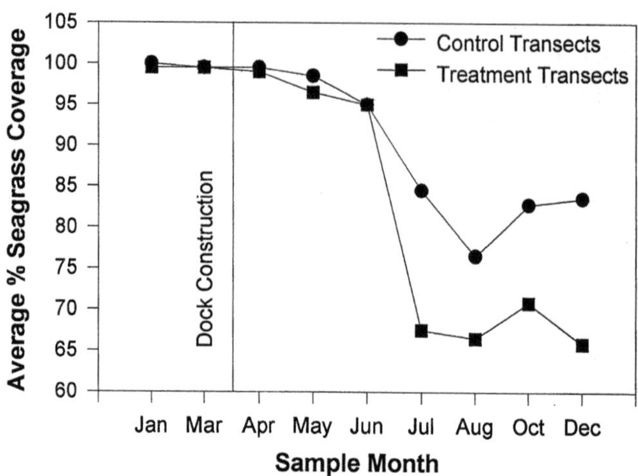

FIGURE 4.2 Average decline in seagrass coverage (*Syringodium filiforme* and/or *Halodule wrightii*) at all sites for 1998. Values represent coverage (number of occupied 100 cm² squares per 2 m²) along entire 12 m transects (control n = 36; treatment n = 108).

Shoot count data analyzed for all dates and sites showed significance for all factors and interaction terms except location (Table 4.7). Analyses of treatment alone ($p < 0.001$) and date alone ($p < 0.001$) were also significant. Tukey's multiple comparisons test of date alone revealed that shoot counts showed short-term responses (1 and 5 wk) to dock construction as well as a marked decrease at the 14 wk (July) sample dates (Table 4.8). The following hierarchy was established for shoot count by treatment independently (smallest to greatest decrease): controls > 1.51 m docks > pilings > 0.91 m docks. Transect point 12 was different from transect points 1 and 6 when analyzed independently ($p < 0.001$). Figure 4.3 shows the average change in shoot count

TABLE 4.3
Tukey's multiple comparisons test for transect points alone in 2-m² areas (number of squares with seagrass present out of 200) shown as individual 95% confidence intervals for mean based on pooled standard deviation.

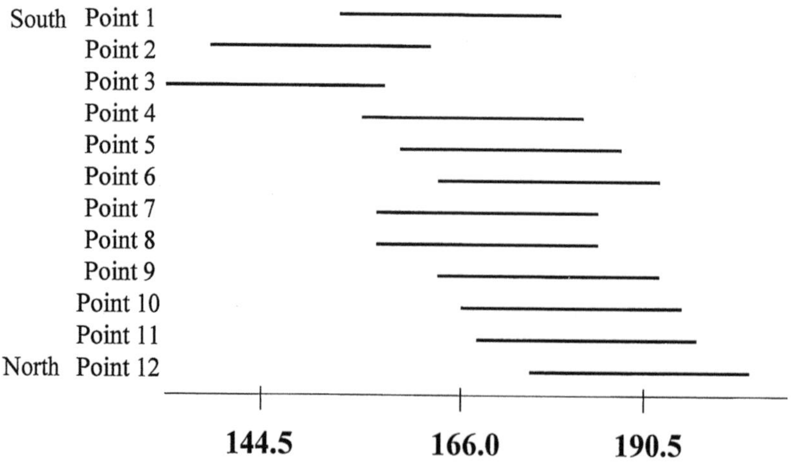

TABLE 4.4
Tukey's multiple comparisons test for transect points alone at Middle Point in 2-m² areas (number of squares with seagrass present out of 200) shown as individual 95% confidence intervals for mean based on pooled standard deviation.

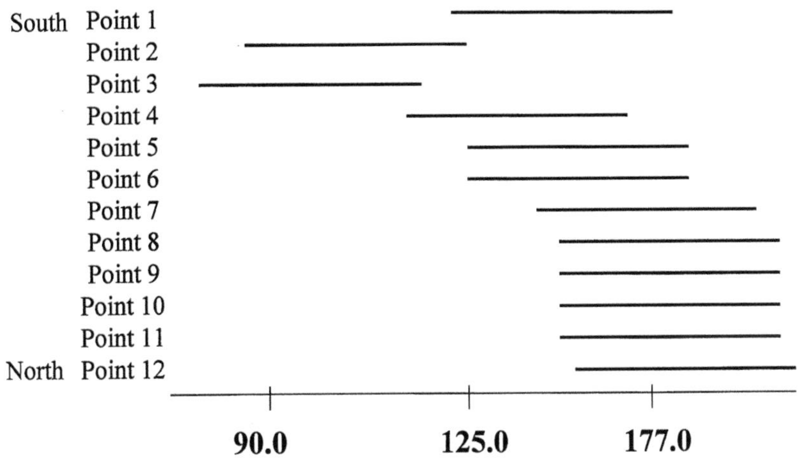

TABLE 4.5
Tukey's multiple comparisons test for transect points alone at Herman's Bay Point in 2-m² areas (number of squares with seagrass present out of 200) shown as individual 95% confidence intervals for mean based on pooled standard deviation.

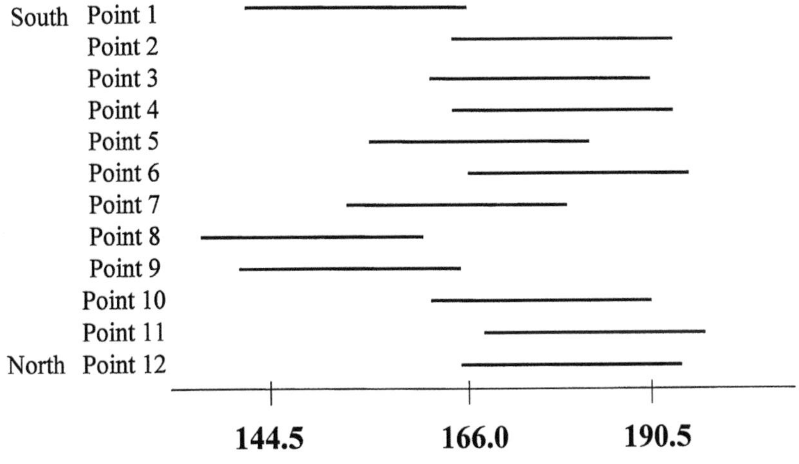

TABLE 4.6
Tukey's multiple comparisons test for transect points alone at Bear Point in 2-m² areas (number of squares with seagrass present out of 200) shown as individual 95% confidence intervals for mean based on pooled standard deviation.

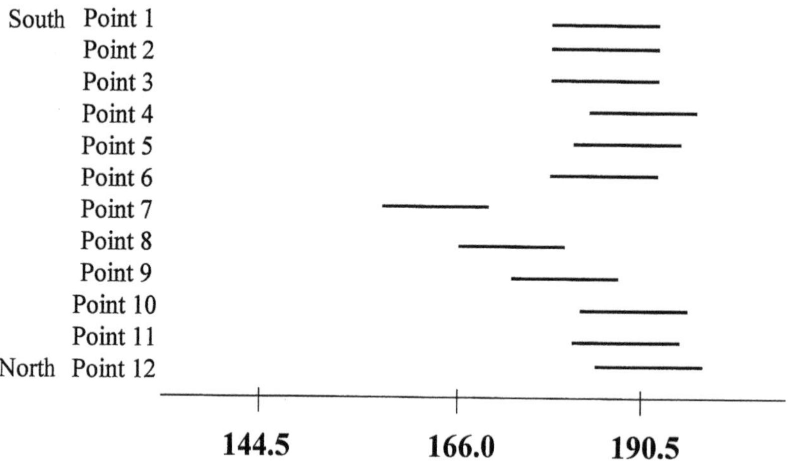

over time for all sites and treatments for transect point 6 only. Analysis of transect point 6 only using ANOVA showed the same hierarchical relationship as the treatment analysis (smallest to greatest decrease): controls > 1.51 m docks > pilings > 0.91 m docks. For all sites and treatments, transect point 1 has decreased an average of 40% (from 8.0 shoots/100cm² to 4.8). Transect point

TABLE 4.7
Results of balanced ANOVA of $\log_{10}$ (x + 1) transformed data testing the effects of location, date, treatment, and transect point on shoot density (number of shoots per 100 cm²)

Source	DF	SS	MS	F Values	P Values
Location	2	0.3694	0.1847	2.54	0.079
Date	8	95.3474	11.9184	163.80	0.000***
Treatment	3	3.9897	1.3299	18.28	0.000***
Meter	2	11.4205	5.7102	78.48	0.000***
Location · Date	16	30.6147	1.9134	26.30	0.000***
Location · Treatment	6	13.9487	2.3248	31.95	0.000***
Location · Tr. Point	4	1.6886	0.4221	5.80	0.000***
Date · Treatment	24	4.7063	0.1961	2.70	0.000***
Date · Tr. Point	16	11.2158	0.7010	9.63	0.000***
Treatment · Tr. Point	6	10.5259	1.7543	24.11	0.000***
Error	4772	347.2164	0.0728		
Total	4859	531.0433			

Note: *** $p < 0.001$, DF = degrees of freedom, SS = sum of squares, MS = mean squares.

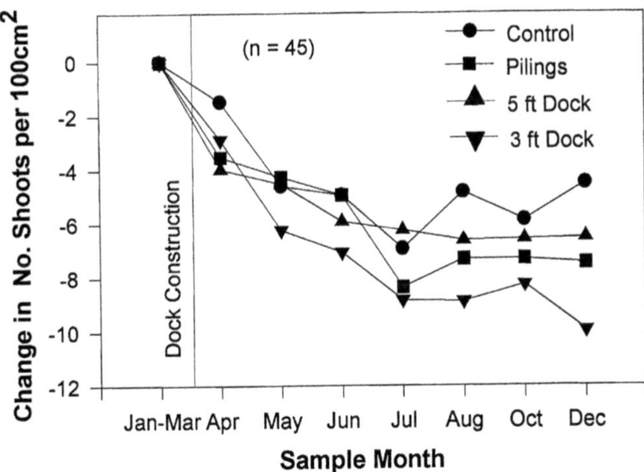

FIGURE 4.3 Average change in transect point 6 shoot counts (number of shoots *Syringodium filiforme* and/or *Halodule wrightii* per 100 cm^2) at all sites.

12 has also decreased an average of 40% (from 10.2 to 6.1) whereas transect point 6 has decreased an average of 56% (from 9.3 to 4.0).

PAR Measurements

Submarine light profiles for ten sample dates between December 1997 and November 1998 were used to calculate mean annual k (m^{-1}) and minimum light requirement for each site. Mean annual attenuation coefficients for Herman's Bay Point, Middle Point, and Bear Point were 1.00, 1.17, and 0.87, respectively. Mean edge of bed depths were 1.10 m at Herman's Bay Point, 1.07 m at Middle Point, and 1.52 m at Bear Point. Minimum light requirements were calculated as 33.3% I_o (529 µmol m^{-2} sec^{-1}), 28.6% I_o (520 µmol m^{-2} sec^{-1}), and 26.7% I_o (491 µmol m^{-2} sec^{-1}), respectively.

Daylong light collections revealed that treatments with decking were different from those without. Figures 4.4 and 4.5 show daylong light surveys of Middle Point in July and November as typical examples. Critical PAR on the figures is the minimum light requirement calculated for each site. Analysis of variance revealed no difference among treatments before 10:00 a.m. and after 4:30 p.m. in July. Therefore, November light data samplings began 2 hr after sunrise and ended 2 hr before sunset. ANOVA revealed that most of the variation stemmed from the treatment factor. The 0.91 and 1.51 m docks were similar and the control and pilings were similar in July at each site. These two groupings are different from each other ($p < 0.001$) at each site. Location is also not significant when analyzed independently for the November data. Treatment analysis ($p < 0.001$) shows the 0.91 m docks as different from the control and piling transects. Also, the 1.51 m docks are different from the pilings. Herman's Bay Point showed all treatments averaged below critical PAR for that site. Middle Point showed the 0.91 m and 1.51 m docks below critical PAR all day. Bear Point also showed the 0.91 and 1.51 m docks with PAR values below the critical value all day, and the control transect was below critical for most of the day.

DISCUSSION

Seagrass Cover and Shoot Count Trends

Structures built over submerged resources are presumed to decrease light below levels suitable for the survival of submerged aquatic vegetation. In the past, docks of differing heights have been constructed based on certain rule criteria for the issuance of State of Florida permits (e.g., presence/absence of natural resources, construction within aquatic preserves). Presently, the guidelines for constructing docks within marine and estuarine aquatic preserves (Chapter 258, Florida Statutes; Chapters 18–20, Florida Administrative Code) include access piers 1.51 m above MHW with 1.27 cm deck spacing. The 1.51 m height requirement has met with opposition from private applicants and marine contractors, mainly because of the added expense of additional lumber. A change in the rule to lower the dock height requirement to 0.91 m in freshwater aquatic preserves was proposed during the 1997 legislative session and incorporated into rule in 1998.

Presumably, the dock height and deck spacing requirements, as well as orientation/avoidance of natural resources considerations, are instituted to allow sufficient light to penetrate below docking structures thereby allowing sufficient PAR for submerged aquatic vegetation survival. A Florida Department of Environmental Protection (FDEP) survey of docks within the southern Indian River Lagoon revealed 151 of 300 docks surveyed had seagrass surrounding the structures but no seagrass directly beneath them. Seagrass presence under docks was strongly correlated with dock height ($r^2 = 0.9844$) for the docks studied (built 0–1.66 m above MHW). Loflin (1995), however, found no significant correlation between dock height and seagrass presence at docks surveyed at Charlotte Harbor estuary, Florida. It must be noted, however, that the docks surveyed exhibited relatively narrow ranges for the parameters studied. In the present study, preliminary results have shown that docks cause sufficient shading to impact seagrass cover and shoot density and that 0.91 and 1.51 m structures built over seagrasses cause unique effects.

Seagrass coverage responses to decreased light have been detected over periods of months (Gordon et al. 1994; Czerny and Dunton 1995; Fitzpatrick and Kirkman 1995) and years (Onuf 1996). The present study shows measurable decreases in cover beginning 14 weeks after commencement of treatment effect. The short-term impacts of piling installation, e.g., halo effect, were minimal and do not appear to have induced variation beyond natural fluctuation in coverage. Sample dates 1, 5, and 10 weeks post-construction were similar to pre-construction samplings (Table 4.2). Pilings were installed using a high pressure water jet and had minimal effects on seagrass in the short term. The halo created around the pilings ranged from 0–.5 m (average ~20 cm). The halos have changed slightly in shape over time and have become deeper than the surrounding substrate through scouring and presence of bioturbators. These depressions tend to trap drift algae and hinder seagrass recolonization.

Both seagrass cover and shoot counts have decreased for all sites and treatments, including controls (Figures 4.2 and 4.3). The unusually wet winter and subsequent nutrient load into the lagoon created a substantial drift algae bloom that persisted into June. The algal coverage in the lagoon shallows was nearly contiguous and as much as 1 m deep. The mats were composed primarily of *Gracilaria* spp., *Dictyota* spp., and *Acanthophora spicifera*. Algal blooms have been documented as major contributors to seagrass decline in Texas *H. wrightii* beds (Dunton 1994; Onuf 1996). The bloom that occurred during the present study served to mask the treatment effect somewhat. However, treatment transects clearly declined to a greater degree than controls (Figure 4.2).

The data confirm our conclusion that the sites were similar prior to dock construction. In general, the three sites have responded to treatment effect in varying degrees with Middle Point showing the greatest impact, followed by Herman's Bay Point and Bear Point. The average decrease in cover for all sites by treatment describes predicted results. The experimental design allowed for a quantification of seagrass decline along a spatial gradient bisecting structures. However, the southern ends of several transects contributed to an overall appearance of treatment effect concentrated at those ends (Table 4.3). Middle Point (Table 4.4), Herman's Bay Point (Table 4.5), and some control samplings (at transect point 1) consistently showed low values for cover at the south ends. Middle Point transect points have also been on average lower in seagrass cover and shoot density than the other two sites, especially at the southern end of the 1.51 m dock transect. Outside of these apparently stochastic effects, the general trend at two of the three sites was steady decline

TABLE 4.8
Tukey's multiple comparisons test for shoot density by date alone (number of shoots per 100 cm²) shown as individual 95% confidence intervals for mean based on pooled standard deviation.

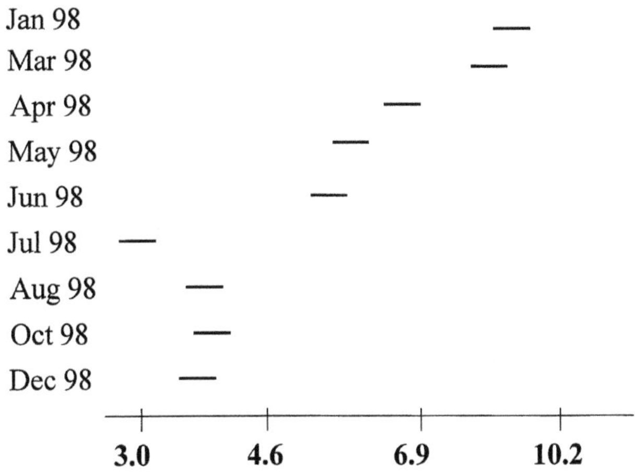

in coverage at the middle of the transects where the treatment effect should be greatest. This trend was realized even when data from controls were included (Tables 4.5 and 4.6). The shadow of all the docks was predominant at transect points 6 and 7 throughout the year and the data confirm its presence. Average percent decreases between 14 weeks pre-construction and 33 weeks post-construction showed predicted trends: cover at controls declined 12.3%, cover at pilings declined 21.3%, cover at 1.51 m docks declined 23.8%, and cover at 0.91 m docks declined 28.4%. However, ANOVA of cover by treatment revealed that 1.51 m docks showed a greater decrease than 0.91m docks. Perhaps ANOVA amplified an apparent stochastic effect occurring at the southern end of the 1.51 m transect at Middle Point. Future observations might reveal a recolonization of the seagrass into the relatively bare patch.

Shoot density has shown trends similar to those seen in seagrass coverage. The most dramatic decreases were noted at the 14 wk post-construction sampling event (Table 4.8). The 1 and 5 wk samples were different from the pre-construction data and different from one another. This might be a real effect of construction or an artifact of small sample size (relative to cover data) and decreases caused by the algae bloom. Presumably, seagrass response to decreased PAR should be manifested initially as diminishing biomass parameters (such as shoot count density) followed by changes in distribution (Onuf 1996). Over time, location became an insignificant factor in shoot density (Table 4.7). Treatment analyzed separately for all sites together also parallels coverage trends somewhat (smallest to greatest decrease): controls > 1.51 m docks > pilings > 0.91 m docks. Declines in coverage at the south end of several transects have also been manifested as decreases in shoot densities at meter 1. However, in terms of real changes over time, the end transect points (1 and 12) have decreased to the same degree and transect point 6 has shown the most dramatic decreases. Transect point 1 was relatively low prior to construction and is therefore unlike transect point 6 data when shoot counts are analyzed together. When transect point 6 is analyzed separately, the trend parallels total shoot density data (smallest to greatest decrease): controls > 1.51 m docks > pilings > 0.91 m docks.

The seagrass coverage had increased at the 25 wk sampling for treatment and control transects (Figure 4.2), perhaps in response to the subsidence of the algae bloom and subsequent lag time.

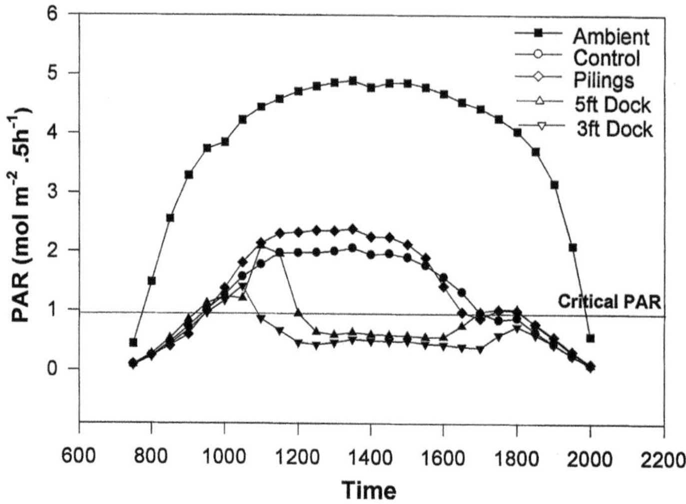

FIGURE 4.4 Daylong PAR profile for Middle Point 7/30/98. Critical PAR line is equivalent to 0.94 mol $m^{-2} \cdot 5\ h^{-1}$.

Also, shoot counts showed a slight increase between the 14 and 19 wk sample dates. The final sample data showed the controls and treatments diverging for both shoot density at transect point 6 and seagrass cover (Figures 4.2 and 4.3). The trend is predicted to continue considering the PAR data collected under the structures.

Worldwide, seagrasses have shown differential responses to experimentally and naturally induced, e.g., algal blooms, light reduction. These responses include decline in above- and below-ground (Dunton 1994) biomass; decline in various leaf floristics, e.g., leaf/shoot density; leaf length (Gordon et al. 1994); changes in distribution (Onuf 1996); and changes in relative and total amounts of blade chlorophyll (Wiginton and McMillan 1979; Czerny and Dunton 1995). In the present study, blade height of *S. filiforme* was noticeably greater (~15cm taller than ambient blades around the docks) under the 0.91 and 1.51 m docks at one site at the 10, 14, and 19 wk sample dates. The blade height was not different at the 25 or 33 wk sample dates. This phenomenon coincides with leaf elongation results documented by Czerny and Dunton (1995), showing that shade effects on blade length are most pronounced in summer when water temperatures are highest. It does appear that the seagrasses have responded to reduced light conditions by decreased shoot density under the structures. Increased bioturbation and changes in hydrology caused by piling presence have also contributed to decreases in cover and shoot density. Known bioturbators such as fishes (*Archosargus probatocephalus*, *Opsanus tau*, *Dasyatis sabina*) and invertebrates (*Panuliris argus*) have been consistently documented around the free-standing pilings and dock pilings. The decrease in grass cover and shoot density along the pilings transects probably cannot be attributed solely to the minimal shading produced by the pilings (Figures 4.4 and 4.5).

PAR and Minimum Light Requirement

At a dock elevated 1.66 m above MHW at Indian River Lagoon, Florida, FDEP staff found significant differences ($p < 0.01$) among PAR collected in open sunlight, and 0.91 and 1.51 m beneath the dock. Four sampling days over 1 year (1 day for each season) revealed a range of 11–25% difference in total daily PAR collected between the 0.91 and 1.51 m-placed sensors. For the present study, peak solar irradiance period sampling (10:00 a.m. to 4:30 p.m.) in July showed that the 0.91 and 1.51 m docks were statistically similar in PAR reduction. However, the difference between the docks might be biologically significant. The 0.91 m docks induced PAR values consistently lower than the 1.51 m docks during the 10:00 a.m. to 4:30 p.m. window: an average

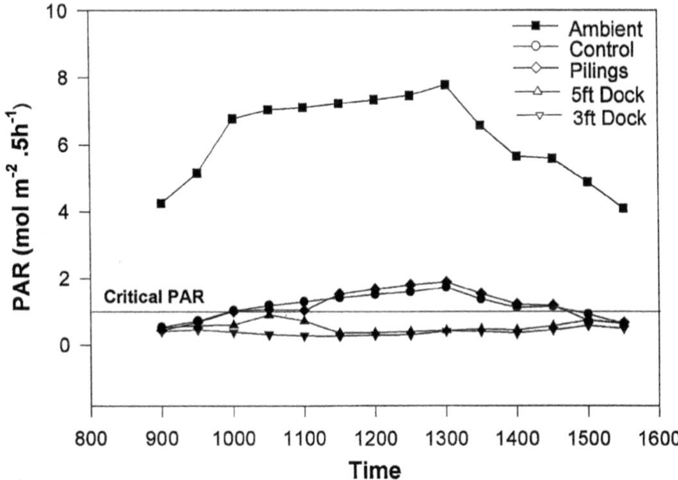

FIGURE 4.5 Daylong PAR profile for Middle Point 11/17/98. Critical PAR line is equivalent to 0.94 mol $m^{-2} \cdot 5\ h^{-1}$.

of 20.1% lower at Herman's Bay Point (range, 5–62%), 30.6% lower at Middle Point (range, 6–66%), and 23.1% lower at Bear Point (range, 3–49%). Average daily PAR values clearly reveal the difference between the docks. In July, at Herman's Bay Point, the 0.91 and 1.51 m docks averaged 376 μmol m^{-2} sec^{-1} and 462 μmol m^{-2} sec^{-1}, respectively. At Middle Point, the values were 301 μmol m^{-2} sec^{-1} and 438 μmol m^{-2} sec^{-1}, respectively. At Bear Point, they were 298 μmol m^{-2} sec^{-1} and 405 μmol m^{-2} sec^{-1}, respectively. The control and pilings values were very similar and together averaged 750 μmol m^{-2} sec^{-1} at Herman's Bay Point, 708 μmol m^{-2} sec^{-1} at Middle Point, and 559 μmol m^{-2} sec^{-1} at Bear Point in July.

In the November surveys, the average difference among treatments was noticeably lower than in July. Weather patterns apparently play a large role in determining light attenuation into the water column during the fall when the east coast of Florida experiences the highest tides of the year, potentially significant rainfall events (28 cm in 24 hr on 11/5/98), consistent 10+ knot winds, and continuous cloud cover. When present, these patterns decrease the effect of dock shading and, at times for brief periods, even control areas do not receive light above calculated minimum light requirement for seagrass survival. If the July and November sample dates are representative of summer and fall days, then the seagrass is not anticipated to survive under the 0.91 m docks at the very least because shading reduces the light underneath decking below the calculated ecological compensation point. The 1.51 m docks might receive enough solar irradiance to maintain the seagrass in a reduced coverage and shoot density state. Because of their small size, the experimental docks only shadow the transects for a portion of the day, whereas typical east–west oriented docks in the Indian River Lagoon shadow seagrasses most of the day. As a result, the experimental design underestimates the effects of docks on underlying seagrasses.

Historical calculations of ecological compensation depth (ECD) for submerged aquatic vegetation probably underestimate seagrass requirements, and site-specific water quality and substrate parameters affect ECD for a given species (Kenworthy 1994). Early studies of *H. wrightii* and *S. filiforme*, seagrasses common to the southeastern U.S. and the Caribbean, reveal that these two species have similar minimum light requirements (15–20% of average annual I_0) (Onuf 1991; Kenworthy et al. 1991; Morris and Tomasko 1993). Recent calculations of minimum light requirements for *H. wrightii* in southern Texas grass beds are 18% I_0 (Dunton 1994), and >16% I_0 (Czerny and Dunton 1995; Onuf 1996). Minimum light requirements for these two species have been calculated for some southern Indian River Lagoon beds (24–37% I_0 from Kenworthy and Fonseca 1996) and their maximum depth limits are similar (Gallegos and Kenworthy 1996; Kenworthy and

Fonseca 1996). Williams and McRoy (1982) showed that *H. wrightii* (a pioneer species) and *S. filiforme* (a climax species) have similar saturation- and half-saturation irradiance constants with respect to carbon uptake for Texas and Puerto Rico samples. The present study determines minimum light requirements for *H. wrightii* and *S. filiforme* as 33.3% I_0 (529 μmol m^{-2} sec^{-1}), 28.6% I_0 (520 μmol m^{-2} sec^{-1}), and 26.7% I_0 (491 μmol m^{-2} sec^{-1}) (an average of 29.5% I_0) at the three sites studied, which are comparable to values calculated by other researchers. It should be noted, however, that the predominance of *S. filiforme* at the sites (including bed edges) dictates that the present study results should be weighted toward that species. It should also be noted that light reduction due to epiphytic attenuation was not accounted for in the studies documented above or in the present study. Therefore, the calculated minimum light requirements must be regarded as approximate estimates only. Based on these parameters, seagrass cover and shoot density are expected to continue to decline under the 0.91 and 1.51 m docks and perhaps be extirpated entirely within specific zones around the structures.

ACKNOWLEDGMENTS

We would like to thank Kent Smith, Florida Department of Environmental Protection (FDEP), Bureau of Protected Species Management, for funding for this project. Special thanks to our reviewers for constructive comment and assistance with calculations. We would also like to thank FDEP employees Lauren Hall, Patricia Adams, Brian Proctor, Chris Bergh, and Georgia Vince for field support.

REFERENCES

Czerny, A. B. and K. H. Dunton. 1995. The effects of *in situ* light reduction on the growth of two subtropical seagrasses, *Thalassia testudinum* and *Halodule wrightii. Estuaries* 18(2):418–427.

Dennison, W. C. 1987. Effects of light on seagrass photosynthesis, growth and depth distribution. *Aquatic Botany* 27:15–26.

Dennison, W. C., R. J. Orth, K. A. Moore, J. C. Stevenson, V. Carter, S. Kollar, P. W. Bergstrom, and R. A. Batuik. 1993. Assessing water quality with submersed aquatic vegetation. *Bioscience* 43:86–94.

Dunton, K. H. 1994. Seasonal growth and biomass of the subtropical seagrass *Halodule wrightii* in relation to continuous measurements of underwater irradiance. *Marine Biology* 120:479–489.

Fitzpatrick, J. R. and H. Kirkman. 1995. Effects of prolonged shading stress on growth and survival of the seagrass *Posidonia australis* in Jervis Bay, New South Wales, Australia. *Marine Ecology Progress Series* 127:279–289.

Fonseca, M. S., G. W. Thayer, and W. J. Kenworthy. 1987. The use of ecological data in the implementation of management of seagrass restorations. In *Proceedings of the Symposium on Subtropical-Tropical Seagrasses of the Southeastern U.S.,* M. J. Durako, R. C. Phillips, and R .R. Lewis (eds.). Florida Marine Research Publications 42, pp. 175–188.

Gallegos, C. L. and W. J. Kenworthy. 1996. Seagrass depth limits in the Indian River Lagoon (Florida, U.S.A.): application of an optical water quality model. *Estuarine Coastal Shelf Science* 42:267–288.

Gordon, D. M., K. A. Grey, S. C. Chase, and C. J. Thompson. 1994. Changes to the structure and productivity of a *Posidonia sinuosa* meadow during and after imposed shading. *Aquatic Botany* 47:265–275.

Heidelbaugh, W. S. and W. G. Nelson. 1995. A power analysis of methods for assessment of change in seagrass cover. *Aquatic Botany* 53:227–233.

Kenworthy, W. J. 1994. Conservation and restoration of the seagrasses of the Gulf of Mexico through a better understanding of their minimum light requirements and factors controlling water transparency. In *Indicator Development: Seagrass Monitoring and Research in the Gulf of Mexico,* H. A. Neckles (ed.). Environmental Protection Agency/620/R–94/029, pp. 17–31.

Kenworthy, W. J. and D. E. Haunert (eds.). 1991. *The Light Requirements of Seagrasses: Proceedings of a Workshop to Examine the Capability of Water Quality Criteria, Standards, and Monitoring Programs to Protect Seagrasses,* NOAA Technical Memorandum, NMFS-SEFC-287.

Kenworthy, W. J. and M. S. Fonseca. 1996. Light requirements of seagrasses *Halodule wrightii* and *Syringodium filiforme* derived from the relationship between diffuse light attenuation and maximum depth distribution. *Estuaries* 19(3):740–750.

Kenworthy, W. J., M. S. Fonseca, D. E. McIvor, and G. W. Thayer. 1989. A comparison of wind-wave and boat wake-wave energy in Hobe Sound: implications for seagrass growth. Draft Report, Beaufort Laboratory, National Marine Fisheries Service, NOAA, Beaufort, NC.

Kenworthy, W. J., M. S. Fonseca, and S. D. Dipiero. 1991. Defining the ecological light compensation point for seagrasses *Halodule wrightii* and *Syringodium filiforme* from long-term submarine light regime monitoring in the southern Indian River. In *The Light Requirements of Seagrasses: Proceedings of a Workshop to Examine the Capability of Water Quality Criteria, Standards, and Monitoring Programs to Protect Seagrasses,* W. J. Kenworthy and D. E. Haunert (eds.). NOAA Technical Memorandum, NMFS-SEFC-287, pp. 67–70.

Kruczynski, W. L. and D. A. Flemer. 1994. Submerged aquatic vegetation research needs. In *Indicator Development: Seagrass Monitoring and Research in the Gulf of Mexico,* H. A. Neckles (ed.). Environmental Protection Agency/620/R-94/029, pp. 51–57.

Livingston, R. J. 1987. Historic trends of human impacts on seagrass meadows in Florida. In *Proceedings of the Symposium on Subtropical-Tropical Seagrasses of the Southeastern U.S.,* M. J. Durako, R. C. Phillips, and R. R. Lewis (eds.). Florida Marine Research Publications 42, pp. 139–151.

Loflin, R. K. 1995. The effects of docks on seagrass beds in the Charlotte Harbor estuary. *Florida Scientist* 58(2):198–205.

Morris, L. J. and D. A. Tomasko (eds.). 1993. *Proceedings and Conclusions of Workshops on Submerged Aquatic Vegetation and Photosynthetically Active Radiation.* Special Publication of the St. Johns River Water Management District, SJ93-SP13, Palatka, FL.

Neckles, H. A. 1994. Ecological indicators. In *Indicator Development: Seagrass Monitoring and Research in the Gulf of Mexico,* H. A. Neckles (ed.). Environmental Protection Agency/620/R-94/029, pp. 43–50.

Onuf, C. P. 1991. Light requirements of *Halodule wrightii, Syringodium filiforme,* and *Halophila engelmanni* in a heterogeneous and variable environment inferred from long-term monitoring. In *The Light Requirements of Seagrasses: Proceedings of a Workshop to Examine the Capability of Water Quality Criteria, Standards, and Monitoring Programs to Protect Seagrasses,* W. J. Kenworthy and D. E. Haunert (eds.). NOAA Technical Memorandum, NMFS-SEFC-287, pp. 59–66.

Onuf, C. P. 1996. Seagrass responses to long-term light reduction by brown tide in upper Laguna Madre, Texas: distribution and biomass patterns. *Marine Ecology Progress Series* 138:219–231.

Sargent, F. J., T. J. Leary, D. W. Crewz, and C. R. Kruer. 1995. Scarring of Florida's seagrasses: assessment and management options. Florida Marine Research Institute Tech. Rep. TR-1, St. Petersburg, FL.

Sokal, R. P. and F. J. Rohlf. 1981. *Biometry: The Perspective and Practice of Statistics in Biological Research.* 2nd Ed., W.H. Freeman, San Francisco.

Wiginton, J. R. and C. McMillan. 1979. Chlorophyll composition under controlled light conditions as related to the distribution of seagrasses in Texas and the U.S. Virgin Islands. *Aquatic Botany* 6:171–184.

Williams, S. L. and C. P. McRoy. 1982. Seagrass productivity: the effect of light on carbon uptake. *Aquatic Botany* 12:321–344.

Winer, B. J. 1971. *Statistical Principles in Experimental Design.* McGraw-Hill Inc., New York.

Wood, E. J. F., W. E. Odum, and J. C. Zeiman. 1969. Influence of seagrasses on the productivity of coastal lagoons. In *Coastal Lagoons, a Symposium.* UNAM-UNESCO, Universidad Nacional Autonoma de Mexico, Mexico, D.F., pp. 495–502.

Zieman, J. C. and R. G. Wetzel. 1980. Productivity in seagrasses: methods and rates. In *Handbook of Seagrass Biology: An Ecosystem Perspective,* R. C. Phillips and C. P. McRoy (eds.). Garland STPM Press, New York, pp. 87–116.

5 Tape Grass Life History Metrics Associated with Environmental Variables in a Controlled Estuary

Stephen A. Bortone and Robert K. Turpin

Abstract Twenty samples of tape grass were removed from four locations along a salinity gradient in the Caloosahatchee River in Lee County, Florida each month in 1998. Examination of the environmental, independent variables indicates a strong seasonal cycle for temperature and a trend toward increasing chlorophyll levels during the year. Dependent response variables recorded for tape grass also indicated a seasonal pattern that mimicked the temperature cycle. There was a time lag in maximal life history attributes. Number of shoots per sample, number of blades per sample, and number of blades per shoot had highest values during the warmer months, i.e., May–August. Blade length, blade width, and biomass were higher during the later part of the summer and early fall. Reproductive attributes of the plants, i.e., number of male and female flowers, were highest during the fall. The salinity gradient that was part of the study design was weak and accounted for only a small part of the variation observed between locations along the river. Typically, the end of the year parameter levels were higher than the beginning of the parameter levels for all response variables among plants. It is suspected that this is due to the inordinately heavy rains during early 1998 that led to lower salinities at all locations. These normally freshwater plants were apparently less stressed because of the lower salinity conditions in the estuary. This situation may have provided a "boost" to their growth that helped expand the extent, size, and fitness of tape grass within the estuarine system. A paradox was revealed in that higher plant growth parameters were recorded among plants from the higher salinity portions of the river. During 1998, salinities were low at the most seaward location, but water clarity was greater, thus providing conditions that may have facilitated growth of this normally freshwater plant in an estuarine ecosystem.

INTRODUCTION

Monitoring seagrasses (including all forms of submerged aquatic vegetation [SAV]) is rapidly becoming one of the foremost methods to determine the overall health and condition of the aquatic environment (Dennison et al. 1993; Stevenson et al. 1993). Seagrasses have shown particular promise in detecting specific factors that may influence both short- and long-term changes to nearshore aquatic ecosystems. More recently, this has been especially true for evaluating the relative health and condition of estuaries (Johannson 1991).

Tape grass, *Vallisneria americana* Michx. (also known as water celery, eel grass, American eel grass, and American wild celery), is widely distributed in nearshore aquatic habitats (see Bortone et al. 1998 for an annotated bibliography on the life history of this species). Tape grass is generally a freshwater species, but it is also an important component of the oligohaline estuarine SAV community (Twilley and Barko 1990; Adair et al. 1994; Kraemer et al. 1999). Because it inhabits the upper portions of estuaries, it is often subjected to wide fluctuations in salinity due to intense

0-8493-2045-3/99/$0.00+$.50
© 2000 by CRC Press LLC

freshwater runoff events after having been subjected to higher salinity exposures due to periods of little runoff or rainfall (Zieman and Zieman 1989).

The Caloosahatchee River in southwest Florida serves as a conduit for fresh water from Lake Okeechobee to the Gulf of Mexico, especially during times of excessively heavy rain events. Tape grass occurs in well-defined beds in shallow waters below a water control structure (Franklin Lock and Dam). Discharge profiles indicate that freshwater releases normally occur during the summer months (June, July, and August) when thunderstorms drop excessive rain onto the surrounding area (Chamberlain and Doering 1998). Also, in anticipation of extra-heavy rain events (such as those resulting from a hurricane), water management authorities release water from Lake Okeechobee during the late spring and summer to allow for situations when excessive storm water may have to be retained in the lake to prevent flooding. Because of this scenario, managing authorities need to know how the retention and release of fresh water affect the contiguous estuarine ecosystem. Monitoring tape grass life history features is one way to examine the impact of these freshwater releases.

Presented here are the results of an initial, baseline monitoring assessment study to establish the life history parameters of tape grass relative to salinity variations in the Caloosahatchee River. While the study is in its infancy, the basic methodology and general results serve to guide the development and implementation of future studies that seek to use seagrasses to assess estuarine conditions. The study was designed to obtain information on both life history attributes of tape grass inhabiting the upper estuarine portions of the Caloosahatchee River and to draw inferences on the association these dependent variables have with the environmental, independent variables. The utility in knowing the species life history parameters and their relationship with the environmental variables will facilitate future management decisions regarding options to release or retain fresh water to the Caloosahatchee River. Ultimately, these data will serve as a basis to maintain the ecological integrity of this ecosystem and may serve as a model for regulation and control of other estuarine areas.

MATERIALS AND METHODS

Description of the Study Area

The area of study is located within the Caloosahatchee River (Figure 5.1) on the southwest coast of Florida, in the southeastern U.S. The study area was chosen because tape grass beds occur in the river along a distance exposed to a broad range of salinity that varies during the year. As tidal flux is usually less than a meter and winds are not consistently strong in any direction, salinity is strongly influenced by the amount of rainfall in the immediate area and the overall drainage basin that extends as far east as Lake Okeechobee.

Study Design

At each of four locations (Figure 5.1) one pair of 100-m transects (one perpendicular to shore and one parallel) was established at two sites (A, downstream; B, upstream). The intersection of the transects for each site was selected *in situ* by visually detecting the presence of tape grass (*V. americana*). Each perpendicular transect began at the shoreline, thus each transect intersection, i.e., midpoint, was located 50 m from shore. Each parallel transect had the midpoint of the perpendicular transect as its midpoint.

Each transect was divided into five intervals (1–20 m, 21–40 m, 41–60 m, 61–80 m, 81–100 m); monthly, a different specific distance within each interval was selected using a random number generator without replacement. Thus, at each of four locations during each month beginning in January 1998, a total of 20 discrete samples for tape grass was collected: 2 sites, 2 transects, 5 intervals.

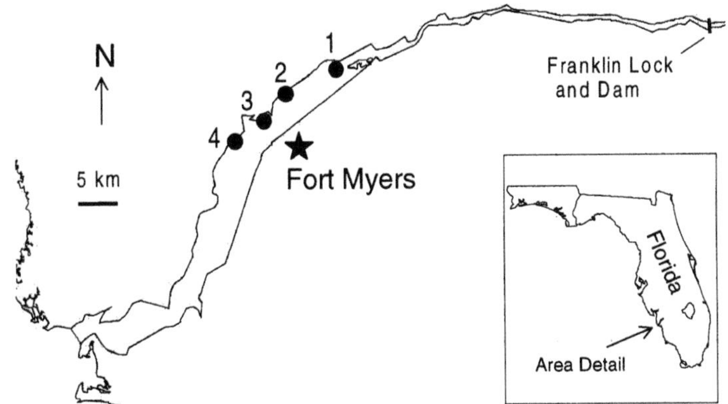

FIGURE 5.1 A map of the study area indicating each of the four sampling locations (filled circles) within the Caloosahatchee River. The insert indicates the study area within the State of Florida.

Sampling Methods

Monthly during 1998, tape grasses were sampled at a randomly selected location along each transect by placing a 0.1-m^2 (0.33 cm × 0.33 cm) weighted PVC quadrat frame on the river bottom and subsequently removing all tape grass plants within the quadrat. Samples were placed in plastic bags, labeled, and stored on ice in an insulated dark box; returned to the laboratory; and, within 24 hr, measured for the dependent variables. Only tape grasses were collected. If the sample contained epiphytes or attached vegetation such as *Ruppia maritima,* notes were taken as to their presence and relative abundance.

Below are each of the independent and dependent variables measured during this study with a brief definition and description of the method used to assess each variable.

Independent Variables

The independent, environmental variables were recorded monthly at each sample location. Most variables were recorded *in situ.* Additionally, water samples were collected 0.5 m below the surface to assess total suspended solids, chlorophyll a, and color. The water samples were placed on ice for up to five hours and taken to the Lee County Environmental Laboratory for detailed water quality analysis.

Date — Recorded by Year Month Day, e.g., 19990131 = January 31, 1999, as a single number to facilitate logical sorting of the database.

Location — Four sample locations on the north side of the Caloosahatchee River (numbered 1, 2, 3, and 4):

- Location 1, 3.5 km west of the I-75 bridge (26°41′23″ N, 81°49′48″ W)
- Location 2, 2 km east of the (Business) US-41 bridge (26°40′21″ N, 81°51′52″ W)
- Location 3, between the (Business) US-41 and US-41 bridges (26°39′18″ N, 81°52′48″ W)
- Location 4, 3 km west of the US-41 bridge (26°38′37″ N, 81°54′8″ W).

Site — Two sample sites (A, downstream; B, upstream) per location separated from each other by 100 to 200 m.

Direction — Direction or orientation of sample transect (either perpendicular or parallel to shore at each site).

Distance — Distance along the 100-m transects. Each transect was divided into five, 20-m intervals. Samples were collected within each 20-m interval at distances selected from a random number generator without replacement. Distances increased going offshore (perpendicular transects) or upstream (parallel transects).

Time — Local time of day (EST or EDT) at the beginning of sampling at each location recorded as military time.

Depth — Water depth to the nearest centimeter at the center of each 20-m sample interval along each transect. Water depths were measured on only one occasion throughout the entire study area (all within 1 hour) to reflect the relative depths among samples.

Tidal stage — Category of tidal stage (L = low, H = high, E = ebbing, F = flooding) at time of sampling; determined from the relative position of Fort Myers using a tidal projection software program (Nautical Software Inc., Beaverton, OR).

Temperature — Surface water temperature (measured to the nearest degree Celsius).

Salinity — Estimated to the nearest part per thousand (ppt) using a temperature corrected refractometer.

Secchi depth — Vertical Secchi disk (20 cm diameter) depth; measured to nearest centimeter as an indicator of water clarity.

Total suspended solids (TSS) — measured as mg/L (detection limit = 1 mg/L).

Chlorophyll a — Measured as mg/m^3 (detection limit = 0.5 mg/m^3).

Color — Measured in color units (detection limit = 1 cu) at 465 nm, platinum cobalt standard of 500 APHA (Eaton et al. 1995), measured at 500 cu undiluted stock; 300 cu, 30 ml diluted to 50 ml; 100 cu, 10 ml diluted to 50 ml; 50 cu, 5 ml diluted to 50 ml with de-ionized H_20.

Dependent Variables

All dependent variables were measured in the laboratory on samples placed on ice for 24 hr, except where noted.

Number of Shoots — Number of *V. americana* shoots counted from each sample. Individual shoots, i.e., plants, may have occurred singly or attached via underground rhizomes. When attached to other plants, each shoot was counted separately.

Number of Blades — Number of *V. americana* blades (leaves) counted from each sample. Where the number of shoots was > 30, the number of blades was counted from a subsample of 30 shoots; the number of blades was counted in the subsample and the total number of blades was calculated for the sample.

Number of Male and Female Flowers — Counted from the entire sample collected at each location.

Blade Length — Mean blade length measured in mm and calculated to nearest 0.1 mm from the five longest blades.

Blade Width — Mean blade width measured in mm and calculated to nearest 0.1 mm from the five widest blades.

Weight — Dry weight in grams (to nearest 0.001 g) of the entire sample. Each sample was dried for 5 days in an oven at 80°C. Where number of shoots in a sample exceeded 30, a subsample of 30 shoots was dried and weighed; dry weight for the entire sample was then calculated.

Analyses

Data were analyzed using the SAS® statistical package (SAS User's Guide 1995). Pearson and Spearman Rank correlation coefficients were used to determine the associations between all variables. Stepwise regression was used to assess the significance and degree each variable contributed to explain the variation observed for the dependent variables. Graphs were prepared using SigmaPlot® software.

RESULTS

Independent Variables

Temperature — Temperature generally followed a pattern of increase and decrease typical for the seasonal changes in southwest Florida (Figure 5.2). Lowest temperatures were in December and January (17–18°C) and warmest during June (32–34°C). Monthly temperature profiles at all four locations followed a seasonal pattern with the exception of November when all locations were abruptly higher. This was due to warm weather conditions.

Salinity — Variation in salinity can most often be attributed to tidal effects as well as local runoff and rains. As salinity was not measured on a consistent lunar cycle, it is not surprising that it showed little evidence of a seasonal pattern. However, the annual profile does indicate evidence of the inordinately heavy rains that occurred January–March with residual runoff occurring until May in 1998. Salinities ranged from 0 to 10 ppt, and most were at or below 5 ppt. Locations were chosen so that lower salinities would be likely upstream (Location 1), and higher salinities would be furthest downstream (Location 4) with Locations 2 and 3 being intermediate. Figure 2 indicates that this was generally true as the lowest salinities were recorded at Location 1 (mean = 0.5 ppt/month) and highest from Location 4 (4.4 ppt). However, considerable variability was noted at the intermediate distance locations which were nearly identical with regard to salinity (2.2 and 2.6 ppt for Locations 2 and 3, respectively). Lack of an obvious upstream-downstream salinity gradient was probably due to local mixing and runoff.

Secchi depth — Water clarity, as measured by Secchi depth, was highly variable (Figure 5.2) with little evidence of a seasonal pattern except that water clarity was greater during the spring and early summer and poorest during the winter and early fall. The high variability and difference in water clarity between the other sites and Location 1 were probably due to the influence of variable winds and boat traffic proximate to this location.

Total suspended solids (TSS) — Inspection of Figure 5.3 indicates that measures of this variable did not display a pattern, but that differences between locations were relatively small. Notable, however, were the two peaks in TSS at Location 4 in July and September.

Chlorophyll a — This variable was generally low early during the sampling period, but gradually increased toward the late summer and fall (Figure 5.3). Generally, Locations 2 and 4 had higher levels during the period of increased chlorophyll levels.

Color — Color information can prove an important indicator of the influence of freshwater runoff because, as runoff increases (particularly from natural, tannic-stained areas such as woodlands), water becomes darker in color. Figure 5.3 indicates that the color pattern was similar to the rainfall pattern in the area during 1998.

Dependent Variables

Number of samples with shoots (Figure 5.4) indicated an overall annual trend to increase during the sampling period among all locations except Location 1, the most upstream sample. Locations 2 and 3 displayed the most dramatic increase in the number of samples with shoots. While the number of samples with shoots at Location 4 increased during 1998, it was always highest among all locations. The number of samples with shoots displayed no seasonal trends at Location 1.

Monthly patterns were similar among the dependent variables: shoots per sample, blades per sample, and blades per shoot (Figure 5.5). These variable parameters were higher during the warmer months of the year (May–August) and lower during months when the water temperature was cooler

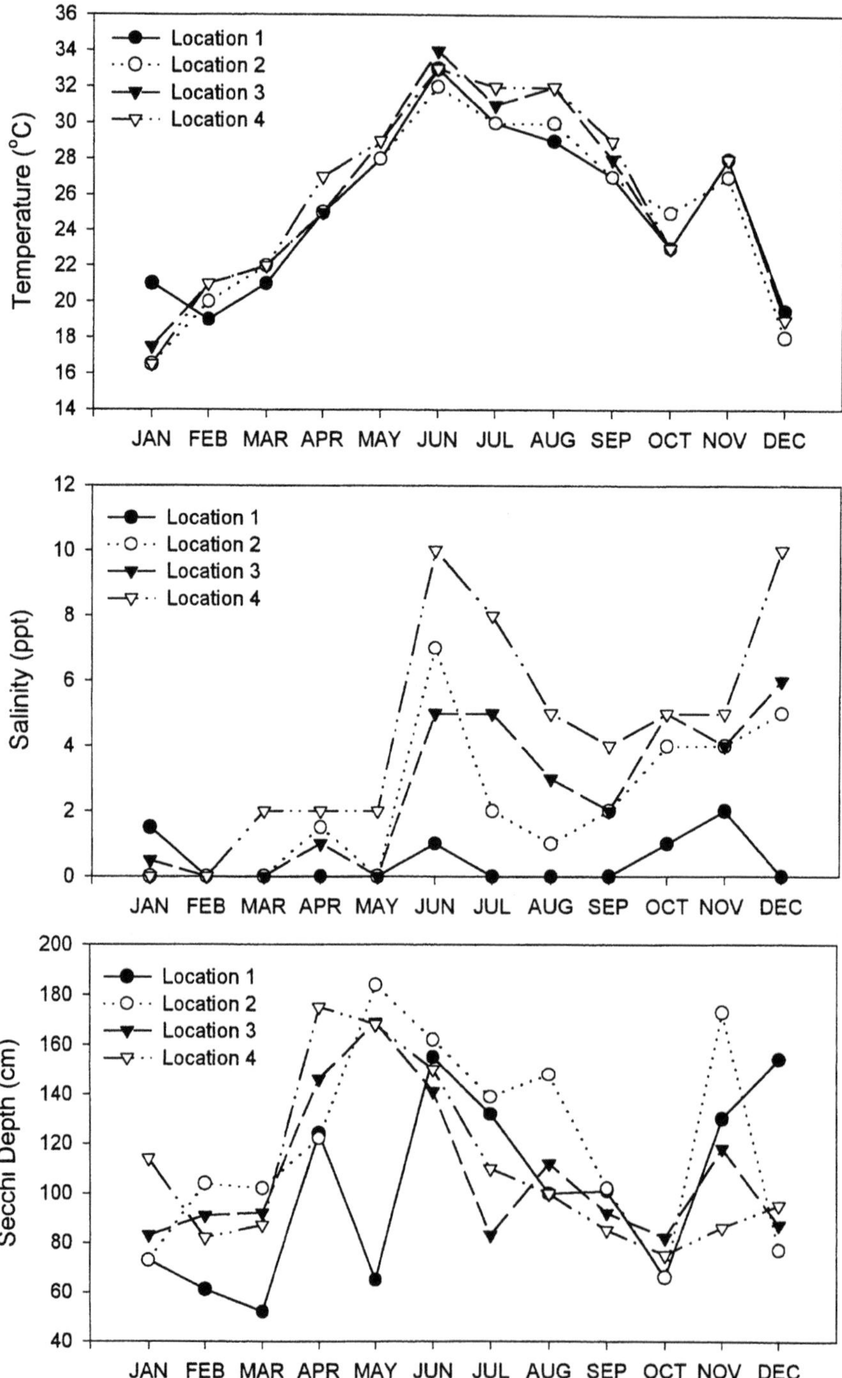

FIGURE 5.2 Monthly temperature (upper graph), salinity (middle graph), and Secchi depth (lower graph) recorded at each of the four sample locations.

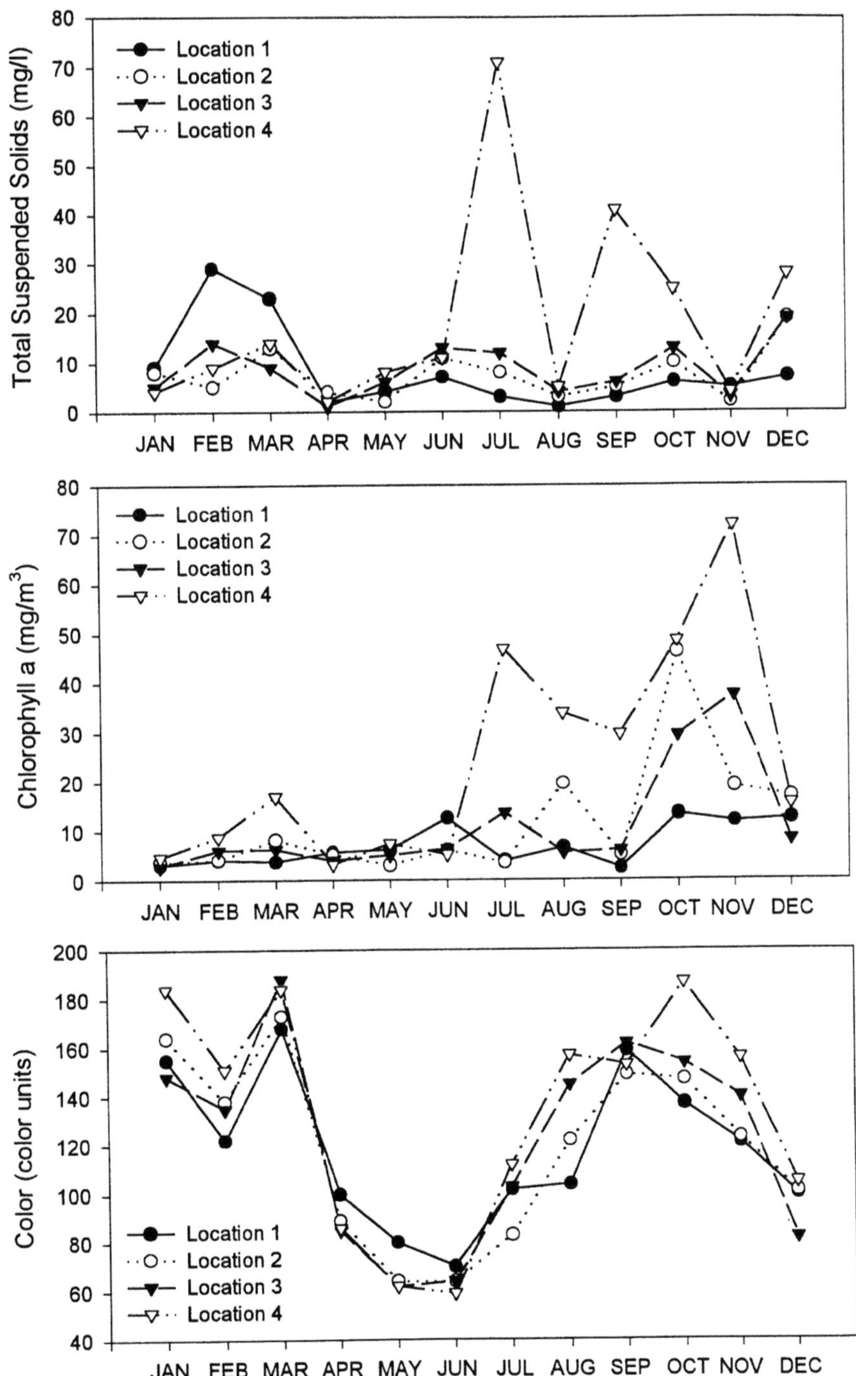

FIGURE 5.3 Monthly total suspended solids (upper graph), chlorophyll a (middle graph), and color (lower graph) recorded at each of the four sampling locations.

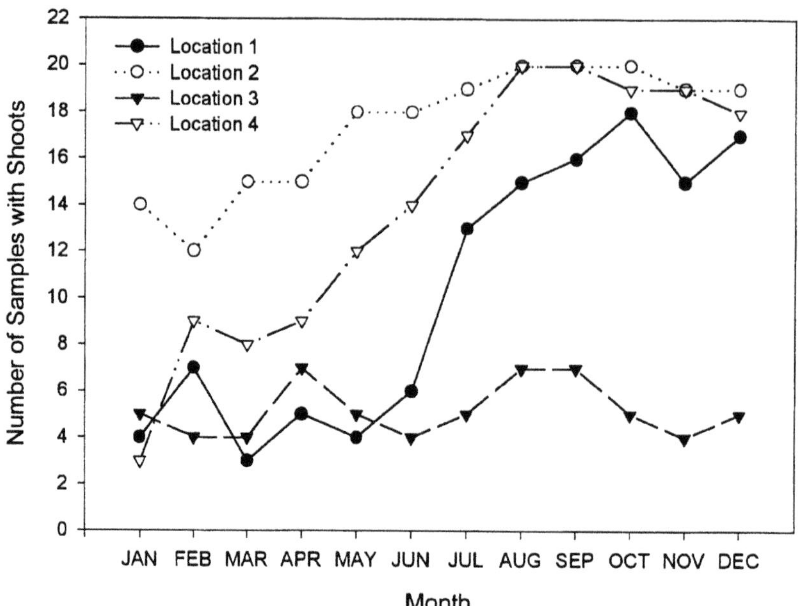

FIGURE 5.4 Monthly trend among locations for the number of samples with shoots.

(January–March and September–December) and the total amount of daylight was reduced. The number of shoots and blades per sample was higher at Location 4.

Blade lengths and blade widths increased toward the late summer and early fall at all locations (Figure 5.6). In contrast, the blade width/blade length ratio was lowest during this period. While a distinct trend seems lacking among these variables, blade length was typically longer at Location 4.

Just as the blade lengths and widths increased during the late summer and early fall, so did the number of female and male flowers (Figure 5.7). It appears, however, that the peak flower months (highest at Location 4) were maximal during August–October. The maximum values for blade lengths and widths were high for an expanded season, but were highest during July–September.

Biomass (as measured by dry weight per sample and per shoot) was low during the early part of the year (January–May), but increased during June and July and remained relatively high until the late fall (Figure 5.8).

Variable Associations

To determine the potential relationships among and between both independent and dependent variables, a Pearson Product correlation coefficient (r) was calculated between all possible pairs of variables. Significant ($p < 0.05$) and strong ($r > \pm 0.50$) correlations among independent variables were noted: date with salinity (+ 0.52) and chlorophyll (+ 0.51), location with salinity (+ 0.51), and Secchi depth with color (– 0.64). Among the dependent variables, significant ($p < 0.05$) and strong ($r > \pm 0.75$) relationships were observed: number of shoots with number of blades (+ 0.96), blade length with blade width (+ 0.88) and weight (+ 0.77).

A comparison of Spearman Rank correlation coefficients (a non-parametric measure of association) indicated that significant and strong associations occurred between salinity and chlorophyll (+ 0.55), as well as between Secchi depth and color (– 0.57). Among the dependent variables, there were several significant and strong Spearman Rank correlations: number of shoots with blade length (+ 0.88), blade width (+ 0.88), and weight (+ 0.95); number of blades with blade length (+ 0.89), blade width (+ 0.89), and weight (+ 0.96); blade length with blade width (+ 0.97) and weight (+ 0.96); blade width with weight (+ 0.95).

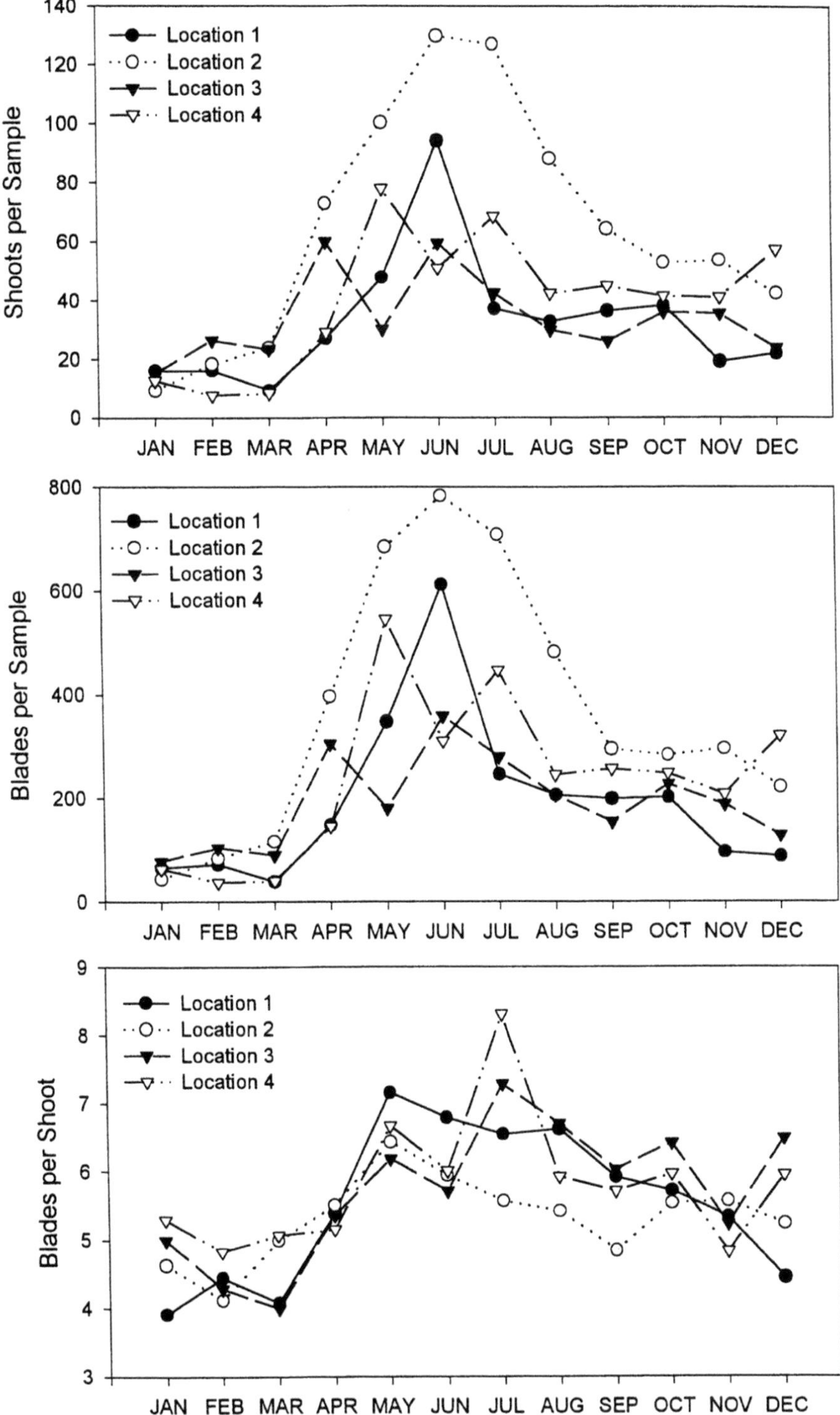

FIGURE 5.5 Monthly trend among locations for the number of shoots per sample (upper graph), number of blades per sample (middle graph), and the number of blades per shoot (lower graph).

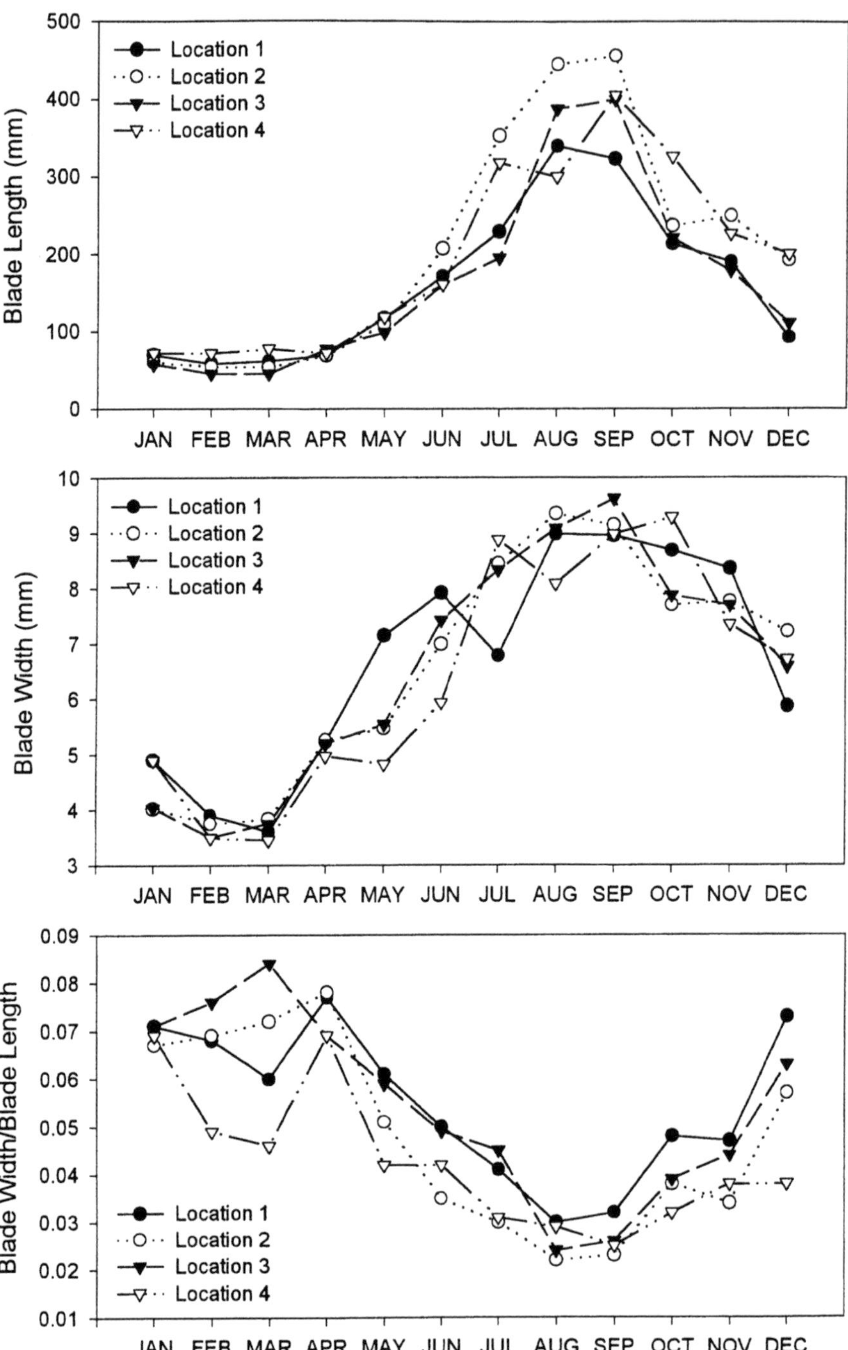

FIGURE 5.6 Monthly trend among locations for the mean blade length (upper graph), mean blade width (middle graph), and blade width/blade length ratio (lower graph).

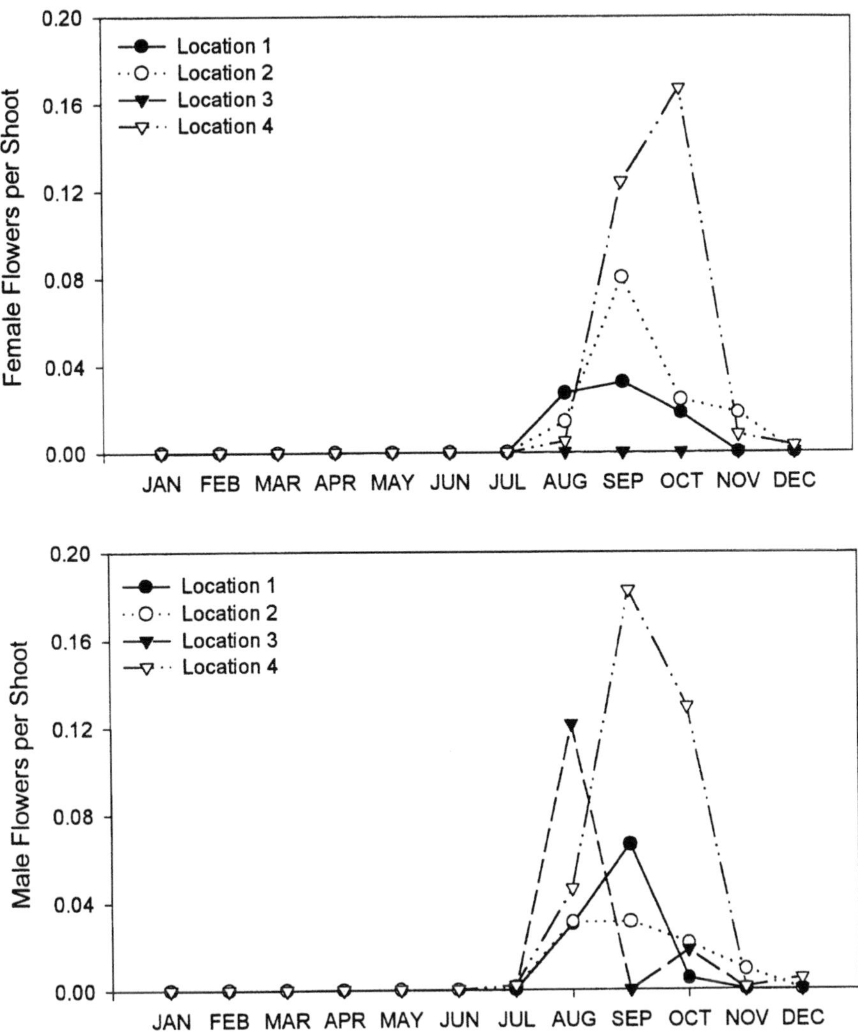

FIGURE 5.7 Monthly trend among locations for the number of female flowers per shoot (upper graph) and the number of male flowers per shoot (lower graph).

Interestingly, there were no significant correlations of any independent with any dependent variable.

Stepwise Regression

A stepwise regression procedure was used to elucidate the amount of variation in the dependent variables that could be explained by the data gathered during this study. The results are summarized in Table 5.1.

Inspection of these analyses indicates that only one or two of the alternative dependent variables can explain most of the variation within each of the dependent variables. For example, variability in the number of shoots can be almost completely predicted from the number of blades. Number of blades and number of shoots are clearly co-dependent as are blade length and blade width. Nearly 60% of the variation in shoot weight can be explained by blade length. Less reliable is the prediction of the number of male and female flowers as only 14.3 and 18.7%, respectively, of the variation of these parameters can be explained with the model offered here.

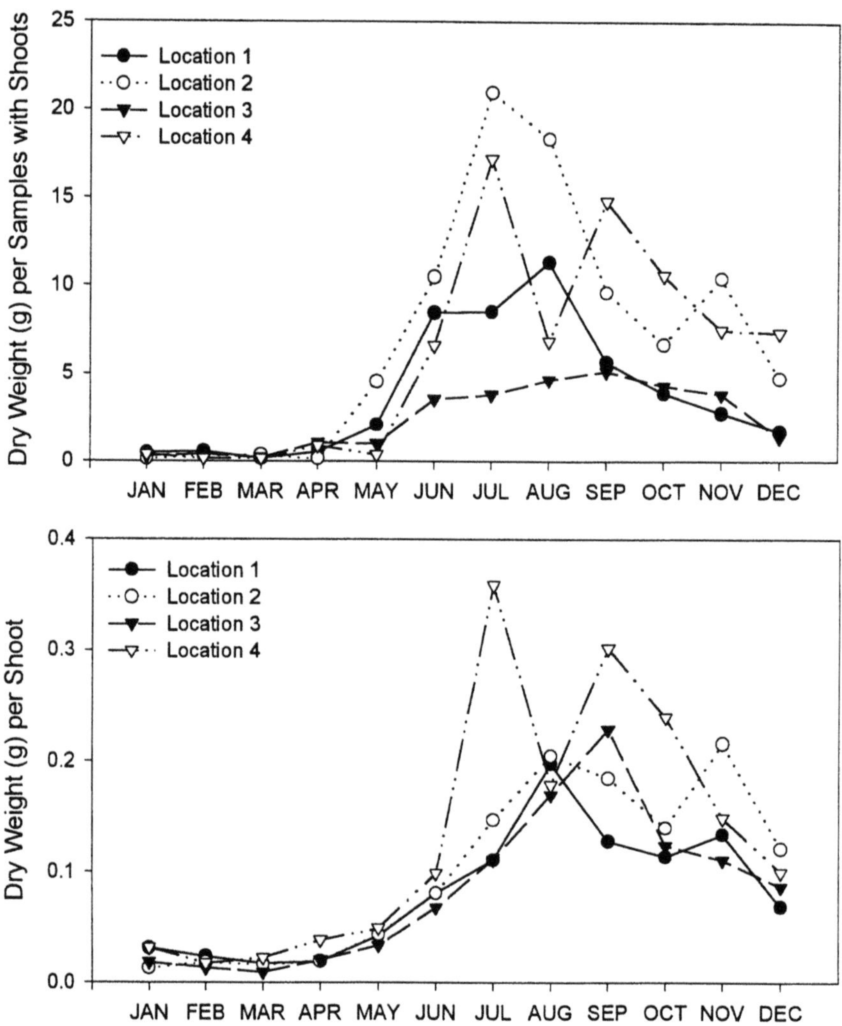

FIGURE 5.8 Monthly trend among locations for the dry weight per samples with shoots (upper graph) and the dry weight per shoot (lower graph).

The main purpose of this analysis, however, was to gain insight into the environmental variables that may be associated with variation in the dependent variables. The number of shoots was only significantly associated with variation in TSS; blade width with temperature; blade length with five factors (location, depth, temperature, Secchi depth, and color); blade width with location, depth, temperature, salinity, Secchi depth, chlorophyll, and color; number of male flowers with depth and color; number of female flowers with location, Secchi depth, and color; and weight with six factors (location, depth, temperature, Secchi depth, TSS, and color).

Of all the factors, location, depth, temperature, Secchi depth, and color are apparently the most influential in affecting (or being associated with) the dependent variables.

DISCUSSION

Donnermeyer and Smart (1985) noted that seasonal growth in tape grass was maximal during mid-to-late July in the freshwaters of Mississippi. The dependent, response variables recorded here for tape grass followed a similar seasonal pattern. There were, however, slight seasonal differences as

TABLE 5.1
Summary of the stepwise procedures to build a predictive model for each of the dependent variables. All values are significant at the 0.15 level.

Variable	No. Shoots	No. Blades	Blade Length	Blade Width	No. Male	No. Female	Weight
Date							
Location			0.001	0.005		0.002	0.001
Distance							
Depth			0.002	0.004	0.003		0.003
Temp		0.001	0.004	0.002			0.001
Salinity				0.001			
Secchi			0.001	0.001		0.002	0.001
TSS	0.001						0.007
Chlor a				0.006			0.002
Color			0.007	0.004	0.016	0.007	
No. Shoots		0.921		0.021	0.007	0.017	0.002
No. Blades	0.921		0.013				0.110
Blade Length		0.001		0.779	0.003	0.135	0.600
Blade Width	0.007	0.001	0.779				0.017
No. Male							0.011
No. Female	0.001		0.003				0.004
Weight	0.001	0.002	0.062	0.013	0.113	0.018	
Total R^2	**0.929**	**0.925**	**0.874**	**0.835**	**0.143**	**0.187**	**0.754**

to when the response variables were maximal. The number of shoots increased in the spring and stayed high in the summer while declining in the fall. This was also true for the number of blades and blade/shoot ratio. The attributes of the plants associated with size, i.e., blade length, blade width, and weight, all displayed greater values in mid-summer and late fall. The reproductive season for tape grass, indicated by the number of male and female flowers, was highest during fall. Tape grasses have seasonally lagged life history features which may each be useful in assessing temporal aspects of stress in estuaries.

The absolute number of plants in the area, as indicated by the number of samples with shoots (Figure 5.4), shows a slightly different seasonal pattern. The number of samples with shoots increased at three of the four locations (Location 1 was the exception). This increase is important because the populations of shoots at these locations were dramatically higher in December than they were during the previous January. In fact, closer inspection of all the study results indicates that monthly parameter levels for virtually all variables recorded here were higher during the end of the year than the beginning of the previous year.

This feature of the tape grass populations for 1998 in the Caloosahatchee River may be attributable to several factors. The levels of all variables are apparently cyclic (probably temperature related). Therefore, they should have declined to lower levels in December, similar to those of the previous January, if the study had been continued. Preliminary evidence indicates that, while parameter levels for all variables continued to decline, they were, nevertheless, higher than during the previous January (Bortone and Turpin, personal observation). A potentially testable hypothesis can be offered to explain the apparent annual increase in plant attributes during 1998. As indicated previously, the amount of annual rainfall during 1998 (especially during the early part of the year) was excessive. Since tape grass is primarily a freshwater plant, it has some growth inhibition when subjected to even moderate levels of salinity (Doering et al. 1999). As the plants were exposed to much lower salinities in 1998, it is reasonable to assume that they responded by becoming more numerous and larger during 1998. Therefore, the 1999 population parameters for beginning the "new" year would be higher than for the previous year.

Accepting this hypothesis leads to a paradox to explain the dependent variable responses. Generally, the highest plant parameters were recorded at Locations 2 and 4. Location 4 is furthest downstream, and plants at this location were thus subjected to the highest salinity. Interestingly, however, water clarity (as measured by Secchi depth) was highest at Locations 2 and 4 (122 and 110 cm, respectively) while water clarity was lowest at Locations 1 and 3 (102 and 108 cm, respectively). Thus, while salinities were higher, plant growth may have been accelerated because of increased light availability.

While an annual mean difference of only a few cm between Locations 3 and 4 may not seem important, it should be noted that the Secchi depths had to be measured some 200–400 m toward the center of the river. A Secchi depth measure could not be obtained at the specific site where plants were located because water depth at both locations was insufficient when water was clearest. Our impression at the time (as indicated in our field notes) was that water clarity was greater at Location 4. Site-specific measures of water clarity would probably have revealed a greater difference in water clarity between the two locations.

Doering et al. (1999) noted in the laboratory the tape grass plants from the Caloosahatchee River curtailed growth at salinities higher than 15 ppt. Also, laboratory (Doering et al. 1999) and confirming field observations (Kraemer et al. 1999) indicated that tape grass growth was inversely associated with salinity. An important balance must therefore be present for tape grasses to thrive in an estuary: enough water clarity to permit light penetration (generally associated with higher salinity water) and enough freshwater to reduce the stress associated with the effects of salinity. Future research efforts should more closely monitor the balance between the stress potential caused by higher salinity levels and the increases in plant growth afforded by higher levels of photosynthetically useful light waves resulting from increased water clarity.

Among the independent variables there was a linkage between location, depth, Secchi depth, and color. These variables are all location specific and should be expected to link together. Among the environmental variables identified as having a significant association with the response variables only temperature is not location dependent. It should be noted that color should be negatively related to Secchi depth because higher color levels interfere with water clarity.

CONCLUSIONS

Establishing a baseline of life history information on tape grass serves to help track changes in the aquatic conditions that lead to favorable SAV growth in the oligohaline portions of the Caloosahatchee River. Our investigation indicates that plant life history attributes vary seasonally and spatially. More important, however, is the realization that plant growth attributes may each respond on a different time scale to the presence of conditions favorable for plant growth. Even more important is the recognition of the need to measure environmental factors with a high degree of specificity with regard to proximity to the sample locations and with a high degree of accuracy, increased frequency, and greater precision. Our study has hinted at the significance of the interactive aspects of temperature, salinity, and water clarity. It behooves estuary managers to specifically determine the interactive effects these factors have on SAV growth in estuaries. Once this determination has been made, then managing estuaries to increase SAV growth (and subsequently provide an abundant, natural habitat for the community of organisms associated with SAV) will become a reality.

ACKNOWLEDGMENTS

We thank D. Haunert, P. Doering, R. Chamberlain, and A. Steinman of the South Florida Water Management District for providing guidance on the study design and for answering numerous questions regarding tape grass. We also thank T. Barnes for laboratory and sampling assistance. M. Giles and J. Evans reviewed an earlier draft of this manuscript. Lee County Environmental Laboratory staff (K. Kibby, Director) conducted the water quality analyses. Funding for this study was provided by the South Florida Water Management District and the Florida Center for Environmental Studies.

REFERENCES

Adair, S. E., J. L. Moore, and C. P. Onuf. 1994. Distribution and status of submerged vegetation in estuaries of the upper Texas coast. *Wetlands* 14(2):110–121.

Bortone, S. A., M. P. Smith, and R. K. Turpin. 1998. Indexed annotated bibliography to The Biology and Life History of Tape Grass *(Vallisneria americana).* Florida Center for Environmental Studies, Tech. Ser. 4:i–ii, Florida Atlantic University (Northern Palm Beach Campus), West Palm Beach, FL, pp. 1–77.

Chamberlain, R. H. and P. H. Doering. 1998. Freshwater inflow to the Caloosahatchee Estuary and the resource-based method for evaluation. In *Proceedings of the Charlotte Harbor Public Conference and Technical Symposium,* S. F. Treat (ed.). Charlotte Harbor National Estuary Program, Tech. Rep. No. 98-02, North Fort Myers, FL, pp. 81–90.

Dennison, W. C., R. J. Orth, K. A. Moore, J. C. Stevenson, V. Carter, S. Kollar, P. W. Bergstrom, and R. A. Batiuk. 1993. Assessing water quality with submerged aquatic vegetation. *Bioscience* 43:86–94.

Doering, P. H., R. H. Chamberlain, K. M. Donahue, and A. D. Steinman. 1999. Effect of salinity on the growth of *Vallisneria americana* Michx. from the Caloosahatchee Estuary, Florida. *Florida Scientist* 62(2):89–105.

Donnermeyer, G. N. and M. M. Smart. 1985. Biomass and nutritive potential of *Vallisneria americana* Michx. in Navigation Pool 9 of the Upper Mississippi River. *Aquatic Botany* 22(1):33–44.

Eaton, A. D., L. S. Clesceri, and A. E. Greenberg (eds.). 1995. *Standard Methods for the Examination of Water and Wastewater.* 19th Ed., American Public Health Association, Washington, D.C.

Johannson, J. O. R. 1991. Long-term trends of nitrogen loading, water quality and biological indicators in Hillsborough Bay, Florida. In *Proceedings, Tampa Bay Area Scientific Information Symposium 2,* S. F. Treat and P. A. Clark (eds.). Tampa, FL, pp. 157–176.

Kraemer, G. P., R. H. Chamberlain, P. H. Doering, A. D. Steinman, and M. D. Hanisak. 1999. Physiological responses of transplants of the freshwater angiosperm *Vallisneria americana* along a salinity gradient in the Caloosahatchee Estuary (Southwestern Florida). *Estuaries* 22(1):138–148.

SAS User's Guide 1995 edition. SAS Institute, Inc. Raleigh, NC.

Stevenson, J. C., L. W. Staver, and K. S. Staver. 1993. Water quality associated with survival of submerged aquatic vegetation along an estuarine gradient. *Estuaries* 16:346–361.

Twilley, R. R. and J. W. Barko. 1990. The growth of submerged macrophytes under experimental salinity and light conditions. *Estuaries* 13(3):311–321.

Zieman, J. C. and R. T. Zieman. 1989. The ecology of sea grass meadows of the west coast of Florida: a community profile. Biological Report 85 (7.25), U.S. Department of the Interior, Fish and Wildlife Service. Washington, D.C., 155 pp.

6 Experimental Studies on the Salinity Tolerance of Turtle Grass, *Thalassia testudinum*

Peter H. Doering and Robert H. Chamberlain

Abstract The effects of salinity on the growth and survival of turtle grass, *Thalassia testudinum*, were examined using laboratory mesocosms. Plants collected from the Caloosahatchee Estuary (southwest Florida) were exposed to five constant salinity treatments (6, 12, 18, 25, or 35‰) for 43 days. Two independent experiments were conducted: one during the winter dry season and the other during the summer wet season. Although there were differences between the two seasonal experiments, these differences were independent of salinity. *T. testudinum* survived exposure to a salinity of 6‰ for 6 weeks in both seasons. After 6 weeks exposure, number of blades, number of blades/shoot, and biomass of blades were similar in the range 12–35‰, but all three parameters exhibited a relative decrease at 6‰. The final length of blades achieved at the end of the experiment increased steadily as salinity increased. In general, production and elongation of new blades were observed at all salinities. Elongation rates of newly formed blades were positively correlated with salinity. The net production of blades was positive at salinities of 12‰ or greater, but ceased at 6‰. The phosphorus and nitrogen content of blades and rhizomes was inversely related to salinity. The biomass of blades in the mesocosm experiments was within the range observed in the Caloosahatchee. Both in the field and in the laboratory, blade biomass increased as salinity increased. Our experimental results suggest that given unlimited light, parameters of growth in *T. testudinum* are negatively impacted at salinities between 6 and 12‰.

INTRODUCTION

The supply of fresh water to estuaries has become an increasingly important issue for water managers (Postel et al. 1996). Anthropogenic modifications of river discharge by diversion, withdrawal, channelization, and damming have altered both the timing and quantity of freshwater delivery to estuaries (Whitfield and Wooldridge 1994; Hopkinson and Vallino 1995; Jassby et al. 1995). Unfortunately, the effects of altered freshwater discharge on downstream estuarine ecosystems and their component communities are not well understood (Hopkinson and Vallino 1995; Livingston et al. 1997). One estuarine ecosystem that is strongly impacted by altered freshwater discharge is the Caloosahatchee Estuary.

The Caloosahatchee Estuary is located on the southwest coast of Florida (Figure 6.1) and is part of the larger Charlotte Harbor system. The Caloosahatchee Estuary has been physically altered at both its freshwater and seaward ends. The major source of fresh water to the estuary is the Caloosahatchee River, which runs 65 km from Lake Okeechobee to the Franklin Lock and Dam (S-79, Figure 6.1). The river was first extended eastward and connected to Lake Okeechobee in 1884. The combination lock and dam structures at the towns of Moore Haven (S-77) and Ortona (S-78) were completed in 1937 to control river flow and discharge from Lake Okeechobee (Figure 6.1). The river was straightened and deepened in 1937, 1941, and again in 1966 to improve

0-8493-2045-3/99/$0.00+$.50
© 2000 by CRC Press LLC

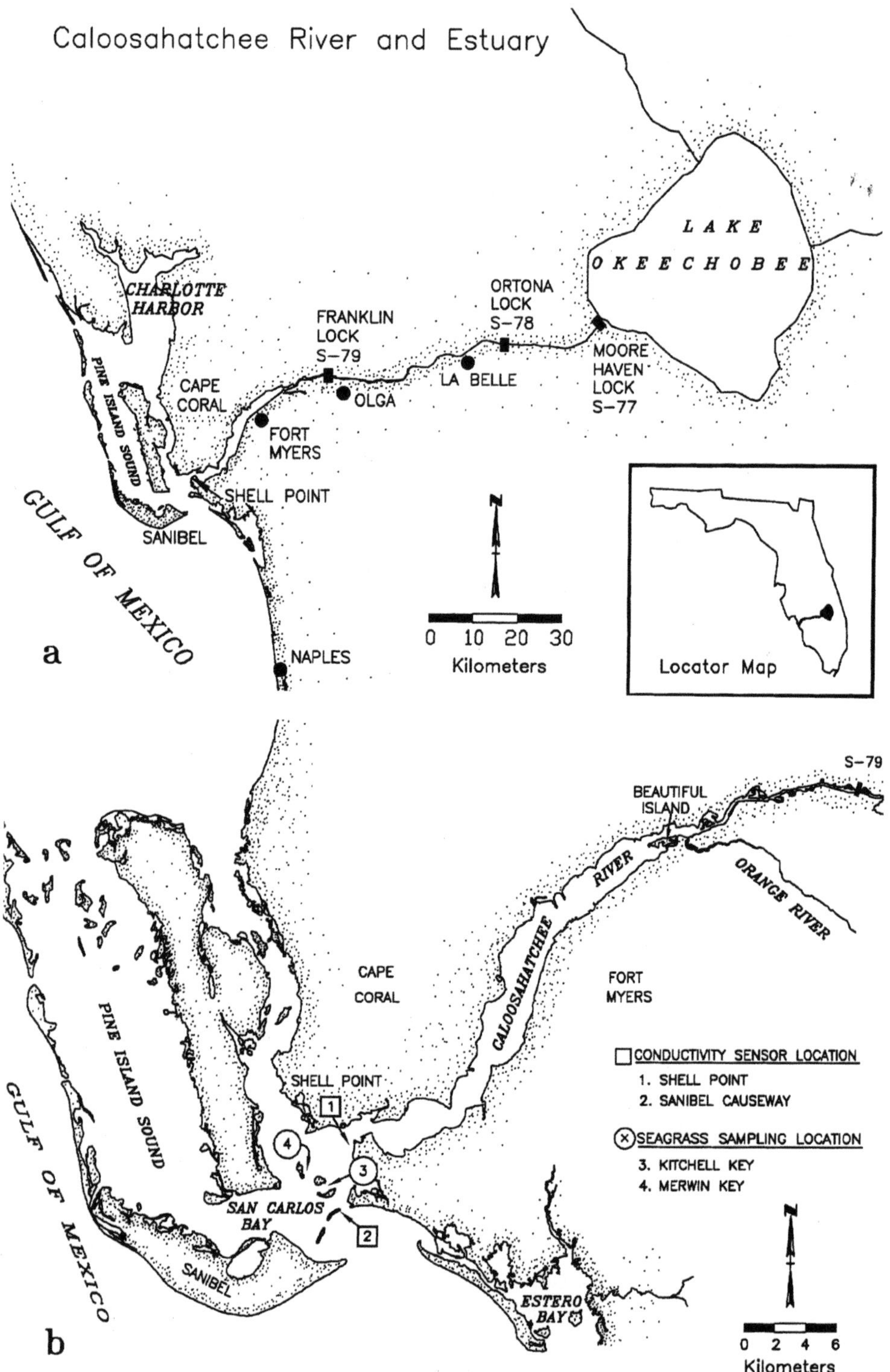

FIGURE 6.1 **a.** Map of the Caloosahatchee River and Estuary showing the connection to Lake Okeechobee and locations of water control structures. **b.** Map of the Caloosahatchee River and Estuary showing the locations of seagrass sampling stations and salinity recorders maintained by the South Florida Water Management District.

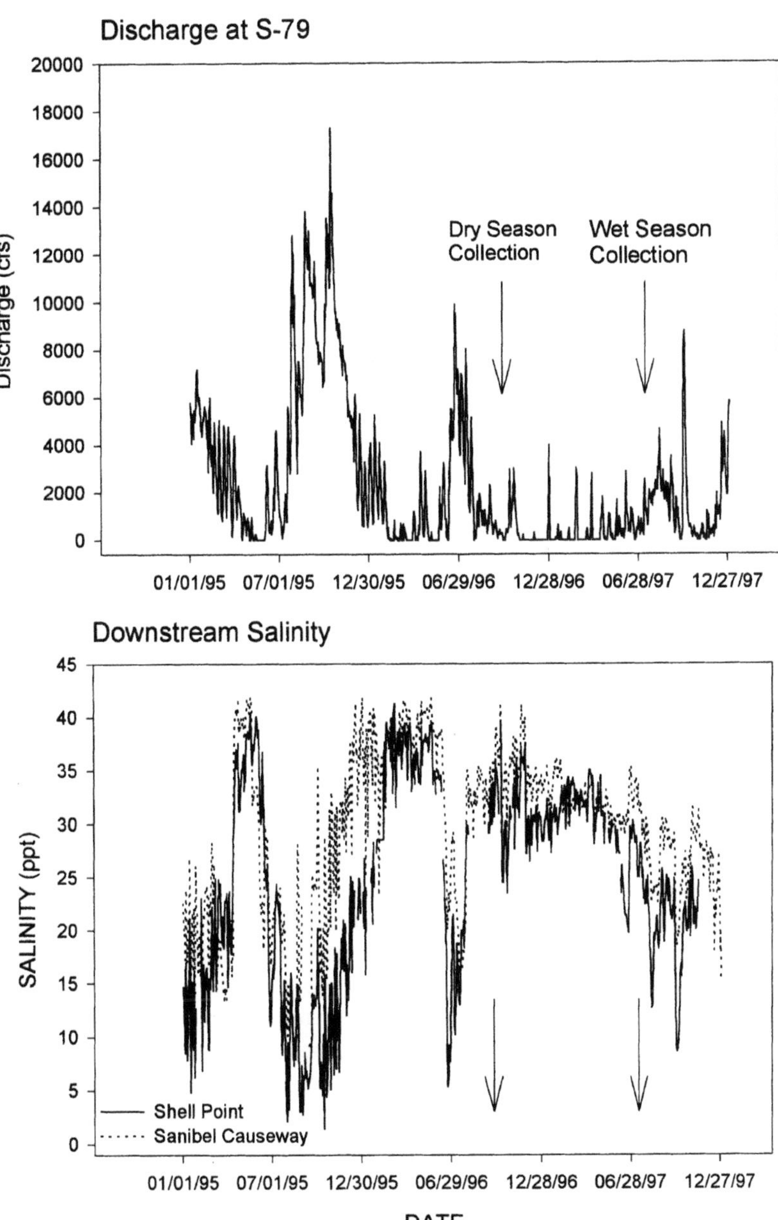

FIGURE 6.2 Daily average discharge (cfs) of fresh water at the Franklin Lock and Dam (S-79), and salinity at Shell Point and the Sanibel Causeway from records maintained by the South Florida Water Management District. Arrows mark the days upon which plants were collected for mesocosm experiments.

navigation and flood control. The final structure, the Franklin Lock and Dam (S-79), was added in 1966 to further control river flow and to act as a salinity barrier (Flaig and Capece 1998). Also in the 1960s, a causeway was constructed across the mouth of San Carlos Bay between Sanibel Island and Punta Rassa. Freshwater discharge at S-79 exhibits substantial temporal variation (Figure 6.2) and this variability drives spatial and temporal fluctuation in salinity in the downstream estuary (Doering et al. 1999; Doering and Chamberlain 1998).

As in other estuaries, seagrasses are a prominent component of the Caloosahatchee estuarine system. Seagrass meadows provide habitat for many benthic and pelagic organisms (Thayer et al.

1984), stabilize sediments (Fonseca and Fisher 1986), and can form the basis of plant-based and detrital-based food chains (Zieman and Zieman 1989). Hence, seagrasses and other submersed aquatic macrophytes often are a focal point of estuarine restoration efforts (SJRWMD and SFWMD 1994; Batiuk et al. 1992). Seagrasses can be a more sensitive barometer of anthropogenic impacts, such as nutrient loading, than traditional indicators of estuarine health. Thus seagrasses are prime candidates for inclusion in monitoring programs (Tomasko et al. 1996). Further, their environmental requirements can form the basis for water quality goals (Dennison et al. 1993; Stevenson et al. 1993).

In the Caloosahatchee system, beds of *Thalassia testudinum* occur downstream of Shell Point in San Carlos Bay and Pine Island Sound (Figure 6.1). Analysis of historical aerial photographs led Harris et al. (1983) to conclude that a 30% loss of seagrass occurred in the Charlotte Harbor system between 1945 and 1982. The authors attributed this loss to several factors, but chief among them were reduced salinities and light penetration attributed to alterations to the Caloosahatchee River and retention of fresh water behind the causeway. Variation in freshwater discharge at S-79 also has the potential to affect *T. testudinum* in San Carlos Bay. As data from continuous salinity recorders show (sites shown in Figure 6.1), during periods of high discharge (Figure 6.2), salinity falls well below optimum values (17–36‰, Zimmerman and Livingston 1976) and approaches lower tolerance limits for death (5‰, McMillan 1974).

This study was undertaken to (1) determine the influence of salinity on the growth and survival of *T. testudinum*; (2) better understand the effects of freshwater discharge and attendant salinity variation on the distribution and abundance of *T. testudinum*; and (3) help identify the maximum level of fresh water that could be discharged at S-79 without adverse impact on seagrass beds in San Carlos Bay.

MATERIALS AND METHODS

An experimental mesocosm approach was used to examine the influence of salinity on the growth and survival of *T. testudinum*. Plants were grown in ten cylindrical mesocosms (1.3 m in diameter × 1 m deep) filled with water to a depth of 60 cm (volume = 800 l). The mesocosms were located indoors at an aquarium facility at the Gumbo Limbo Nature Center in Boca Raton, Florida. One 1000 watt metal halide lamp, kept on a L:D photoperiod of 12:12 h, was suspended over each tank to provide light. A given salinity was maintained by mixing appropriate volumes of fresh and salt water (total volume = 114 l) from each of two head tanks located above each mesocosm. Head tanks were alternately filled and emptied into the mesocosms using solenoid valves controlled by timers. Thus, water was delivered to the mesocosms in a series of 114-liter pulses. Water in the mesocosms was replaced three times daily. Salt water was pumped from the Atlantic Ocean. Tap water, passed through a series of activated charcoal towers and filters (20 micron pore size) to remove chlorine, was used as a source of fresh water. Seasonal variation in salinity tolerance was examined by conducting two experiments: one during the dry season (November–April) and one during the wet season (May–October). Similar methods, summarized below, were employed during each experiment.

COLLECTION OF PLANTS

Plants and associated sediments were collected from a site adjacent to Merwin Key in San Carlos Bay (Figure 6.1). Plastic, rectangular tubs (18 cm H × 40.5 cm L × 27 cm W) were filled with sediment in the field. Sediment was collected with a shovel and passed through a 0.25 cm^2 mesh screen to remove shells, pebbles, and large infauna. Rhizomes, bearing 2–4 shoots and an apical meristem, were gently removed from the sediment by hand. Plants were placed in ice chests, covered with water from the site, and transported back to the laboratory on the day of their collection. Plants for the dry season experiment were collected on September 24, 1996 about a month before the end of the wet season (Figure 6.2). The major discharges of the wet season had subsided and salinity was already climbing in San Carlos Bay (Figure 6.2). Plants for the wet season experiment were collected on July 14, 1997, as seasonal discharges were beginning and salinity was falling (Figure 6.2).

Salinity Experiments

Upon return to the laboratory, plants were held overnight in seawater and planted in the tubs on the following day. Two or three rhizomes, with associated shoots, were planted to yield 4–7 shoots/tub. Tubs were randomly distributed among the 10 mesocosms with 4 tubs/mesocosm in the dry season experiment and 5 tubs/mesocosm in the wet season experiment.

Plants were allowed to acclimate in seawater (35‰) for 4 weeks (wet season experiment) and 7 weeks (dry season experiment). The dry season experiment had a longer acclimation period to ensure that the experiment was conducted during the dry season (November–April). Following acclimation, salinity in all treatments, except the 35‰ treatment, were adjusted to 25‰ and then lowered (1.5‰/day) as needed to final treatment salinities of 35, 25, 18, 12, and 6‰. These treatment levels were chosen to encompass the range of salinities observed in the field (Figure 6.2). Two mesocosms were randomly assigned to each salinity treatment. Plants were held at treatment salinities for 43 days. The dry season experiment commenced on November 21, 1996, and the wet season experiment on August 21, 1997. Tanks were scrubbed as necessary to remove wall growth. Once a week, epiphytic growth was gently removed by hand from plant blades. Salinity and temperature were monitored daily with a YSI 600XL data sonde. PAR was checked weekly with a LI-COR spherical quantum sensor and data logger and adjusted to 475–525 μmol photons/m^2/sec at the bottom of a tank by raising or lowering the lamp. Water samples were analyzed weekly for dissolved inorganic nitrogen and phosphorus on an Alpkem Auto Analyzer (APHA 1985; SFWMD 1994).

Plant Response Variables

The following non-destructive measurements were made weekly during each experiment: number of shoots per tub, number of blades per shoot, and blade length of 15 randomly selected blades per tub. In addition, one newly formed blade in each tub was marked each week with a piece of thread sewn through the blade. Length was measured at the time of marking and after 1 week. This procedure yielded an elongation rate for each tub during each week of the experiment.

At the end of each experiment, all plants were harvested and plant material was separated into blades, roots, and rhizomes and dried to constant weight at 80°C for biomass determination. Tissue nitrogen was determined by elemental analysis on a Fissons CNS Analyzer Series 2 against a BBOT ($C_{26}H_{26}N_2O_2S$) standard. Analytical replicability was ±5%. Recovery of National Institute of Standards and Technology reference material averaged 97.6 and 99.9%, respectively, for nitrogen in peach and spinach leaves. Tissue P was analyzed by a method modified from Solorzano and Sharp (1980). Ground 50 mg samples were ashed (550°C) and hydrolyzed at 80°C with 0.025 *N* HCl. Hydrolyzed samples were diluted 1:100 with distilled water and analyzed for soluble reactive phosphorus on an Alpkem Auto Analyzer. Analytical replicability was ±5%, and recovery of P from NIS reference material averaged 92.7 and 93.7%, respectively, for peach and spinach leaves.

Statistical Analysis

Data were analyzed using a three-factor-nested analysis of variance (ANOVA) design. Season (wet or dry) and salinity (6, 12, 18, 25, and 35‰) were considered fixed factors, while mesocosms (the two mesocosms at each salinity) were considered random and nested within the salinity factor. To avoid pseudoreplication, data taken on a given day in each mesocosm were averaged across tubs. This procedure yielded one observation from each of the two mesocosms assigned to each season × salinity combination (n = 2). In such a design, the replicate mesocosm is the experimental unit.

Rules for determining F-tests in nested ANOVA designs are summarized in Winer (1971). Differences between main effect and interaction means were evaluated with the Student–Newman–Keuls (SNK) test (Winer 1971). The ANOVA was used to evaluate relative treatment differences in plant biomass, chemical composition, and morphology, measured at the end (day 43) of the experiment. Non-destructive measurements, taken on day 1 of the experiment, were similarly analyzed.

As applied here, the ANOVA was used to examine differences between treatments at the beginning and end of the experiments. To determine if salinity affected the net production of shoots and blades, net growth rates were calculated for each tub by difference between the end (day 43) and beginning (day 1) of each experiment. Average net changes were calculated for each salinity treatment and tested for a difference from zero using the *t*-test (two-tailed).

Elongation rates of newly formed blades, measured by marking with a thread, were analyzed by repeated measures ANOVA. Factors in the design were season, salinity, and time (the weekly measurements) being the repeated measures factor. To avoid pseudoreplication, rates for each week were averaged across tubs in a mesocosm. All statistical analyses were accomplished using SAS Version 6 (SAS 1989).

Field Observations

Above-ground (blade) biomass samples (grams dry wt/m^2) of *T. testudinum* were collected from the Merwin Key site and a site near Kitchell Key at monthly intervals during two periods: February 1986–May 1989, when discharges at S-79 were low to moderate, and November 1994–December 1995, when discharges were often high. An additional sampling was conducted in April 1998 after a period of very high discharge. At each sampling, above ground biomass was harvested from six randomly placed 0.1 m^2 quadrats. Material was oven-dried (90°C) to determine dry weight biomass. Data were averaged by sampling date. Salinity on the day of sampling was measured with a Hydrolab Surveyor III or YSI 600 XL data sonde.

RESULTS

With minor exceptions, measured salinities over the courses of both experiments were maintained within ±2‰ of nominal treatment salinities (Figure 6.3, Table 6.1). Temperature in the ten mesocosms averaged between 24.5 and 25.5°C during the dry season and between 25 and 28°C during the wet season (Table 6.1). Three-way ANOVA (season, salinity, and mesocosms) revealed that mean dissolved inorganic nitrogen (DIN) and dissolved inorganic phosphorus (DIP) concentrations decreased with increasing salinity treatment in both seasons (6 > 12 > 18 > 25 > 35‰, SNK test $p < 0.05$). Mean concentrations of both nutrients also were significantly ($p < 0.05$) higher in the wet season than in the dry season (Table 6.1).

Morphometric Measurements

Statistically significant differences (ANOVA, $p < 0.05$) were found in the initial conditions of the two experiments (Figure 6.4). There were more shoots/tub and longer blades at the beginning of the wet season experiment, but more blades/shoot and more blades/tub at the beginning of the dry season experiment (Figure 6.4). Because season and salinity were completely crossed in the experimental design, these seasonal differences were independent of salinity. In each experiment, initial conditions were similar among salinity treatments (main effect of salinity, $p > 0.05$; season × salinity interaction, $p > 0.05$).

After 43 days, one or both of the main effects of season and salinity were statistically significant for blades/tub, shoots/tub, blades/shoot, and blade length. No statistically significant interactions (season × salinity) were detected for any parameter, signifying the independence of the effects of salinity and season. The seasonal differences present at the beginning of the experiments persisted

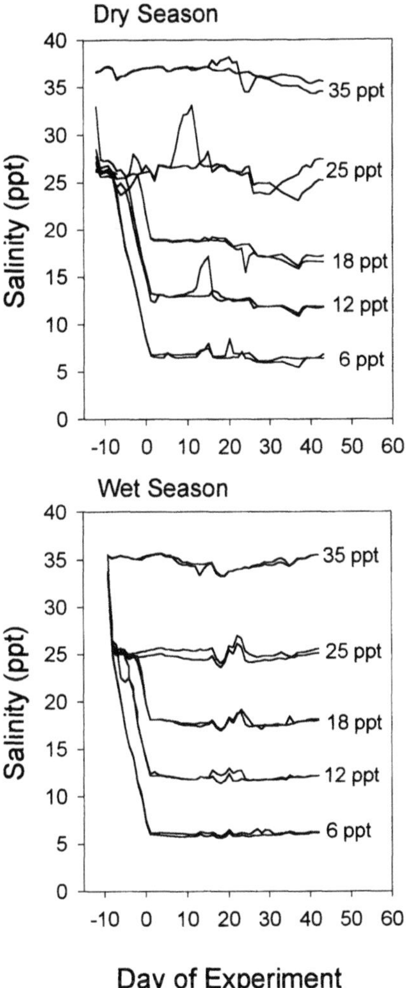

FIGURE 6.3 Time course of salinity in the mesocosms during the two seasonal experiments. The two lines per treatment represent the replicate mesocosms.

at the end and were significant at $p < 0.05$, except for the difference in shoots which was significant at $p < 0.10$ (Figure 6.4).

Salinity effects also were detected for the average number of blades/tub and the number of blades/shoot and confined to the 6‰ treatment (Figure 6.5). Results of the SNK test ($p < 0.05$) revealed the following grouping of treatment means: 6‰ < (12 = 18 = 25 = 35‰). Number of shoots/tub showed the same pattern but no statistical differences were found (Figure 6.5). Blade length increased as salinity increased, with blades being the shortest in the 6 and 12‰ treatments, of intermediate length in the 18 and 25‰ treatments, and longest in the 35‰ treatment (Figure 6.6).

Rates of Growth

Elongation rates (cm/day) of newly formed blades were significantly influenced by the main effect of salinity ($p < 0.05$) with growth rates increasing as salinity increased (Figure 6.6). No significant difference existed between seasons, nor was the season × salinity interaction statistically significant at $p < 0.05$. There were significant interactions between salinity, season, and time, indicating that elongation rates changed from week to week during the experiments, depending on salinity and

TABLE 6.1
Average (std) salinity (‰, n = 43), temperature (°C, n = 43), and concentrations (mg/l, n = 6) of dissolved inorganic nitrogen (DIN = NH_4 + NO_2 + NO_3) and dissolved inorganic phosphorus (DIP) in the two mesocosms assigned to each salinity.

	Dry Season				Wet Season			
Treat	Salinity	Temp	DIN	DIP	Salinity	Temp	DIN	DIP
6‰	6.5(0.4)	25.2(0.3)	0.23(0.13)	0.009(0.003)	6.2(0.2)	26.6(0.3)	0.33(0.05)	0.032(0.041)
	6.8(0.4)	25.3(0.3)	0.24(0.17)	0.010(0.003)	6.0(0.1)	26.9(0.2)	0.34(0.05)	0.033(0.044)
12‰	12.6(0.6)	25.1(0.3)	0.21(0.13)	0.005(0.003)	12.0(0.2)	26.6(0.3)	0.24(0.05)	0.022(0.032)
	12.8(1.3)	25.0(0.3)	0.19(0.12)	0.006(0.004)	12.2(0.3)	26.5(0.3)	0.26(0.06)	0.020(0.031)
18‰	18.2(0.9)	25.2(0.4)	0.13(0.07)	0.004(0.002)	17.2(1.1)	25.6(3.9)	0.15(0.06)	0.012(0.02)
	18.1(1.0)	25.2(0.4)	0.12(0.08)	0.004(0.003)	17.8(0.4)	27.3(0.3)	0.17(0.04)	0.022(0.05)
25‰	26.2(1.1)	24.2(0.4)	0.07(0.06)	0.003(0.002)	25.3(0.5)	27.2(0.4)	0.13(0.07)	0.009(0.009)
	26.6(2.3)	24.7(0.5)	0.10(0.07)	0.003(0.002)	24.7(0.5)	27.7(0.4)	0.11(0.03)	0.010(0.007)
35‰	36.4(0.9)	24.7(0.5)	0.03(0.05)	0.002(0.0)	34.6(0.7)	26.6(0.6)	0.08(0.11)	0.004(0.004)
	36.6(0.9)	24.9(0.5)	0.03(0.04)	0.002(0.001)	34.8(0.6)	28.1(0.6)	0.04(0.05)	0.005(0.003)

season. Graphical inspection of the data (not shown) suggested that rates tended to decrease during the experiment in both seasons at 6‰ while remaining constant at 25 and 35‰. At 18‰, rates remained constant (wet season) or increased (dry season), while at 12‰, rates remained constant (wet season) or decreased (dry season).

Absolute net changes in number of shoots per tub (Table 6.2) achieved over the 43-day experiments were small (0–1.5 shoots/tub). In the dry season, there were significant (non-zero) changes, but the relationship to salinity was not linear. In the wet season, positive growth occurred at salinities of 18‰ or above (Table 6.2). The number of blades/tub increased at salinities of 12‰ or above. Net production of blades ceased at 6‰.

Biomass

Total dry weight biomass was higher in the dry season (4.4 g dry wt/tub) than in the wet season (3.8 g dry wt/tub, $p < 0.05$), but did not vary with salinity in either season. Above-ground biomass of blades did not vary seasonally, but was significantly lower at 6‰ (0.56 g dry wt/tub, $p < 0.05$) than at higher salinities, which were not statistically different from one another (range: 0.89–1.08 g dry wt/tub). Below-ground tissues (roots + rhizomes) accounted for most of the total biomass. Like the total, they were higher in the dry season (3.5 g dry wt/tub) than in the wet season (2.9 g dry wt/tub, $p < 0.05$) and did not vary with salinity.

Elemental Composition

The concentration of phosphorus in blade and rhizome tissue varied both with season and salinity (main effects, $p < 0.05$, no interaction). Concentrations in both tissues were higher in the wet season (2.5 and 3.8 μg P/mg dry wt for blades and rhizomes, respectively) than in the dry season (2.1 and 3.1 μg P/mg dry wt for blades and rhizomes, respectively). In general the concentration of P in rhizome tissue was higher than in blades (Figure 6.7). The phosphorus concentration decreased in both tissues as salinity treatment increased (Figure 6.7). The concentration of P in roots also was higher ($p < 0.05$) in the wet season (0.86 μg P/mg dry wt) than in the dry season (0.68 μg P/mg dry wt) and lower

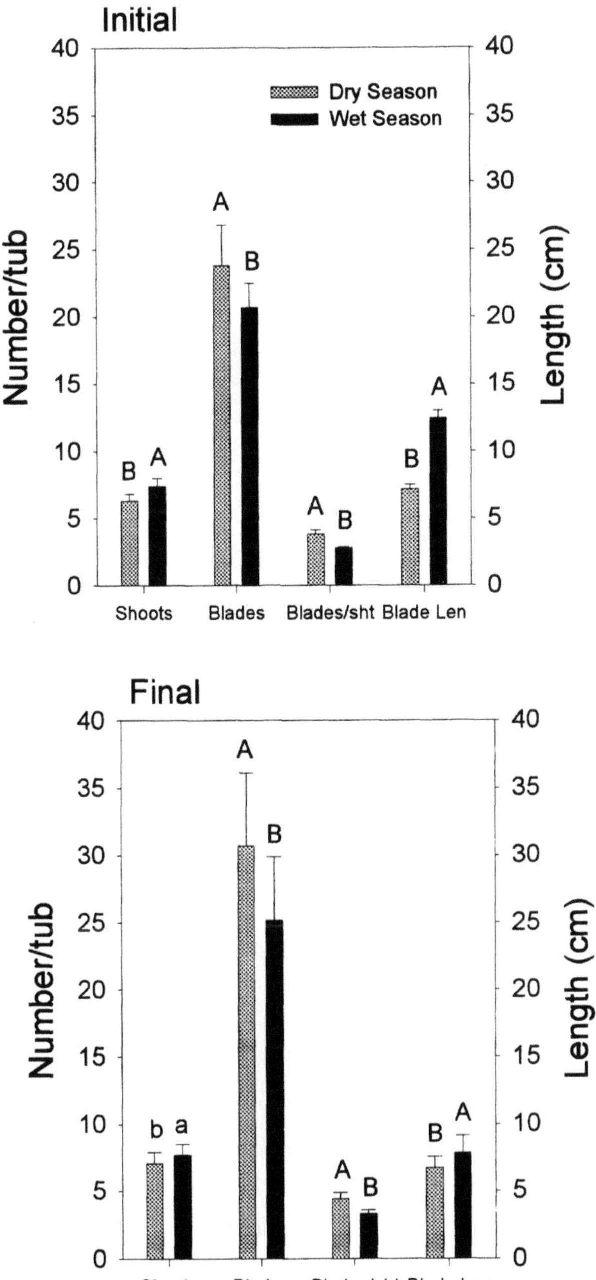

FIGURE 6.4 Comparison of initial conditions of the two seasonal experiments and of final conditions of the two seasonal experiments. Bars are seasonal means ± standard deviation (n = 10). Pairs of bars with different letters indicate a statistical difference between the two experiments. Upper case letters indicate $p < 0.05$, lowercase letters indicate $p < 0.10$.

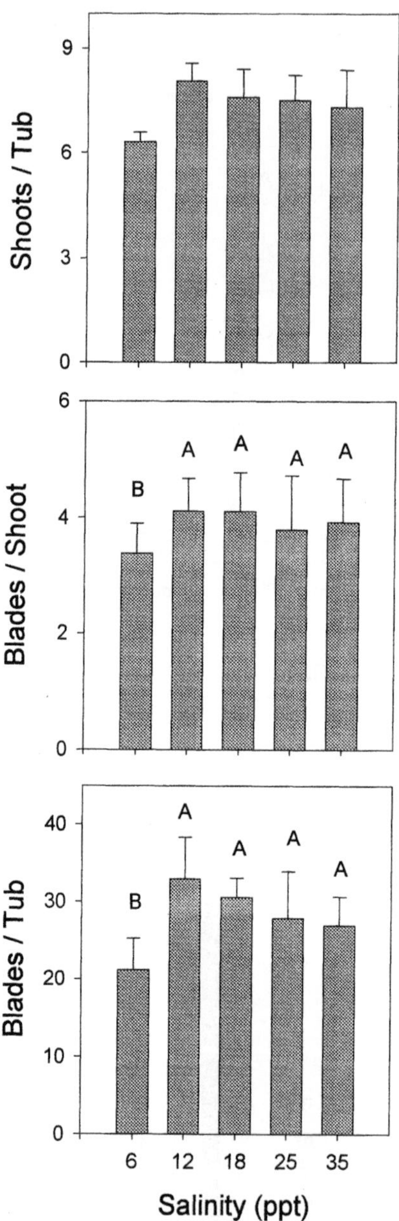

FIGURE 6.5 Combined average (± standard deviation, n = 4) number of shoots/tub, number of blades/shoot, and number of blades/tub, for both experiments, as a function of salinity treatment after 43 days of exposure. Letters indicate statistical differences between salinity treatments. Means with different letters are statistically different at $p < 0.05$.

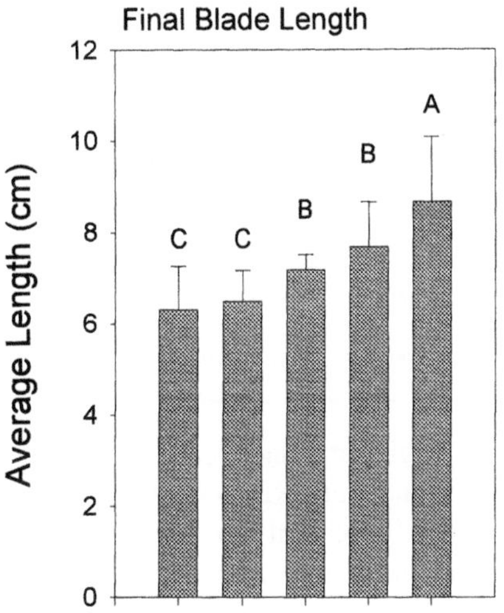

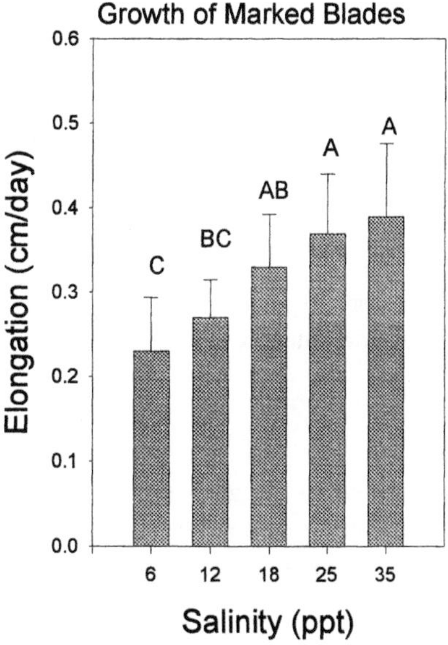

FIGURE 6.6 Average length of blades (± standard deviation, n = 4) achieved in the salinity treatments after 43 days of exposure, and average elongation rate (± standard deviation, n = 24) of newly formed blades. Averages include both experiments. Letters indicate statistical differences between salinity treatments. Means with different letters are statistically different at $p < 0.05$.

TABLE 6.2
Average (std) net change in number of shoots/tub and blades/tub in each salinity treatment during two seasonal experiments.

		Salinity		Treatment		
Attribute	Season	6‰	12‰	18‰	25‰	35‰
Shoots	Dry	0.0 (0.93)ns	1.3 (1.2)	0.8 (1.0)ns	1.1 (0.8)	0.5 (1.4)ns
	Wet	−1.4 (1.5)	0.1 (1.4)ns	1.0 (1.2)	0.8 (1.0)	1.0 (0.8)
Blades	Dry	2.0 (3.8)ns	9.4 (4.8)	7.3 (3.7)	9.4 (2.1)	6.3 (2.7)
	Wet	−3.7 (6.4)ns	7.1 (6.5)	9.1 (5.2)	3.2 (3.1)	6.4 (5.3)

Note: n = 8 tubs for dry season, n = 10 for wet season, ns = not statistically

than in the other two tissues. Although some statistical differences between salinity treatments were detected in both seasons, these did not vary with salinity in a regular manner.

The concentration of nitrogen in blades and rhizomes varied with season and salinity in patterns similar to phosphorus. Concentrations were higher in the wet season (28.3 and 18.0 μg N/mg dry wt for blades and rhizomes, respectively) compared with the dry season (26.3 and 15.0 μg N/mg dry wt for blades and rhizomes, respectively). The nitrogen concentration also decreased in both tissues as salinity treatment increased (Figure 6.7). By contrast, the concentration of N was lower in rhizome tissue than in blade tissue. Thus, molar N:P ratios were higher in blades (26–28) than in rhizomes (9–12). Nitrogen in root tissue was higher in the wet season (12.4 μg N/mg dry wt) than in the dry season (9.5 μg N/mg dry wt), but did not vary with salinity.

Field Observations

Above-ground biomass of *T. testudinum* blades in the field was significantly correlated (non-parametric Spearman's rank correlation) with salinity measured on the day of collection (Figure 6.8). Biomass data from the laboratory experiments described here fell within the range observed in the field and also exhibited a significant, positive correlation with salinity (Figure 6.8).

DISCUSSION

Submersed aquatic angiosperms found in estuaries are either freshwater species, with some salt tolerance, or marine species, with some tolerance to salinities below oceanic levels (Kemp et al. 1984). Typically, marine species dominate higher salinity regions (> 20‰) while salt-tolerant freshwater species inhabit the lower salinity regions (Kemp et al. 1984). Establishing the effects of salinity on seagrasses and other submerged aquatic angiosperms is key to understanding their occurrence, distribution, abundance, and performance in estuarine systems.

Our results demonstrate that *T. testudinum* can survive a salinity of 6‰ for at least 6 weeks. However, growth is curtailed or arrested at this salinity and the magnitude of suppression appears sensitive to specific growth measures. After 6 weeks of exposure, the number of blades/tub, number of blades/shoot, and biomass of blades were similar in the range of 12–35‰ but all three parameters exhibited a relative decrease at 6‰. The final length of blades achieved varied across the entire range of treatment salinities, increasing steadily as salinity increased. Blade length may be more sensitive to salinity variation than the other morphometric parameters that were measured.

Changes in number of shoots over the course of the experiment were small and did not show a clear response to salinity. Production and elongation of newly formed blades were observed at

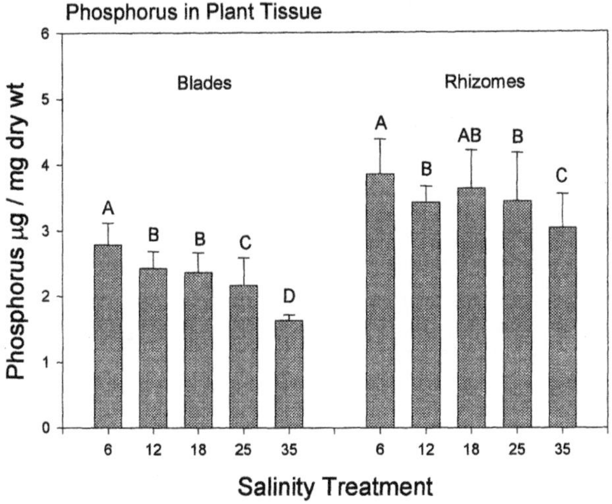

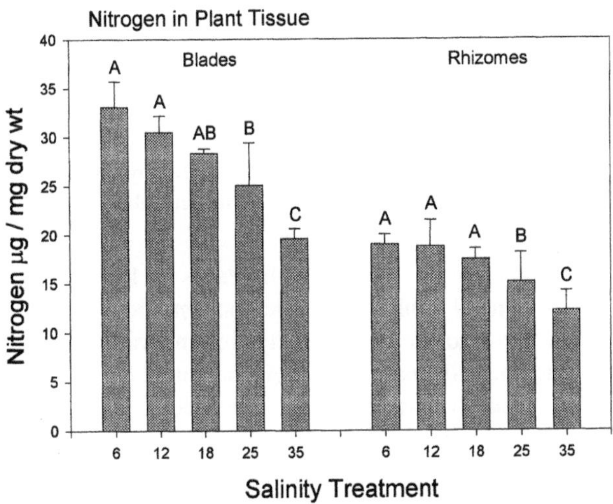

FIGURE 6.7 Combined average concentration (± standard deviation, n = 4) of nitrogen and phosphorus in plant tissue, for both experiments, as a function of salinity treatment, after 43 days of exposure. Letters indicate statistical differences between salinity treatments. Means with different letters are statistically different at $p < 0.05$.

all salinities. Elongation rates increased as salinity increased. Elongation rates of individual blades were within the range reported for *T. testudinum* in Florida (0.2–0.5 cm/day, Dawes 1987). The net production of blades was positive at salinities of 12‰ or greater, but ceased at 6‰. Thus, at 6‰, the production of new blades was offset by loss of older blades.

This experiment was designed to test the effects of salinity on the growth and survival of *T. testudinum*. The impacts of salinity are difficult to assess because other factors, such as nutrient concentrations, can co-vary with salinity even under the controlled conditions employed here. Whether availability of nutrients accounted for the differences between salinity treatments is difficult to assess. Critical tissue nutrient concentrations are often used to indicate potential nutrient limitation in aquatic angiosperms (Gerloff and Krombholz 1966). There is some debate concerning the validity of the approach for seagrasses (Udy and Dennison 1997). Nevertheless, Duarte (1990)

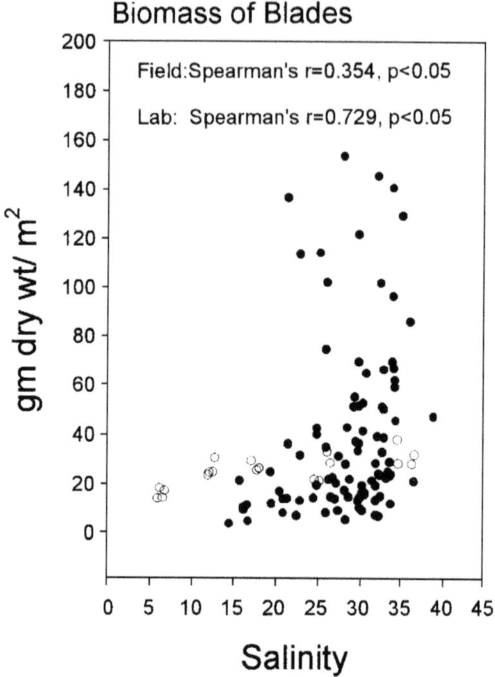

FIGURE 6.8 Biomass of blades at two field sites and salinity on the day of collection (filled circles). Average biomass of blades in the mesocosms and average salinity during the laboratory experiments (open circles).

identified potential critical nutrient concentrations in leaves of 1.8% dry weight N and 0.2% dry weight P for seagrasses in general. Only leaf tissue concentrations in the 35‰ treatment, where growth was highest, were close to these potentially limiting values (Figure 6.7). Rather, tissue concentrations in this study usually exceeded these potentially limiting values. Light was not confounded with salinity, since PAR was maintained between 475 and 525 μmol photons/m²/sec in all treatments. This intensity is well above that required for a positive carbon balance for the whole plant (40 μmol photons/m²/sec, Fourqurean and Zieman 1991) and saturation intensities for leaves (66–163 μmol photons/m²/sec, Dawes 1998) or whole plants (290 μmol photons/m²/sec, Herzka and Dunton 1997).

Elemental Composition

The concentrations and molar ratios of N and P that we observed in leaves of *T. testudinum* from the Caloosahatchee were within the range reported for this species by other workers (Jensen et al. 1998; Fourqurean et al. 1992: Duarte 1990). In our experiment, the nutrient content of blade and rhizome tissue decreased as salinity increased (Figure 6.7). The observed patterns of tissue nutrient concentrations may reflect the fact that growth rates increased with increasing salinity. As growth rate increased, internal nutrient pools may have been diluted by growth, resulting in lower tissue concentrations (Jensen et al. 1998). Water column concentrations of DIN and DIP also decreased as salinity increased (Table 6.1), and tissue concentrations may have reflected these conditions in the overlying water. Without measurements of nutrients in the fresh and salt water sources, it is not possible to determine (1) whether the decrease in water column nutrient concentrations with increasing salinity arose from mixing of the two sources, representing a true exposure gradient or (2) whether it was due to assimilation by the plants.

TABLE 6.3
Summary of literature concerning the salinity tolerance of turtle grass, *Thalassia testudinum*.

Tolerance	Conditions	Location	Reference
Optimum 28–36 ppt Lowest 10 ppt[b]	Field Observations	Florida	Phillips 1960
Upper Limit 60 ppt	Indoor Tanks, Outdoor Ponds, 55-day exposure[a]	Texas	McMillan and Moseley 1967
Died at 5 ppt Survival 10–50 ppt Upper 60 ppt	Laboratory, 2-week constant exposure	Texas	McMillan 1974
Optimum 17–36 ppt Lowest 6 ppt[b]	Field Observations	Florida	Zimmerman and Livingston 1976
Optimum 24–35 ppt	Literature Summary	Florida	Zieman and Zieman 1989 Zieman 1982
Occurrence 30–40 ppt	Field Observations	Texas	Adair et al. 1994
Optimum 22–36 ppt	Field Observations	Florida	This study
Survival 6–35 ppt	Indoor Tanks	Florida	This study

[a] Increased salinity from 32 to 67 ppt during experiment.
[b] One sampling date.

Comparison with Other Studies

Most information concerning the salinity tolerance of *T. testudinum* derives from field observations which suggest an optimum salinity range of 17–36‰ (Table 6.3). Our field data, which show high blade biomass in the range of 22–36‰, are in general agreement. Field observations also indicate that *T. testudinum* can withstand low salinities in the range of 6–10‰ (Table 6.3). Since such observations are invariably limited to one sampling date, the length of time *T. testudinum* can tolerate low salinity remains unknown.

We found only two manipulative experiments that addressed the salinity tolerance of *T. testudinum* (Table 6.3). One examined tolerance to high salinity, raising the salinity from 32 to 67‰ over a 55-day period (McMillan and Moseley 1967). The other also addressed low salinity and results indicated that *T. testudinum* did not retain green tissue after more than a few days at 5‰, but survived at 10‰ for at least 2 weeks (McMillan 1974). Having survived exposure to 6‰ for 6 weeks, our experiment affirms that *T. testudinum* is tolerant to low salinity.

Comparison with Field Data

Our results support the contention that the correlation between salinity and biomass, observed in the field is, at least in part, the consequence of a cause and effect relationship. Although we did not produce high biomass at salinities of 22–36‰, laboratory data fell within the range of observations from the field (Figure 6.8). Both sets of data exhibit a statistically significant correlation with salinity. This correspondence suggests that salinity can be an important determinant of biomass in the field.

The effect of salinity on the biomass of *T. testudinum* in San Carlos Bay appears most likely to be mediated through its effects on growth, length, and number of blades. In the lab, effects on number of blades were detected only at 6‰. By contrast, effects on growth and length of blades were evident even as salinity decreased from 35 to 25‰. All these effects could lead to lower biomass.

The role of mortality or complete defoliation of leaves induced by salinity fluctuation in San Carlos Bay is less certain simply because salinity does not appear to reach critical levels either very frequently or for very long. At Sanibel Causeway (the downstream boundary of San Carlos Bay), salinity did not fall below 8‰ during the 3-year record between 1995 and 1997 (Figure 6.2). At the upstream boundary (Shell Point), salinity fell below 6‰ only 12 times in 3 years with a maximum duration of 7 days and a median duration of 2 days. For one period, salinity remained below 7‰ for 22 days. In San Carlos Bay, salinity does not appear to often fall below the 6‰ that *T. testudinum* survived in the lab.

ACKNOWLEDGMENTS

We thank Keith Donohue and Mike McMunigal of the Harbor Branch Oceanographic Institution and Callie Buchanan of Florida Atlantic University for technical assistance in the laboratory. We thank Dan Crean and Kathy Haunert of the South Florida Water Management District for their help in collecting and processing field samples.

REFERENCES

Adair, S. E., J. L. Moore, and C. P Onuf. 1994. Distribution and status of submerged vegetation in estuaries of the Upper Texas Coast. *Wetlands* 14:110–121.

American Public Health Association, American Water Works Association, and Water Pollution Control Federation. 1985. *Standard Methods for Examination of Waste Water.* 16th Ed., American Public Health Association, Washington, D.C.

Batiuk, R. A., R. J. Orth, K. A. Moore, W. C. Dennison, J. C. Stevenson, L. W. Staver, V. Carter, N. B. Rybicki, R. E. Hickman, S. Kollar, S. Bieber, and P. Heasly. 1992. Chesapeake Bay submerged aquatic vegetation habitat requirements and restoration targets: a technical synthesis. Chesapeake Bay Program CBP/TRS 83/92, 186 pp.

Dawes, C. J. 1987. The dynamic seagrasses of the Gulf of Mexico and Florida Coasts. Florida Department of Natural Resources, Bureau of Marine Research, St. Petersburg, FL. Florida Marine Research Publications 42:25–52.

Dawes, C. J. 1998. Biomass and photosynthetic responses to irradiance by a shallow and a deep water population of *Thalassia testudinum* on the west coast of Florida. *Bulletin of Marine Science* 62:89–96.

Dennison, W. C., R. J. Orth, K. A. Moore, J. C. Stevenson, V. Carter, S. Kollar, P. W. Bergstrom, and R. A. Batiuk. 1993. Assessing water quality with submersed aquatic vegetation. *Bioscience* 43(2):86–94.

Doering, P. H. and R. H. Chamberlain. 1998. Water quality in the Caloosahatchee Estuary, San Carlos Bay and Pine Island Sound. In *Proceedings of the Charlotte Harbor Public Conference and Technical Symposium, March 15–16, 1997,* Punta Gorda, FL. Charlotte Harbor National Estuary Program Technical Report No. 98-02, pp. 229–240.

Doering, P. H., R. H. Chamberlain, K. M. Donohue, and A. D. Steinman. 1999. Effect of salinity on the growth of *Vallisneria americana* Michx. from the Caloosahatchee Estuary, Florida. *Florida Scientist* 62(2):89–105.

Duarte, C. M. 1990. Seagrass nutrient content. *Marine Ecology Progress Series* 67:201–207.

Flaig, E. G. and J. Capece. 1998. Water use and runoff in the Caloosahatchee watershed. In *Proceedings of the Charlotte Harbor Public Conference and Technical Symposium, March 15–16, 1997,* Punta Gorda, FL. Charlotte Harbor National Estuary Program Technical Report No. 98–02, pp. 73–80.

Fonseca, M. S. and J. S. Fisher. 1986. A comparison of canopy friction and sediment movement between four species of seagrass and with reference to their ecology and restoration. *Marine Ecology Progress Series* 29:15–22.

Fourqurean, J. W. and J. C. Zieman. 1991. Photosynthesis, respiration and whole plant carbon budgets of *Thalassia testudinum, Halodule wrightii* and *Syringodium filiforme*. In *The Light Requirements of Seagrasses: Proceedings of a Workshop to Examine the Capability of Water Quality Criteria, Standards, and Monitoring Programs to Protect Seagrasses,* W. J. Kenworthy and D. E. Haunert (eds.). NOAA Technical Memorandum, NMFS-SEFC-287, pp. 39–44.

Fourqurean, J. W., J. C. Zieman, and G. V. N. Powell. 1992. Phosphorus limitation of primary production in Florida Bay: evidence from C:N:P ratios of the dominant seagrass *Thalassia testudinum. Limnology and Oceanography* 37:162–171.

Gerloff, G. C. and P. H. Krombholz. 1966. Tissue analysis as a measure of nutrient availability for the growth of angiosperm aquatic plants. *Limnology and Oceanography* 11:529–537.

Harris, B. A., K. D. Haddad, K. A. Steidinger, and J. A. Huff. 1983. Assessment of fisheries habitat: Charlotte Harbor and Lake Worth, Florida. Final Report, Florida Department of Natural Resources, Bureau of Marine Research, Marine Researh Laboratories, St. Petersburg, FL, 211 pp.

Herzka, S. Z. and K. H. Dunton. 1997. Seasonal photosynthetic patterns of the seagrass *Thalassia testudinum* in the western Gulf of Mexico. *Marine Ecology Progress Series* 152:103–117.

Hopkinson, C. S., Jr. and J. J. Vallino. 1995. The relationships among man's activities in watersheds and estuaries: a model of runoff effects on patterns of estuarine community metabolism. *Estuaries* 18(4):598–621.

Jassby, A. D., W. J. Kimmerer, S. G. Monismith, C. Armor, J. E. Cloern, T. M. Powell, J. R. Schubel and T. J. Vendlinski. 1995. Isohaline position as a habitat indicator for estuarine populations. *Ecological Applications* 5:272–289.

Jensen, H. S., K. J. McGlathery, R. Marino, and R. W. Howarth. 1998. Forms and availability of sediment phosphorus in carbonate sand of Bermuda seagrass beds. *Limnology and Oceanography* 43:799–810.

Kemp, W. M., W. R. Boynton, R. R. Twilley, J. C. Stevenson, and L. G. Ward. 1984. Influences of submersed vascular plants on ecological processes in upper Chesapeake Bay. In *The Estuary as a Filter,* V. S. Kennedy (ed.). Academic Press, New York, pp. 367–394.

Livingston, R. J., X. Niu, F. G. Lewis, III, and G. C. Woodsum. 1997. Freshwater input to a Gulf estuary: long-term control of trophic organization. *Ecological Applications* 7:277–299.

McMillan, C. 1974. Salt tolerance of mangroves and submerged aquatic plants. In *Ecology of Halophytes,* R. J. Reimold and W. H. Queen (eds.). Academic Press, New York, pp. 379–399.

McMillan, C. and F. N. Moseley. 1967. Salinity tolerances of five marine spermatophytes of Redfish Bay, Texas. *Ecology* 28:503–505.

Phillips, R. C. 1960. Observations on the ecology and distribution of the Florida seagrasses. Professional papers series, No. 2, Contribution No. 44, Florida State Board of Conservation Marine Laboratory, St. Petersburg, FL, 72 pp.

Postel, S. L., G. C. Daily, and P. R. Ehrlich. 1996. Human appropriation of renewable fresh water. *Science* 271:785–788.

SAS Institute Inc. 1989. SAS/STAT User's guide, Version 6, Cary, NC, 1674 pp.

Solorzano, L. and J. Sharp. 1980. Determination of total dissolved phosphorus and particulate phosphorus in natural waters. *Limnology and Oceanography* 25:754–758.

South Florida Water Management District. 1994. SFWMD comprehensive quality assurance plan, DRE Inventory Control No. 318.

St. Johns River Water Management District and South Florida Water Management District. 1994. Indian River Lagoon surface water improvement and management (SWIM) plan, Palatka and West Palm Beach, FL, 120 pp.

Stevenson, J. C., L. W. Staver, and K. W. Staver. 1993. Water quality associated with survival of submersed aquatic vegetation along an estuarine gradient. *Estuaries* 16:346–361.

Tomasko, D. A., C. J. Dawes, and M. O. Hall. 1996. The effects of anthropogenic enrichment on turtle grass (*Thalassia testudinum*) in Sarasota Bay, Florida. *Estuaries* 19(2B):448–456.

Thayer, G. W., W. J. Kenworthy, and M. S. Fonseca. 1984. The ecology of eelgrass meadows of the Atlantic Coast: A community profile. U.S. Fish and Wildlife Service Report No. FWS/OBS-84/02, 147 pp.

Udy, J. W. and W. C. Dennison. 1997. Growth and physiological responses of three seagrass species to elevated sediment nutrients in Moreton Bay, Australia. *Journal of Experimental Marine Biology and Ecology* 217:253–277.

Whitfield, A. K. and T. H. Wooldridge. 1994. Changes in freshwater supplies to southern African estuaries: some theoretical and practical considerations. In *Changes in Fluxes in Estuaries: Implications from Science and Management*, K. R. Dyer and R. J. Orth (eds.). Olsen and Olsen, Fredensborg, pp. 41–50.

Winer, B. J. 1971. Statistical principles in experimental design. McGraw-Hill, New York, 907 pp.

Zieman, J. C. 1982. The ecology of the seagrasses of South Florida: a community profile. U.S. Fish and Wildlife Services, Office of Biological Services, Washington, D.C., FWS/OBS-82/25, 158 pp.

Zieman, J. C., and R. T. Zieman. 1989. The ecology of the seagrass meadows of the west coast of Florida: a community profile. U. S. Fish and Wildlife Service Biological Report 85(7.25), 155 pp.

Zimmerman, M. S. and R. J. Livingston. 1976. Seasonality and physio-chemical ranges of benthic macrophytes from a North Florida estuary (Apalachee Bay). *Contributions in Marine Science* 20:33–45.

7 Effects of the Disposal of Reverse Osmosis Seawater Desalination Discharges on a Seagrass Meadow (*Thalassia testudinum*) Offshore of Antigua, West Indies

David A. Tomasko, Norman J. Blake, Craig W. Dye, and Mark A. Hammond

Abstract Seawater desalination using reverse osmosis technology is being investigated as a potential alternative source of potable water in southwest Florida. Although the technology for desalinating brackish groundwater and/or seawater is well established, questions remain concerning the effects of reverse osmosis seawater desalination discharges, i.e., brine. This study examined the response of a meadow of turtle grass *Thalassia testudinum* to the direct discharge of brine from a desalination plant in Antigua, West Indies. Shoot density, areal blade biomass, and areal blade productivity were determined prior to, and then 3 and 6.5 months after, the diversion of a discharge plume onto a previously unimpacted turtle grass meadow. No relationship was found between temporal changes in density, biomass, or productivity, and the degree of exposure to brine discharge as quantified by the change in salinity vs. background values. After 3 months of discharge diversion, a weak yet statistically significant relationship was detected between macroalgal biomass and degree of exposure to brine discharge. The increased macroalgal abundance is believed to be caused by increased availability of nitrogen in the discharge plume. This is most likely due to episodic loading from backflushing of filters, incidental releases of detergents used to eliminate bio-fouling, and combined stormwater discharges.

INTRODUCTION

Southwest Florida is currently faced with potable water supply problems associated with a rapidly expanding population and the overdraft of traditional groundwater resources. Further complicating the issue, most surface-water resources in southwest Florida are undependable as long-term water supplies (Southwest Florida Water Management District — SWFWMD 1992). In this region, alternative sources of water such as seawater desalination are currently being evaluated to determine what future role, if any, might be played by non-traditional sources of water supply (Dye et al. 1995; Hammond et al. 1998).

Few studies have been conducted on the environmental effects of discharges of drinking water byproduct, i.e., brine, from desalination plants. Studies have evaluated the chemical quality of brackish groundwater discharges (Whiting and Wolfe 1996), trends in coral coverage at the discharge point of a thermal distillation plant in the Red Sea (Mabrook 1994), and the response of seagrass meadows to brine discharge from a reverse osmosis plant in Bermuda (Smith 1995).

0-8493-2045-3/99/$0.00+$.50
© 2000 by CRC Press LLC

In earlier efforts, the SWFWMD performed laboratory scale studies of reverse osmosis concentrate effects on standard marine test organisms (mysid shrimp, *Mysidopsis bahia*, and the silverside minnow, *Menidia beryllina*). Results of these studies have shown that Florida water quality standards can be easily met for concentrate discharge (Dye et al. 1995).

For seagrasses and macroalgae, a wealth of information exists on the ability of different species to adjust to fluctuations in salinity, including the increased salinities expected with nearshore disposal of brine discharges (Dawes 1987, 1989; Dawes and Tomasko 1988; Gessner 1971; Jagels 1973, 1983; Koch and Lawrence 1987; Lapointe 1989; Lapointe et al. 1984, 1992, 1994; Lazar and Dawes 1991; McMillan and Moseley 1967; Pulich 1986).

Based on these and other studies on the salinity tolerances of seagrasses and macroalgae, it was anticipated that most species would have little trouble accommodating relatively minor (ca. 2–5 ppt) increases in salinity that might be expected in a "worst case" scenario associated with the discharge of reverse osmosis concentrate from a seawater desalination plant. Nonetheless, in this study, the responses of seagrasses and macroalgae to brine discharge from a reverse osmosis seawater desalination plant are further investigated. The purpose of this investigation was to document the structure and health, under actual field conditions, of this important benthic community within the discharge zone of a seawater reverse osmosis facility. Additional studies on the response of benthic microalgae, foraminifera, infaunal invertebrates, fouling communities, and mobile epifauna are contained within the report compiled by Hammond et al. (1998).

MATERIALS AND METHODS

After initial surveys of potential locations for this study, a site on the north shore of Antigua,West Indies, was chosen (Figure 7.1). This location had a number of features that made it the choice for our investigation: (1) sufficiently high discharge volume (ca. 1.0–1.5 mgd); (2) sufficiently high discharge salinity (ca. 57–65 ppt); (3) the presence of a healthy and diverse biological community close to the discharge point (e.g., soft and hard corals, extensive meadows of the seagrass *Thalassia testudinum*); and (4) a discharge outlet that could be modified to divert the flow to a nearby (ca. 32 m), unaffected location.

The experimental approach decided upon involved studying a healthy and productive seagrass meadow before (March 1997) and then at 3 (June 1997) and 6.5 months (October 1997) after diverting the discharge plume of a reverse osmosis seawater desalination plant onto this location. From previous assessments of the volume and spatial distribution of concentrate discharge, the following experimental outlay was decided upon: six separate transects (at 60° angles from each other) were set in a pattern radiating from a central point termed Ground Zero (GZ). GZ was the location where the concentrate plume was discharged after its diversion (Figure 7.2). At each of the six transects, stations were located at 2 m intervals until a distance of 10 m from GZ (Figure 7.3). Other biological data (summarized in Hammond et al. 1998) were collected at these same stations, as illustrated in Figure 7.3.

The discharge plume was delimited by measuring standard water quality variables (i.e., temperature, pH, salinity, and dissolved oxygen) at the surface, bottom, and mid-depth for each of the 31 sampled locations. Data were collected with a Hydrolab© Surveyor. These variables were measured at all locations on 15 separate occasions, under a variety of tide and wind conditions.

Due to the considerable amount of spatial variation for seagrass parameters prior to the diversion of the discharge plume, it was decided that the most appropriate technique for determining effects of the discharge on this biological community would be (1) to detect the relationships (if any) between changes in community parameters over time (e.g., seagrass density, biomass, and productivity, etc.) and (2) to detect the intensity of discharge effects measured as the salinity change vs. background values.

Short-shoot density, above-ground standing crop, and blade productivity were determined using standard procedures (Tomasko and Lapointe 1991; Lapointe et al. 1994; Tomasko et al. 1996). At

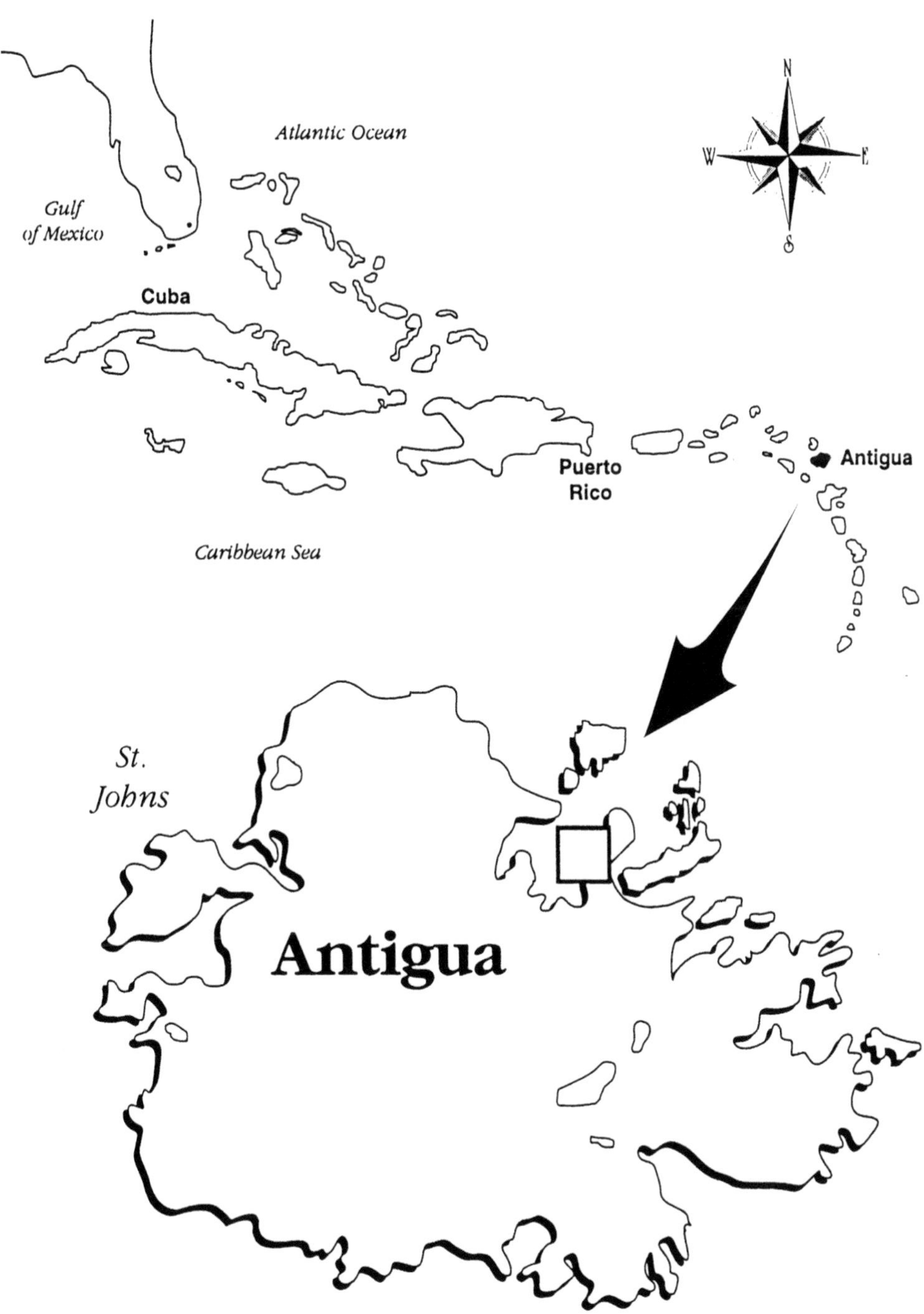

FIGURE 7.1 Location of study site in Antigua, West Indies indicated by the square.

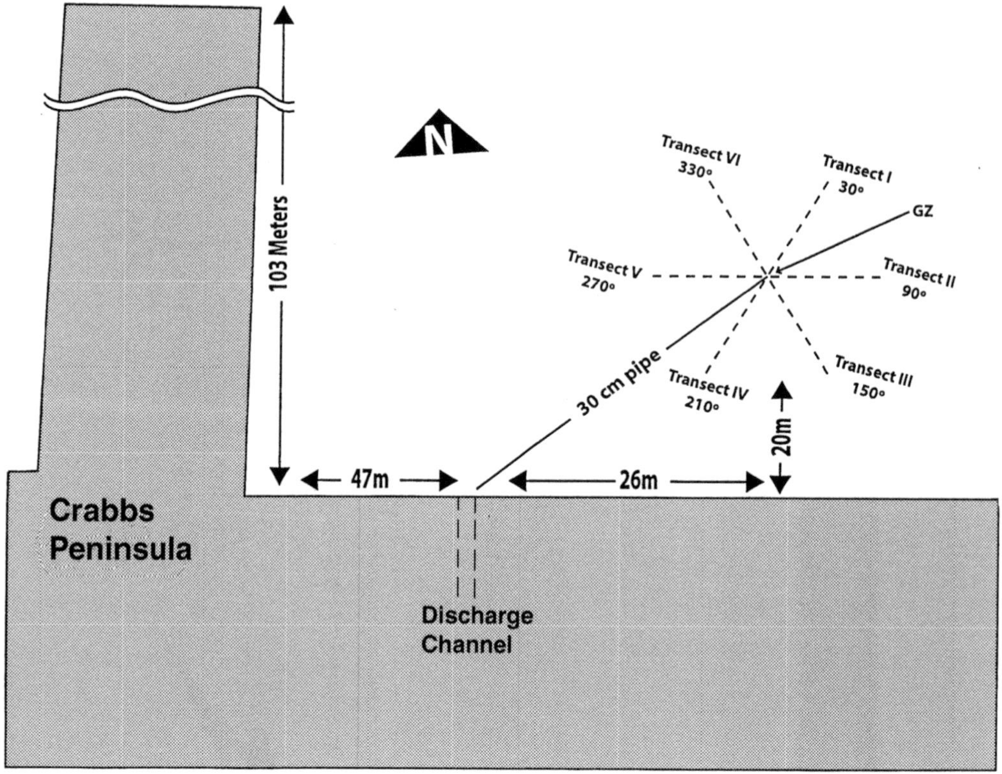

FIGURE 7.2 Overview of study site, showing desalination plant, original outfall location, outfall diversion pipe, and transect lines.

each station, short shoots were enumerated in four replicate quadrats (25 × 25 cm). Additionally, five haphazardly chosen short shoots from within these quadrats were tagged and their blades punched with a 25-gauge hypodermic needle. After 3 to 4 days, the marked short shoots were collected and placed in an ice-filled cooler for transportation back to Florida. Within 12 hours of being harvested, samples were stored in a freezer until later analysis. At that time, blade epiphytes were removed by lightly scraping the leaves with a double-edged razor blade. Newly formed blade material was then separated from old material. Finally, epiphytes, new blade material, and old blade material were dried for at least 24 hr at 65°C. Mean values for biomass and productivity per short-shoot at each station were multiplied by average short shoot densities per m^2 to determine areal blade biomass (g dw m^{-2}) and areal blade productivity (g dw m^{-2} d^{-1}).

Additional efforts focused on determining factors responsible for a seemingly non-random distribution of the brown alga *Dictyota dichotoma* in the area around the discharge zone that was noted on the second trip. Macroalgal coverage was determined by dividing a 25 × 25 cm quadrat into 16 separate sub-quadrats, and "scoring" sub-quadrats where macroalgae comprised more than 50% of the surface area. Four replicate quadrats (each with 16 sub-quadrats) were used at each of the 31 stations. Additional studies involved determining the most likely limiting nutrient for the growth of this species.

On the second trip, four samples were collected from each of the following stations: GZ, Transect 30° at 2 and 4 m (30-2 and 30-4), Transect 90° at 2 and 4 m (90-2 and 90-4), and at locations outside the study area to the east and west of the discharge plume. Samples were immediately placed in ice-filled coolers, transported to Florida, and kept frozen until they could be airmailed in a cooler filled with dry ice for elemental analysis at the University of Texas Marine Science Institute in Port Aransas. Carbon, nitrogen, and phosphorus contents of plant tissue were determined using

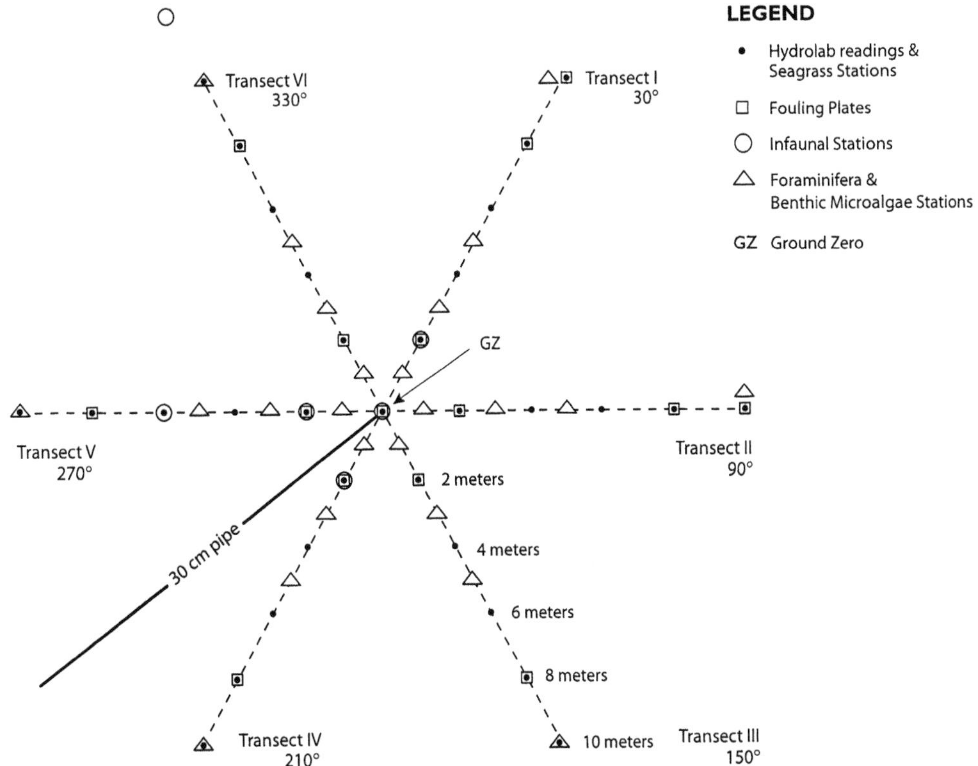

FIGURE 7.3 Diagram of transect lines and stations in relation to Ground Zero, the location where brine discharge was diverted.

standard techniques, i.e., Perkin–Elmer 205C Elemental Analyzer for carbon and nitrogen, acid digestion and spectrophotometry for phosphorus (Ken Dunton, personal communication).

An additional study, completed during the third trip, involved assessing the growth potential of *D. dichotoma* though the use of a 2 × 2 factorial experiment. For this assessment, the experimental design was as follows: plants from an area outside the influence of the discharge plume were collected, weighed, tagged, and placed in modified cages (citrus bags) attached to poles at two locations — outside the plume (the site of collection), and inside the plume (station 30-4). Also, plants from inside the plume (station 30-4) were collected, weighed, tagged, and placed in cages at the same two locations as just described. Three plants each were placed in two bags per treatment for a total of 24 plants.

The experimental design allowed us to compare growth rates of plants outside the plume transplanted into the plume, plants outside the plume transplanted back outside the plume, plants from inside the plume transplanted outside the plume, and plants from inside the plume transplanted back inside the plume. Growth rates (% d^{-1}) were determined after 4 days by measuring wet weights of individual plants.

During the third trip, it was noted that there appeared to be a considerable amount of grazing on the seagrass blades. As samples were processed in the laboratory, shoots with blades that showed evidence of grazing were noted. Based on discussions with Dr. John Ogden, Director of the Florida Institute of Oceanography, it appeared that the most obvious grazing marks were made by the bucktooth parrotfish (*Sparisoma radians*). The bucktooth parrotfish leaves easily identified semi-circular bite marks in grazed blades, with the diameter of these marks between 1 and 10 mm (Ogden and Zieman 1977). Previous studies have shown this species to be a major grazer on turtle grass meadows in the Caribbean (Ogden and Zieman 1977; Ogden 1980). To determine if the discharge plume affected grazing intensity by *S. radians*, a two-way analysis of variance (ANOVA) was performed to determine the effects of transect and distance on grazing rates.

Statistical analyses were performed by examining the relationships, if any, between changes in seagrass growth parameters (short-shoot density, areal blade biomass, and areal blade productivity) over time vs. the intensity of the plume's impact on each station. The intensity of the plume's impact was quantified by determining the increase in salinity at each site vs. the lowest salinity recorded on each of 15 water quality monitoring surveys. The average salinity increase for all surveys was used as the indicator of the average degree of influence of the discharge plume at each station. Significance was set at $p < 0.05$.

RESULTS

Water Quality

The influence of the discharge plume (quantified as the average increase in salinity vs. the background for each of the 15 water quality surveys) ranged from less than 0.20 ppt at sites 330-8, 30-10, 30-8, and 270-8 to values higher than 4.10 ppt at sites 90-2 and GZ. Fittingly, GZ had the highest degree of plume influence, with an average salinity increase of 4.47 ppt vs. the background value. These data were then used to compare the response of seagrass growth parameters to the intensity of the plume by comparing, for example, the change in blade biomass at individual stations over time to the average salinity increase at individual stations.

Seagrass and Macroalgae Growth Parameters

When comparing the change in short-shoot densities for the second trip vs. the first trip (Figure 7.4), no significant relationship was found between the percent change in density and the average salinity increase. Variation in short-shoot densities recorded on the third trip vs. the first trip was not significantly associated with the average salinity increase (Figure 7.5). When shoot densities from before the diversion of the plume are plotted against densities 6.5 months after the diversion, it can be seen that approximately as many stations had increased densities as decreased densities (Figure 7.6). The average shoot density for all stations prior to diversion was 374.6 m^{-2}, while 6.5 months later, the average density was 387.6 m^{-2}, a difference that was not statistically significant (two-tailed Student's *t*-test).

However, there was evidence for a weak but statistically significant relationship between percent algal cover (mainly the brown alga *D. dichotoma*) on the second trip vs. plume influence (Figure 7.7; $p = 0.02$, $r^2 = 0.18$, 30 d.f.). On the third trip, there was no statistical relationship between these variables.

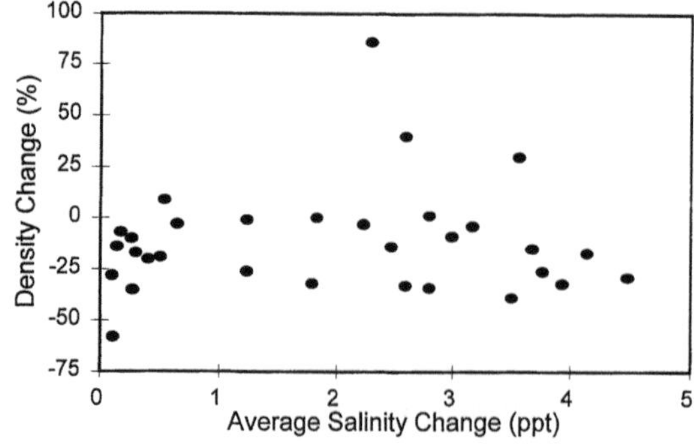

FIGURE 7.4 Change in short shoot-density of *Thalassia testudinum* at all locations (percent of original value) after 3 months of exposure to brine discharge vs. the average salinity change from background (ppt).

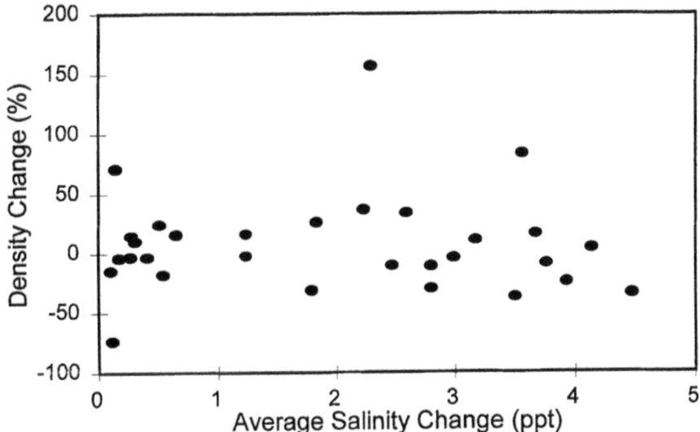

FIGURE 7.5 Change in short-shoot density of *Thalassia testudinum* at all locations (percent of original value) after 6.5 months of exposure to brine discharge vs. the average salinity change from background (ppt).

Values for nitrogen to phosphorus ratios of the tissue of *D. dichotoma* (Figure 7.8) suggest co-limitation by nitrogen and phosphorus, although nitrogen would probably be more strongly limiting than phosphorus (Lapointe et al. 1992). Plants collected from inside the plume had significantly higher tissue nitrogen concentrations than plants collected from areas outside the plume (Figure 7.9) with the highest concentrations occurring at stations 30-4 and 30-2.

Growth rates of *D. dichotoma* from outside the influence of the discharge plume transplanted into the plume (OUT to IN) actually grew more slowly than plants from outside the plume transplanted back outside the plume (OUT to OUT; Figure 7.10). In contrast, plants from inside the plume transplanted back inside the plume (IN to IN) grew faster than all other categories. Plants from inside the plume maintained their high growth rates when transported to the site outside the influence of the plume (IN to OUT).

Prior to diverting the discharge pipe, areal blade biomass values exhibited substantial variation, with values ranging from less than 5 to more than 60 g dw m^{-2} (data not shown). Three months after diversion, values ranged up to more than 30 g dw m^{-2}, and 6.5 months after diversion, values were similar to those from 3 months after diversion. When the percent change in biomass for individual stations on the second trip vs. the first trip was plotted against the average salinity change at these same stations (Figure 7.11), no significant relationship was found. When plotting the change in biomass on the third trip vs. the first one, no significant relationship was found when plotted against the average salinity change (Figure 7.12).

Areal seagrass blade productivity values prior to plume diversion ranged to just more than 1.5 g dw m^{-2} d^{-1} (data not shown). Three months after plume diversion, productivity values ranged up

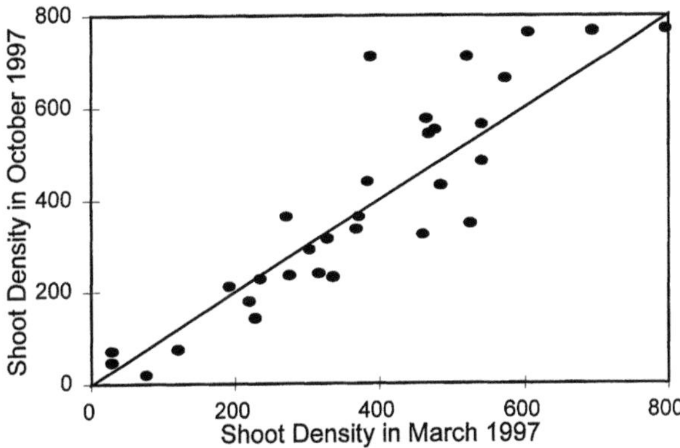

FIGURE 7.6 Short-shoot density of *Thalassia testudinum* (no. m^{-2}; n = 4) in March 1997 plotted against values in October 1997. Line is 1:1 ratio.

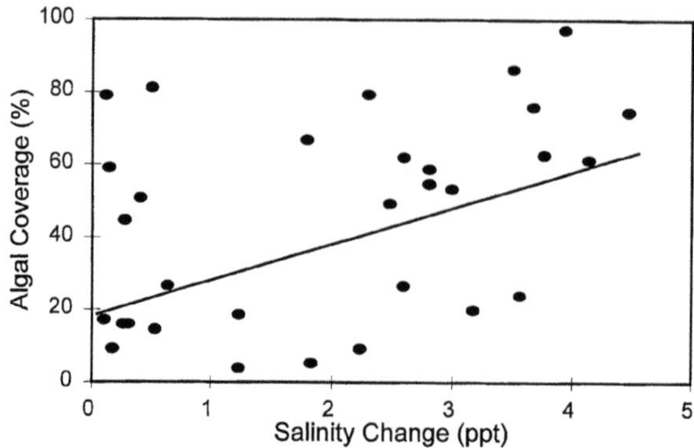

FIGURE 7.7 Algal coverage at all locations (percent of area; n = 4) plotted against the average salinity change from background (ppt). Line is best-fit relationship.

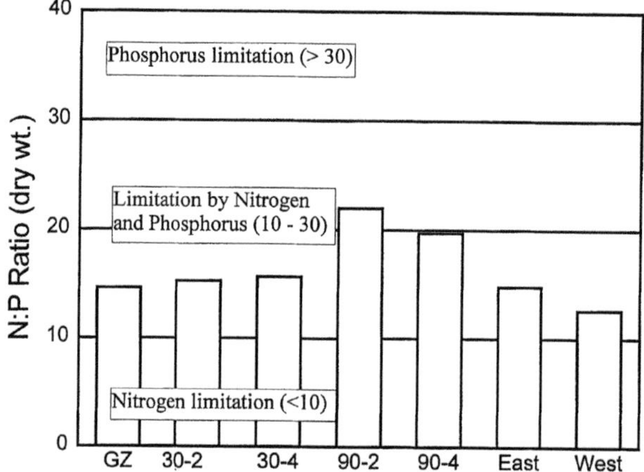

FIGURE 7.8 Nitrogen to phosphorus ratios (by weight) for *Dictyota dichotoma* at ground zero (GZ), stations 30-2, 30-4, 90-2, and 90-4, and sites located east (East) and west (West) of the area influenced by the discharge (n = 4). See text for explanation of station locations.

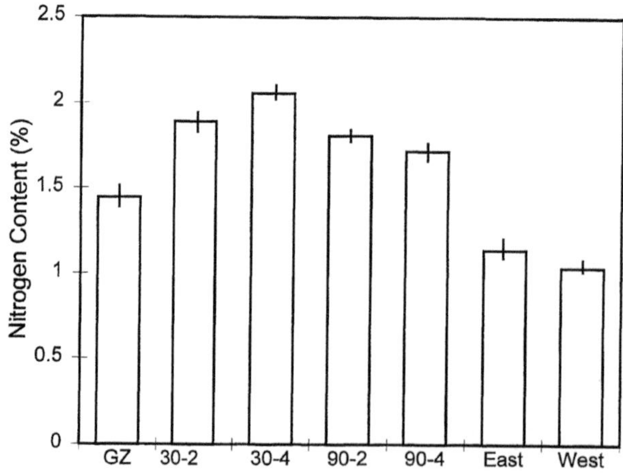

FIGURE 7.9 Nitrogen content (percent of dry weight) for *Dictyota dichotoma* at ground zero (GZ), stations 30-2, 30-4, 90-2, and 90-4, and sites located east (East) and west (West) of the area influenced by the discharge (n = 4). See text for explanation of station locations.

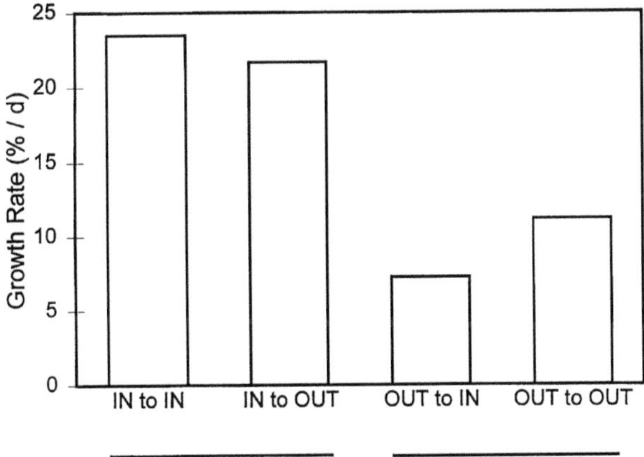

FIGURE 7.10 Growth rates (percent per day) for *Dictyota dichotoma* gathered from inside and outside of the discharge plume and transplanted inside and outside of the discharge plume (see text for details; n = 6). Lines connect values not significantly different from each other.

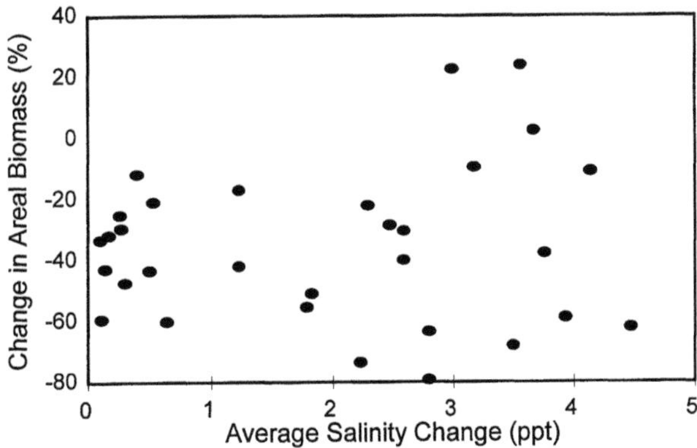

FIGURE 7.11 Change in areal blade biomass of *Thalassia testudinum* at all locations (percent of original value) after 3 months of exposure to brine discharge vs. the average salinity change from background (ppt).

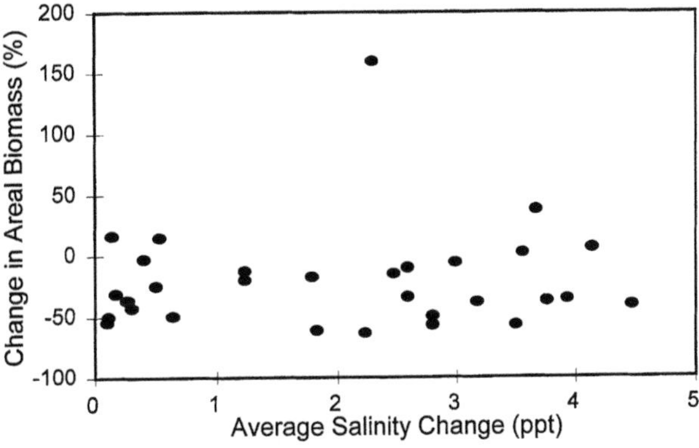

FIGURE 7.12 Change in areal blade biomass of *Thalassia testudinum* at all locations (percent of original value) after 6.5 months of exposure to brine discharge vs. the average salinity change from background (ppt).

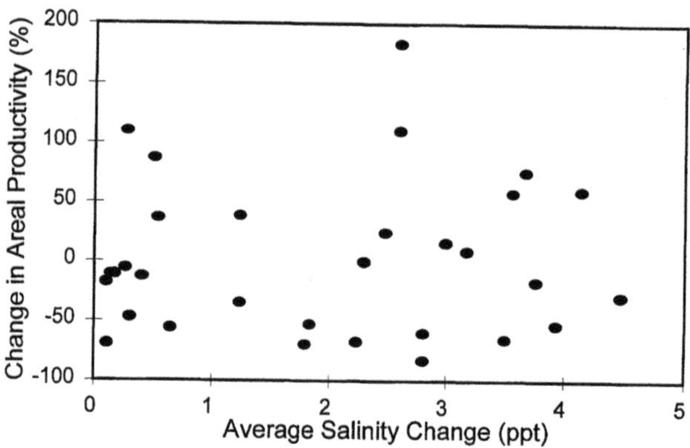

FIGURE 7.13 Change in areal blade productivity of *Thalassia testudinum* at all locations (percent of original value) after 3 months of exposure to brine discharge vs. the average salinity change from background (ppt).

to 1.0 g dw m^{-2} d^{-1}, and 6.5 months after diversion, productivity values ranged up to slightly higher than 1.0 g dw m^{-2} d^{-1}. When the percent change in areal seagrass blade productivity from the second trip vs. the first trip was plotted against the average salinity change (Figure 7.13), no clear relationship was evident. After 6.5 months of plume diversion, no association was found between the percent change in productivity (compared to the first trip) and the average salinity change (Figure 7.14).

The percent of shoots with identified grazing marks from the bucktooth parrotfish, *S. radians*, is shown in Figure 7.15. Two-way ANOVA showed no effect of either transect ($p = 0.229$, 5 d.f.) or distance along transects ($p = 0.608$, 4 d.f.). The percent of shoots grazed by *S. radians* ranged from 90% at 6 meters from GZ to 100% at GZ.

DISCUSSION

For seagrasses, a wealth of information exists on the ability of different species to adjust to fluctuations in salinity. Early work by Jagels (1973, 1983) showed that *T. testudinum*, which is perhaps the most common species in southwest Florida, has specialized cells in the leaf epidermal layer that help this species tolerate salinity stresses. In one of the earliest studies on salinity tolerances in seagrasses, McMillan and Moseley (1967) grew four species (collected in Texas) at salinities up to 74 ppt. The species they examined are all found in southwest Florida. Of the four

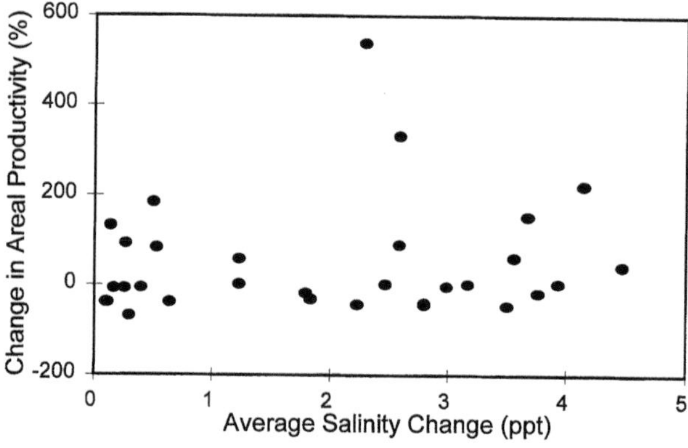

FIGURE 7.14 Change in areal blade productivity of *Thalassia testudinum* at all locations (percent of original value) after 6.5 months of exposure to brine discharge vs. the average salinity change from background (ppt).

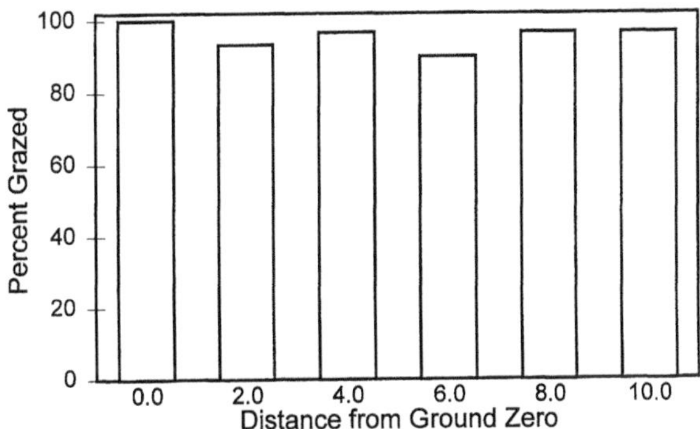

FIGURE 7.15 Percent of blades of *Thalassia testudinum* showing evidence of grazing by the bucktooth parrotfish (*Sparisoma radians*) at 0, 2, 4, 6, 8, and 10 m from Ground Zero (n = 30 for stations 2, 4, 6, 8, and 10 m, n = 5 for station 0 m).

species, *Syringodium filiforme* (manatee grass) showed the lowest upper salinity tolerance limit, as growth stopped when plants were exposed to salinities higher than 45 ppt. The three remaining species, *T. testudinum* (turtle grass), *Halodule wrightii* (shoal grass), and *Halophila engelmannii* (star grass) all continued to grow at salinities up to and beyond 60 ppt.

A study on a Caribbean population of turtle grass by Gessner (1971) showed that while cellular damage occurred at salinities twice that of natural seawater (ca. 70 ppt), no cellular damage occurred at salinities 20% higher than natural seawater (ca. 42 ppt). In another study from Texas, Pulich (1986) found that turtle grass, shoal grass, star grass, and widgeon grass (*Ruppia maritima*, a common species in southwest Florida) all had physiological mechanisms that allowed them to tolerate salinities of up to 47 ppt. Also in Texas, Tomasko and Dunton (1995) documented vigorous growth of shoal grass in the hypersaline Laguna Madre, where salinities ranged from 38 to 48 ppt.

In a study conducted in Tampa Bay, Lazar and Dawes (1991) examined the physiological ecology of two populations of widgeon grass from distinctly different locations. The first population was from Old Tampa Bay, an area where annual average salinities tend to be less than 22 ppt (Boler 1986, as illustrated by Estevez 1989). The second population was from Lower Tampa Bay, where salinities average between 27 and 30 ppt. Both populations showed broad and similar abilities to photosynthesize at the experimental salinities of 0, 17.5, and 35 ppt.

Additionally, a number of studies have examined responses to salinity variation for macroalgae in southwest Florida. For the common brown alga *Sargassum filipendula*, plants collected from a variety of locations in the Tampa Bay area showed similar photosynthetic and respiratory rates when cultured at salinities of 15, 25, and 35 ppt, although plants did not do well when grown at salinities of 5 ppt (Dawes 1987; Dawes and Tomasko 1988; Dawes 1989). For the common red alga *Gracilaria tikvahiae*, Lapointe et al. (1984) found no significant difference in growth rates for plants cultured for 2 weeks at 16 and 26 ppt even though the plants were originally grown at a salinity of 26 ppt. Koch and Lawrence (1987) found that for the related species *G. verrucosa*, plants originally cultured at 10 ppt salinity did not differ in their photosynthetic or respiratory rates when grown 20 days at either 10 or 32 ppt.

These studies indicate that seagrasses and macroalgae tend to have relatively broad salinity tolerances. As such, it was anticipated that the turtle grass meadow at the study site would have little trouble accommodating the salinity increases associated with the seawater desalination plant's discharge. It should be noted that this study investigated a "worst case" scenario, as the desalination discharge was undiluted prior to disposal, a situation unlikely to be allowed should a similar plant be built in southwest Florida.

Considerable spatial variation existed for the seagrass meadow at this site prior to diversion of the discharge. Consequently, seagrass growth parameters were analyzed for potential effects of the discharge by comparing changes over time vs. relative plume effects — which were measured as the average change in salinity vs. background values.

After 3 months of diversion of concentrate discharge, there were no detectable effects of the plume on the density, biomass, and production of the seagrass *T. testudinum*. However, there was a positive, but weak, correlation between the intensity of the plume (measured as the change in salinity vs. background) and the abundance of the alga *D. dichotoma*. Information on N:P ratios suggests that *D. dichotoma* was more likely to be nitrogen limited rather than phosphorus limited. Specimens of the alga *D. dichotoma* from inside the plume had significantly higher nitrogen contents than plants from outside the plume. These data suggest that one of several processes could increase the nutrient loads impacting the areas influenced by the plume, thus increasing macroalgal biomass in those areas.

Processes capable of increasing local nitrogen loads include at least the following: (1) chronic nutrient increases via concentration of nutrient molecules within the discharge volume (i.e., selective removal of water molecules, leaving nutrients behind); (2) episodic nutrient increases associated with backflushing of filters; (3) episodic nutrient increases associated with detergent discharges used to clean biofouled surfaces; and (4) episodic nutrient increases associated with rain events and stormwater discharge from the factory itself (which has a combined stormwater outfall with the discharge pipe).

Macroalgae can store nutrients delivered in pulses, using these nutrients to grow at faster rates when nutrient levels return to normal (Lapointe 1989; Lapointe et al. 1992). In the follow-up study on algal growth rates, *D. dichotoma* from inside the plume maintained its elevated growth rate even when grown in cages outside the plume. In contrast, *D. dichotoma* from outside the plume did not grow faster when placed in cages inside the plume as compared to controls grown in cages outside the plume. These results suggest that the elevated algal abundance seen in the second trip (but not the third trip) was likely caused by an episodic event or a series of such events. However, nutrient-enriched *D. dichotoma* from inside the plume seemed able to retain its elevated growth rate (for a period of several days, at least) even when transferred to areas outside the plume.

It seems probable, when assessing the responses of *D. dichotoma* growth rates, that elevated algal abundances were not due to chronic effects associated with the discharge, but episodic events associated with backflushing, detergent discharges, and/or stormwater runoff from the desalination plant complex. Such phenomena would have to be reduced or eliminated, should a similar plant be built in southwest Florida, lest similar nutrient-induced algal blooms be stimulated.

On the final trip to this site, 6.5 months after diverting the concentrate discharge plume, algal abundance was much lower than on the second trip. In addition, shoot densities of the seagrass meadow showed no significant difference from densities recorded prior to diversion of the discharge plume. No relationship was found between the intensity of the plume's influence (measured as the average increase in salinity vs. background) and changes in shoot density. Similar results were found for data on productivity and biomass.

These results suggest that there is no discernible toxicity to the seagrass *T. testudinum* associated with the discharge from the seawater desalination reverse osmosis plant in Antigua, as changes in density, biomass, and productivity exhibited no relationship with the intensity of the plume's influence. While some locations had reduced biomass and productivity after 3 and 6.5 months of plume influence, other locations showed increases. The period of exposure to brine discharge (6.5 months) was considerably longer than previous salinity manipulation experiments that were able to invoke seagrass responses (Gessner 1971; McMillan and Moseley 1967; Pulich 1986). As such, it is unlikely that the lack of response to salinity increases in this study was due to an insufficient experimental duration. Rather, the lack of response is most likely due to the magnitude of salinity increases being insufficient to invoke such responses.

In addition to the lack of any observed effect on the seagrass meadow itself, it appeared that the discharge plume did not affect the grazing rate of a major seagrass consumer, the bucktooth parrotfish (*S. radians*). Blades of the seagrass *T. testudinum* are a major food item for the bucktooth parrotfish (Ogden and Zieman 1977; Ogden 1980). As such, the lack of impact of the discharge plume on grazing rates suggests that the plume did not affect the feeding behavior of this important herbivorous fish.

These results are consistent with the conclusion reached by Smith (1995), who studied the effects of a discharge plume from a seawater desalination plant in Bermuda, and who stated that, "Overall it appears that, after 10 months of operation (June 1994 to June 1995), the reverse osmosis plant effluent is having little effect on the surrounding seagrass beds."

REFERENCES

Boler, R. 1986. Hillsborough County water quality, 1984–1985. Hillsborough County Environmental Protection Commission, Tampa, FL, 205 pp.

Dawes, C. J. 1987. Physiological ecology of two species of *Sargassum* (Fucales, Phaeophyta) on the west coast of Florida. *Bulletin of Marine Science* 40:198–209.

Dawes, C. J. 1989. Physiological responses of brown seaweeds *Sargassum filipendula* and *S. pteropleuron* before and after transplanting on the west coast of Florida. *Journal of Coastal Research* 5:693–700.

Dawes, C. J. and D. A. Tomasko. 1988. Physiological responses of perennial bases of *Sargassum filipendula* from three sites on the west coast of Florida. *Bulletin of Marine Science* 42:166–173.

Dye, C. W., M. D. Farrell, M. A. Hammond, and Q. D. Wylupek. 1995. Seawater desalination: an investigation of concentrate disposal by means of a coastal ocean outfall. Southwest Florida Water Management District, September 1995, 17 pp.

Estevez, E. 1989. Water quality trends and issues, emphasizing Tampa Bay. In *Tampa and Sarasota Bays: Issues, Resources, Status, and Management.* NOAA Estuary-of-the-Month Seminar Series No. 11, National Oceanic and Atmospheric Administration, Washington, D.C., pp. 65–88.

Gessner. F. 1971. The water economy of the sea grass *Thalassia testudinum. Marine Biology* 10:258–260.

Hammond, M. A., N. J. Blake, C. W. Dye, P. Hallock-Muller, D. A. Tomasko, and G. Vargo. 1998. Effects of the disposal of seawater desalination discharges on nearshore benthic communities. Southwest Florida Water Management District, Brooksville, FL, 250 pp.

Jagels, R. 1973. Studies of a marine grass, *Thalassia testudinum.* I. Ultrastructure of the osmoregulatory leaf cells. *American Journal of Botany* 60:1003–1009.

Jagels, R. 1983. Further evidence for osmoregulation in epidermal leaf cells of seagrasses. *American Journal of Botany* 70:327–333.

Koch, E. W. and J. Lawrence. 1987. Photosynthetic and respiratory responses to salinity changes in the red alga *Gracilaria verrucosa. Botanica Marina* 30:327–329.

Lapointe, B. E. 1989. Macroalgal production and nutrient relations in oligotrophic areas of Florida Bay. *Bulletin of Marine Science* 44:312–323.

Lapointe, B. E., D. L. Rice, and J. M. Lawrence. 1984. Responses of photosynthesis, respiration, growth and cellular constituents to hypo-osmotic shock in the red alga *Gracilaria tikvahiae. Comparative Biochemistry and Physiology* 77A:127–132.

Lapointe, B. E., M. M. Littler, and D. S. Littler. 1992. Nutrient availability to marine macroalgae in siliciclastic versus carbonate-rich coastal waters. *Estuaries* 15:75–82.

Lapointe, B. E., D. A. Tomasko, and W. Matzie. 1994. Trophic structuring of marine plant communities in the Florida Keys. *Bulletin of Marine Science* 54:696–717.

Lazar, A. C. and C. J. Dawes. 1991. A seasonal study of the seagrass *Ruppia maritima* L. in Tampa Bay, Florida: organic constituents and tolerances to salinity and temperature. *Botanica Marina* 34:265–269.

Mabrook, D. 1994. Environmental impact of waste brine disposal of desalination plants, Red Sea, Egypt. *Desalination* 97:453–465.

McMillan, C. and F. N. Moseley. 1967. Salinity tolerances of five marine spermatophytes of Redfish Bay, Texas. *Ecology* 48:503–506.

Ogden, J. C. 1980. Faunal relationships in Caribbean seagrass beds. In *Handbook of Seagrass Biology: An Ecosystem Perspective*, R. C. Phillips and C. P. McRoy (eds.). Garland STPM Press, New York.

Ogden, J. C. and J. C. Zieman. 1977. Ecological aspects of coral reef-seagrass bed contacts in the Caribbean. In *Proceedings, Third International Coral Reef Symposium*. University of Miami, Miami, FL, pp. 377–382.

Pulich, W. M., Jr. 1986. Variations in leaf soluble amino acids and ammonium content in subtropical seagrasses related to salinity stress. *Plant Physiology* 80:283–286.

Southwest Florida Water Management District. 1992. Water supply needs and sources, 1990–2020. Southwest Florida Water Management District, Brooksville, FL.

Smith, S. R. 1995. Final report on the marine environmental impact of the Watlington Water Works reverse osmosis plant. Bermuda Biological Station for Research, Inc., Ferry Reach, Bermuda.

Tomasko, D. A. and B. E. Lapointe. 1991. Productivity and biomass of *Thalassia testudinum* as related to water column nutrient availability and epiphyte levels: field observations and experimental studies. *Marine Ecology Progress Series* 75:9–17.

Tomasko, D. A. and K. H. Dunton. 1995. Primary productivity in *Halodule wrightii*: a comparison of techniques based on daily carbon budgets. *Estuaries* 18:271–278.

Tomasko, D. A., C. J. Dawes, and M. O. Hall. 1996. The effects of anthropogenic nutrient enrichment on turtle grass (*Thalassia testudinum*) in Sarasota Bay, Florida. *Estuaries* 19:448–456.

Whiting, D. and S. Wolfe. October 1996. Toxicity from major seawater ion "imbalance." Florida Department of Environmental Protection, 8 pp.

Section II

Monitoring and Trends

8 Development and Use of an Epiphyte Photo-Index (EPI) for Assessing Epiphyte Loadings on the Seagrass *Halodule wrightii*

Robbyn Miller-Myers and Robert W. Virnstein

Abstract A visual, nondestructive technique for measuring epiphyte abundance was developed and evaluated. The technique provides a tool for rapid estimates of *in situ* epiphyte biomass on *Halodule wrightii*, to be used in association with a large-scale seagrass monitoring program. Tool development included quantifying and documenting, with photographs, a representative distribution of epiphyte loadings (epiphyte biomass to seagrass biomass) within the Indian River Lagoon, Florida. This tool, the Epiphyte Photo-Index (EPI), may be taken into the field to quantify epiphyte loads by matching observed, *in situ*, epiphyte loads with the most representative EPI photograph. Evaluation tests found the accuracy of the EPI estimate increased as the actual *in situ* loadings increased. The EPI explained 56% of the variability with a mean coefficient of variability of 14%. The EPI consists of 25 representative photographs and is utilized during summer sampling of 81 fixed seagrass transects throughout the lagoon. The EPI was implemented with minimal training. Further testing to assess precision and repeatability of this sampling tool is warranted.

INTRODUCTION

Seagrass beds have been defined in the Indian River Lagoon Surface Water Improvement and Management (SWIM) Plan as the most critical habitat in the lagoon (Steward et al. 1994). Therefore, the management goal of the St. Johns River Water Management District for the lagoon system is to protect and restore the integrity and functionality of seagrass habitats within the estuarine system (Virnstein and Morris 1996). This protection will be accomplished, in part, by providing sediment and water quality supportive of healthy seagrass. Status of seagrass is assessed biennially by mapping from aerial photographs and semi-annually (summer and winter) by monitoring 81 fixed transects to detect changes in both density and species composition throughout the lagoon.

Seagrass blades provide structure for the attachment and growth of epiphytic algae, which are major food sources for invertebrates (Virnstein 1987). Nutrient enrichment is suspected as a direct cause of increased epiphytic algae on seagrass blades (Silberstein et al. 1986; Tomasko and Lapointe 1991). Although epiphytes provide food for invertebrate communities, too much epiphytic coverage inhibits light availability for seagrass photosynthesis. Light attenuation by epiphytes has been found to be directly correlated with epiphyte biomass (Harden 1994). Reduction of available light creates stress on seagrass and is a probable cause of seagrass habitat decline (Backman and Barilotti 1976; Czerny and Dunton 1995).

As a complement to the seagrass monitoring program in the lagoon, the ability to quantify epiphyte biomass loads on seagrass was deemed beneficial. Quantification, however, is time

0-8493-2045-3/99/$0.00+$.50
© 2000 by CRC Press LLC

consuming and tedious. To meet this challenge, we developed a method to estimate the abundance of epiphytic algae with a nondestructive sampling technique (the Epiphyte Photo-Index (EPI)), thus reducing the time and cost of laboratory analyses. A visual estimate of epiphyte loads was considered possible because epiphytic algae are discernible without magnification.

The EPI is composed of reference photographs which represent a wide range of epiphyte loadings (ratio of epiphyte biomass to seagrass biomass) present on seagrass in the lagoon. The EPI, when laminated, is waterproof so that it may be readily used in the field. Estimates of epiphyte loads can be made quickly and easily by selecting the best match between the reference photographs and *in situ* epiphyte loadings. This chapter evaluates EPI's accuracy and precision, recommends changes, and describes procedures for its utilization.

METHODS

DEVELOPMENT OF THE EPI

Development of the EPI began by documenting the range of epiphyte loadings that occur throughout the lagoon (Miller-Myers 1997). Shoal grass, *Halodule wrightii* Aschers., was chosen because it is the most widely distributed and most commonly occurring species in the lagoon (Woodward-Clyde 1994; Dawes et al. 1995; Virnstein 1995; Virnstein et al. 1997; Morris et al. 2000).

Seagrass/epiphyte loading samples were collected lagoon-wide at 54 seagrass transects to represent a range of loading conditions. Collections were made at the shore, mid-, and end regions of each transect. During sampling it was observed that epiphyte loading was low in most collections. Therefore, we had to seek areas of observably heavy loading to meet the objectives of the EPI development. An epiphyte loading sample from *H. wrightii* consisted of three samples of three shoots each.

After samples were brought to the lab, they were frozen for storage. Just prior to analysis the samples were placed in a photodeveloping tray, flooded with distilled water, and photographed. The analysis that followed included scraping the epiphytes from the leaves and measuring both leaf area and dry weight of epiphytes and seagrass. Epiphytes were removed by scraping with a single-edged razor blade. Scraped leaves were then rinsed with deionized water. The epiphytic slurry was filtered onto a 47 mm-diameter, glass fiber filter (0.5 μm pore size), dried at 80°C, and weighed. Samples were then ashed at 360°C to obtain ash-free dry weight.

Surface area of the seagrass blades was measured in the laboratory using a LI-COR Leaf Area Meter (Model 3100). Measurements were made in triplicate and averaged. Epiphyte biomass per seagrass surface area was calculated as milligrams ash-free dry weight of epiphytes per unit surface area of seagrass (mg/cm^2). Seagrass blades were then dried at 80°C and weighed. Epiphyte biomass loadings per seagrass biomass were calculated as grams of ash-free dry weight of epiphytes per gram dry weight of seagrass (grams per gram). A high positive correlation was found between epiphyte biomass per seagrass shoot surface area and epiphyte biomass per seagrass biomass ($mg/cm^2 = (10.98)\ g/g - 0.05$, $R = 0.94$) (Miller-Myers 1997). Such a high correlation indicates that either measure could be used to predict the other with a high degree of accuracy; therefore, only the latter are presented here.

PRELIMINARY EVALUATION: MATCHING SAMPLE PHOTOGRAPHS TO EPI PHOTOGRAPHS

A first evaluation consisted of matching test photographs to categories represented by photographs of known epiphyte loading. Participants (n = 58) at the International Estuarine Research Federation Conference (ERF 1995) were asked to match nine photographs to one of three epiphyte loading categories, each represented by two photographs per category.

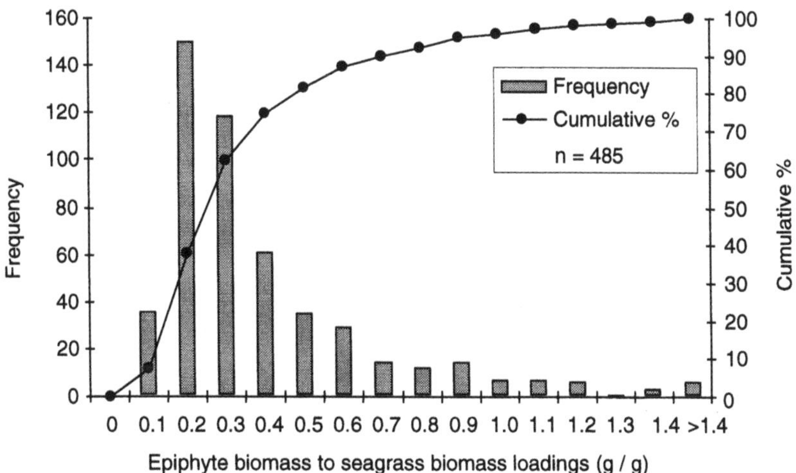

FIGURE 8.1 Frequency distribution of epiphyte loadings from the initial survey collections. Few samples had epiphyte loading > 0.6 g epiphyte/g seagrass (approximately 10%) although some loadings were more than twice that.

Quantitative Field Evaluation: Matching *In Situ* Loadings to EPI Photographs

The field evaluation utilized an earlier draft of the EPI that consisted of the three categories: epiphyte biomass to seagrass biomass ratio less than 1, equal to 1, and greater than 1. Each category contained six representatives, for a total of 18 photographs.

Field evaluation began by haphazardly placing a 1-m^2 quadrat within a seagrass bed. Participants used the 18 EPI reference photographs to compare to the *in situ* epiphyte loadings. Participants selected three shoots from the quadrat and matched them to their estimate of the loadings based on the three EPI categories. "*In situ*" is defined here and throughout this chapter as being the quantification or "actual" epiphyte loading measured by field sampling and laboratory analyses. Therefore, quantification of *in situ* epiphyte abundance began prior to participants making their selections, by haphazardly choosing three samples of three shoots from within the quadrat. Samples were brought to the laboratory in a cooler for later analysis (as above).

RESULTS

The field population sampled for the construction of the EPI had epiphyte loadings in which 90% of the epiphyte loadings were less than 0.6 g/g (grams of epiphyte per gram of seagrass) (Figure 8.1). This distribution is important considering that respondents at the ERF conference matched the evaluation photographs to the correct category more often (81%) when epiphyte loading was high than when loadings were low (51%) ($R = 0.83$) (Figure 8.2).

Results from the field evaluation of the EPI showed that participants typically underestimated epiphyte loads on seagrass (Figure 8.3). Participants' estimates were lower than the *in situ* loads in seven of the nine tests (Figure 8.3). The test indicates that even if the number of estimates is increased, the accuracy may not increase. For example, Tests 4 and 6 had the same number of participant estimates ($n = 6$), yet the results from the two tests were quite different. The mean for Test 4 was well within the standard error of the actual *in situ* epiphyte loading whereas, in Test 6, participant estimates were only half the actual *in situ* loading evaluation (Figure 8.3).

The field evaluation indicated that 56% of the variability can be explained by use of the EPI (Figure 8.4). The accuracy of the estimate increased as *in situ* loadings increased. Natural variability

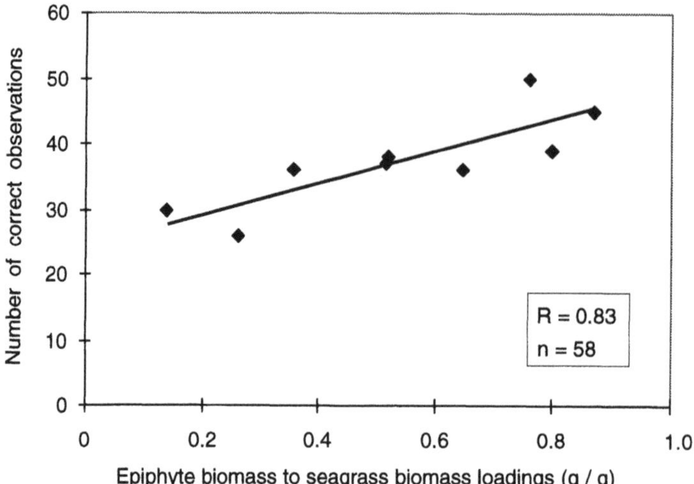

FIGURE 8.2 Number of correct matches (out of 58) for nine test photographs (n = 58 per photograph). Accuracy averaged 68% and increased as loading increased.

of the *in situ* loads was estimated for the nine field tests. Mean coefficient of variability of the *in situ* samples was 14%.

DISCUSSION

CURRENT USE OF THE EPI

Based on the results from the quantitative field evaluation, the number of reference photographs on the EPI was increased to attempt to decrease the variability of visual estimates. Therefore, the current EPI contains five categories of five photographs each. The new 5 × 5 EPI has not yet been field tested for repeatability and accuracy.

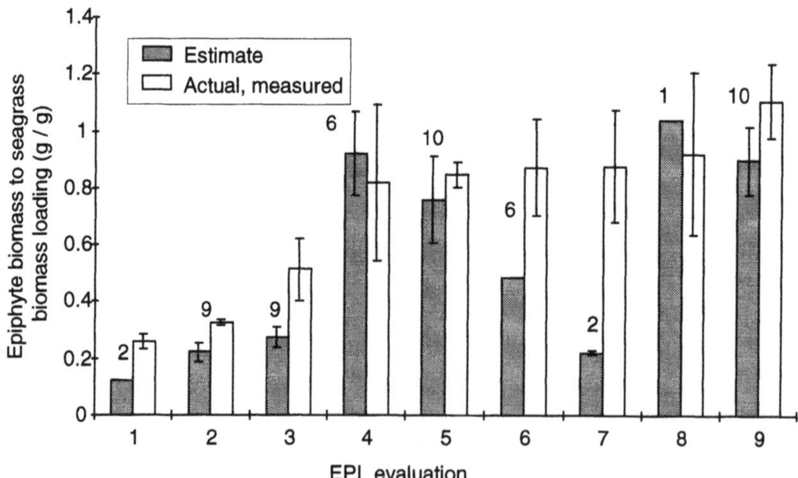

FIGURE 8.3 Comparison of estimated epiphyte loading (gray bars), using the draft EPI reference photographs, to *in situ* epiphyte loading measurements (white bars). Note that most visual estimates underestimate *in situ* loading measurements. Number of estimates per test = n.

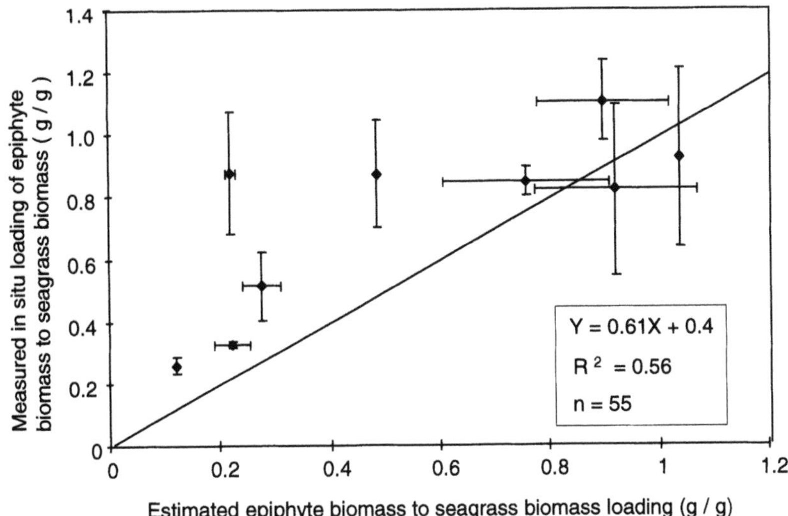

FIGURE 8.4 Regression of actual measured *in situ* epiphyte loadings on loadings estimated by using the EPI (mean + S.E.). Estimates are more accurate at higher epiphyte loading. Diagonal line represents 1:1.

The intended use of the new EPI is to provide a tool for quick visual assessment and categorization of epiphyte loadings. The new EPI can be used to assign epiphyte loadings into one of five categories (Figure 8.5). Each of the five categories contains 5 reference photographs, providing 25 total reference photographs to best match the *in situ* loadings. Two levels of loading estimates can be obtained with the EPI: (1) choosing one of five general categories (1 through 5); and (2) more specifically, selecting one of five sub-categories (a through e) within each category. The actual measures of epiphyte loadings for each photograph are provided (Table 8.1). The data are used to visually calibrate the epiphyte biomass to seagrass biomass loadings represented on the EPI and subsequently applied in the field when making an estimate.

Steps in Using the Current EPI

Four steps are involved in optimizing the new EPI to estimate *in situ* epiphyte loadings on the seagrass *H. wrightii*. These are described below:

1. Define the area
   Defining an area consists of selecting boundaries around the seagrass area for which the epiphyte load is to be estimated — either visually or with quadrat(s). To ensure that only one area is sampled, the area should have no dimension greater than a few meters.
2. Collect random samples from within the area
   To properly estimate *in situ* loads with the EPI, a sample of three shoots should be randomly collected from the area in question. Shoots are best collected by clipping or pinching off the shoot at the sediment surface and gently lifting the shoot by its base. For best comparison to the EPI, shoots should be observed in a shallow, white pan containing water.
3. Compare sample to the EPI to assign it to a category
   A. Select an overall category (1–5), based on its best match to the EPI, comparing adjacent categories to reduce inadvertent matching errors.
   B. If able, select a sub-category (a–e) based on its best match to the EPI reference photographs and table. Before a final assignment is made, check the following considerations (Step 4).

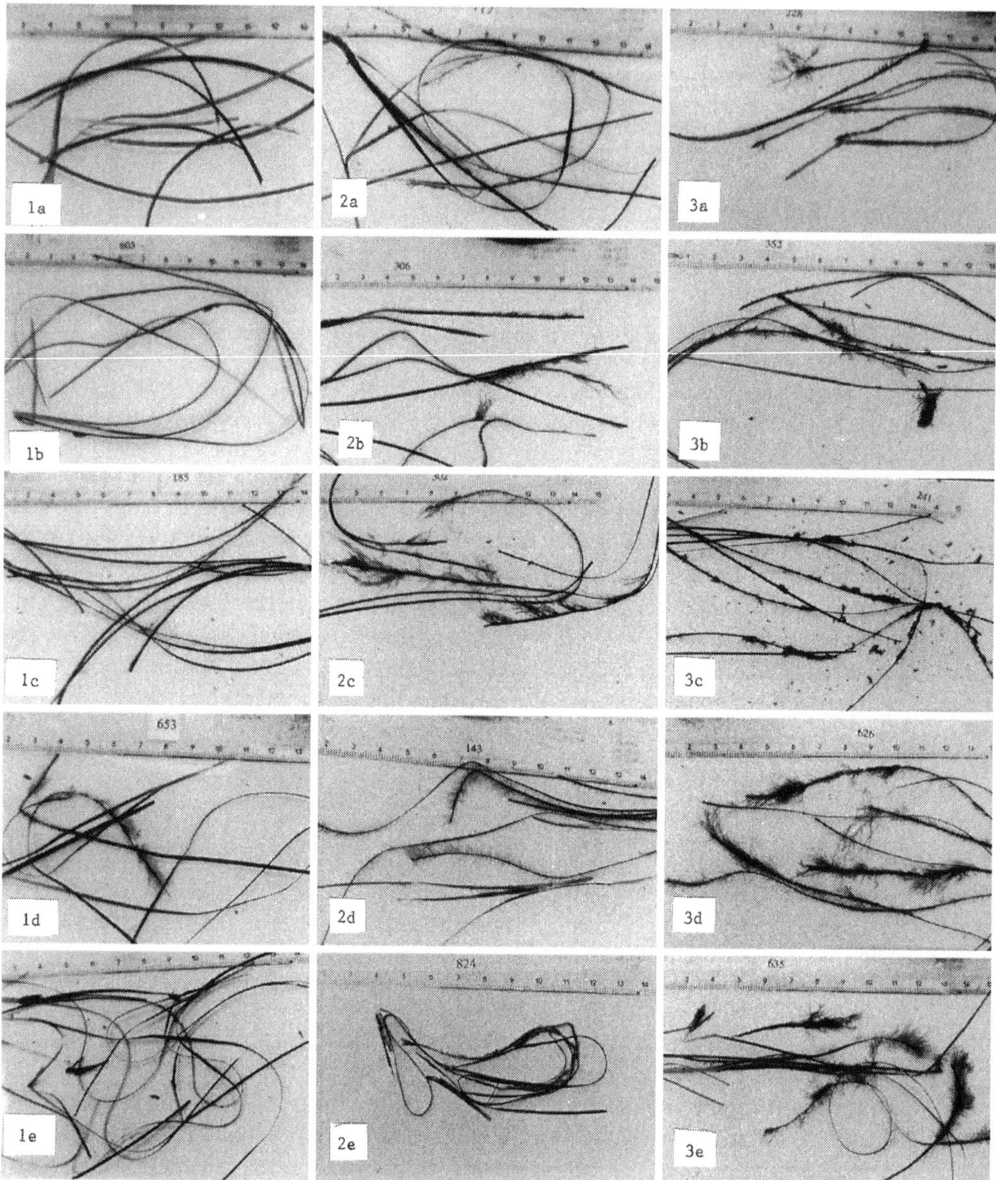

FIGURE 8.5 The EPI, showing five categories (1–5) of epiphyte loadings, each illustrated by five examples (a–e). Epiphyte loading increases between and within each category (values given in Table 8.1).

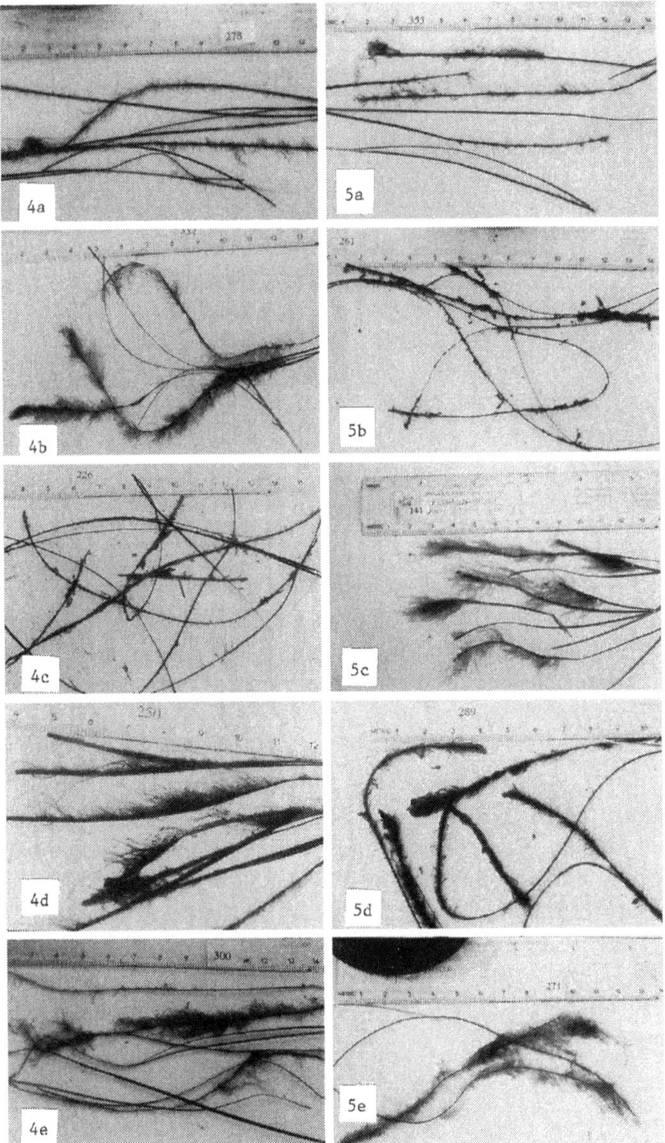

FIGURE 8.5 *Continued.*

4. Refine the selection based on considerations below
   A. The length and width of the seagrass blades determines seagrass biomass. Because epiphyte loading is a ratio of epiphyte biomass to seagrass biomass, the same amount of epiphytes on short, narrow blades would result in a higher loading than if on long, wide blades.
   B. The distribution of epiphytes along seagrass blades may affect one's first impression of loading. Dense accumulations of epiphytic algae concentrated in one area along a blade give a visual first impression of heavy loading. One must scan the entire lengths of the blades and take into account the number of blades with or without epiphytes attached in order to make a proper judgment of overall loadings. Heavy accumulations of algae and/or organic material may be attached to the sheaths of

TABLE 8.1
Quantitative, actual epiphyte loading of grams of epiphyte per gram of seagrass for EPI photographs (Figure 8.5), categories (1–5), and sub-categories (a–e).

	Category				
Sub-category	1	2	3	4	5
a	0.121	0.264	0.465	0.761	0.910
b	0.177	0.282	0.517	0.799	0.980
c	0.190	0.342	0.544	0.800	1.087
d	0.228	0.390	0.593	0.805	1.120
e	0.232	0.416	0.672	0.870	1.525

Note: Table is located on the EPI to assist the user in visually calibrating reference photographs.

sample blades and should be excluded from the loading estimates. EPI biomass loads account for algae attached to photosynthetically active portions of the plant.

C. The type of epiphytes can affect judgment of loading. Sparse coverage by highly visible filamentous epiphytes may falsely appear to produce a heavy loading.

RECOMMENDATIONS

The EPI is a tool that can be used to rapidly assess epiphyte loadings on the seagrass *H. wrightii* in the Indian River Lagoon. The accuracy and precision of this tool can probably be increased by proper instruction and training of field personnel. The tests reported here were first-time evaluations with only preliminary instruction. Further studies are needed to apply the EPI to other areas or to other seagrass species. To evaluate and better refine the new EPI, further tests of accuracy and repeatability should be performed as were conducted for the draft EPI evaluated in this chapter. Periodic testing and training could help ensure consistency within and among field personnel using the EPI.

ACKNOWLEDGMENTS

The authors are grateful to the scientists who assisted us in our lagoon-wide seagrass program, providing their time to test the validity of this work and continuing to help us understand and monitor the Indian River Lagoon. This work was funded primarily by the St. Johns River Water Management District.

REFERENCES

Backman, T. W. and D. C. Barilotti. 1976. Irradiance reduction: effects on standing crops of the eelgrass *Zostera marina* in a coastal lagoon. *Marine Biology* 34:33–40.

Czerny, A. B. and K. H. Dunton. 1995. The effect of *in situ* light reduction on the growth of two subtropical seagrasses, *Thalassia testudinum* and *Halodule wrightii*. *Estuaries* 18:418–427.

Dawes, C. J., D. Hanisak, and W. J. Kenworthy. 1995. Seagrass biodiversity in the Indian River Lagoon. *Bulletin of Marine Science* 57:59–66.

Harden, S. W. 1994. Light Requirements, Epiphyte Load and Light Reduction for Three Seagrass Species in the Indian River Lagoon, Florida. Master's thesis, Florida Institute of Technology, Melbourne, FL, 130 pp.

Miller-Myers, R. R. 1997. Spatial Distribution of Epiphyte Loadings on Seagrass (*Halodule wrightii*) in the Indian River Lagoon, Florida. Master's thesis, Florida Institute of Technology, Melbourne, FL, 57 pp.

Morris, L. J., R. W. Virnstein, J. D. Miller, and L. M. Hall. 2000. Monitoring seagrass changes in Indian River Lagoon, Florida using fixed transects. In *Seagrasses: Monitoring, Ecology, Physiology, and Management*, S. A. Bortone (ed.). CRC Press, Boca Raton, FL, pp. 167–176.

Silberstein, K., A. W. Chiffings, and A. J. McComb. 1986. The loss of seagrass in Cockburn Sound, Western Australia. III. The effect of epiphytes on productivity of *Posidonia australis* Hook. F. *Aquatic Botany* 24:355–371.

Steward, J., R. W. Virnstein, D. Haunert, and F. Lund. 1994. Surface Water Improvement and Management (SWIM) Plan for the Indian River Lagoon. St. Johns River Water Management District and South Florida Water Management District, Palatka and West Palm Beach, FL, 120 pp.

Tomasko, D. A. and B. E. Lapointe. 1991. Productivity and biomass of *Thalassia testudinum* as related to water column nutrient availability and epiphyte levels: field observations and experimental studies. *Marine Ecology Progress Series* 75:9–17.

Virnstein, R. W. 1987. Seagrass-associated invertebrate communities of the southeastern USA: a review. In *Proceedings of the Symposium on Subtropical-Tropical Seagrasses of the Southeastern United States*, M. J. Durako, R. C. Phillips, and R. R. Lewis, IIII (eds.). Florida Marine Research Institute Publication No. 42, Department of Natural Resources, St. Petersburg, FL, pp. 89–116.

Virnstein, R. W. 1995. Seagrass landscape diversity in the Indian River Lagoon, Florida: the importance of geographic scale and pattern. *Bulletin of Marine Science* 57:67–74.

Virnstein, R. W. and L. J. Morris. 1996. Seagrass preservation and restoration: a diagnostic plan for the Indian River Lagoon. Technical Memorandum No. 14, St. Johns River Water Management District, Palatka, FL, 43 pp.

Virnstein, R. W., L. J. Morris, J. D. Miller, and R. Miller-Myers. 1997. Distribution and abundance of *Halophila johnsonii* in the Indian River Lagoon. Technical Memorandum No. 24, St. Johns River Water Management District, Palatka, FL, 14 pp.

Woodward-Clyde Consultants. 1994. Historical Imagery Inventory and Seagrass Assessment, Indian River Lagoon. Final Technical Report, Indian River Lagoon National Estuary Program, Melbourne, FL, 106 pp.

9 Establishing Baseline Seagrass Parameters in a Small Estuarine Bay

Margaret A. Wilzbach, Kenneth W. Cummins, Lourdes M. Rojas, Paul J. Rudershausen, and James Locascio

Abstract Baseline seagrass parameters were established in autumn-winter 1998 for a small estuarine bay, Tarpon Bay, in the J.N. "Ding" Darling National Wildlife Refuge of Sanibel Island, Florida. Distribution, abundance, and condition of the seagrasses were estimated using the transect method protocol adopted by the Southwest Florida Water Management District's Surface Water Management Program. Above-sediment biomass was estimated by weighing clipped plants. The 23 transects, varying in length from 9 to 200 m, involved the sampling of 80 1-m^2 quadrats that contained one or more of the three seagrass species present in Tarpon Bay, turtle grass (*Thalassia testudinum*), shoal grass (*Halodule wrightii*),and manatee grass (*Syringodium filiforme*).

The seagrasses were evaluated for Blaun-Blanquet cover classes and mean blade length in each m^2 quadrat along with the number of short shoots of each species that were enumerated in three 100-cm^2 frames placed within the quadrat. Length–mass regressions were independently developed to allow conversion of short-shoot densities, together with mean blade lengths, into estimates of standing crop biomass. The calculated and directly measured estimates of grams dry mass m^{-2} of the three species in Tarpon Bay are in the low range of those reported in other studies (*T. testudinum* = 15–119, *H. wrightii* = 2–28, and *S. filiforme* = 66–105). The comparatively low estimates may be attributable to seasonal variation, differences among studies in the portion of the plant measured, variation in the presence of encrusting organisms on the seagrass blades, and the effect of variations in blade width on the calculated biomass estimates.

Of the three species in Tarpon Bay, *T. testudinum* was most ubiquitous, occurring in 66% of the quadrats. All three species showed a preference for sand over muddy sand (the most abundant bottom type, 61%) and sediments with coarse shell. In addition, *H. wrightii* exhibited a very strong affinity for mud sediments. In all Blaun Blanquet cover classes, the three species showed a gradient in depth of occurrence with *H. wrightii* at the shallowest depths, *T. testudinum* at intermediate depths, and *S. filiforme* in the deepest locations. Although light transmission showed an expected negative correlation with depth, sediments were not consistently correlated with depth. This study is the initial phase in the long-term evaluation of seagrass community health in Tarpon Bay as a major reference site for the Caloosahatchee River estuary–bay system.

INTRODUCTION

Tarpon Bay is on Sanibel Island off the lower west coast of south Florida. It is part of the Caloosahatchee River Estuary system that includes San Carlos Bay, Pine Island Sound, Matlacha Pass, plus the lower Caloosahatchee River, down river from the Franklin Lock and Dam to Shell Point. The estuary is generally believed to have been environmentally degraded over the last decade; however, data directly documenting these changes are largely circumstantial and anecdotal (Harris

0-8493-2045-3/99/$0.00+$.50
© 2000 by CRC Press LLC

et al. 1983; Livingston 1987). The impression of local residents is that water clarity, seagrass coverage, and shellfish have declined in the estuary–bay system. This impression is supported by some aerial photographic evidence of seagrass decline (Harris et al. 1983) and some intermittent measurements of water quality, seagrass distribution, and invertebrate populations (Chamberlain and Doering 1998; Doering and Chamberlain 1998). The major environmental stressor to the system is likely to be releases of fresh water to the estuary that exceed the ranges of quantity and timing of natural fluctuations experienced by seagrass beds before the Franklin Lock and Dam was in place. The salt barrier Franklin Lock and Dam, on the lower Caloosahatchee River, was installed in the mid-1960s. Degraded water quality, resulting from agricultural and urban runoff in the watershed, which is exacerbated by the un-natural releases of runoff, is also a major stressor.

The research reported here on seagrass communities is part of a program to establish a database for Tarpon Bay that can serve as a reference site for the estuary–bay system. This database can be used to document changes in ecological health while serving as an early warning of further system degradation. Specifically, in this chapter we describe the relative abundance of seagrass species and their distribution relative to light, depth, and sediment type in Tarpon Bay. Blade length:mass relationships for each species are developed, and standing stock biomass estimates are reported and compared with literature data.

METHODS

Study Site

Tarpon Bay, a 13 km^2 shallow embayment of San Carlos Bay in the Caloosahatchee River Estuary system of lower Charlotte Harbor, lies entirely within the J.N. "Ding" Darling National Wildlife Refuge on Sanibel Island. Sanibel is a subtropical barrier island of sand and shell on a drowned limestone plateau with low sand dunes on the gulf side, and a mangrove fringe on most of the bay side (McNulty et al. 1972). Geological age of the island is 5000 years. Average annual air temperature is 23°C. Annual precipitation is 107 cm with approximately 70% of rainfall occurring during the wet season (June–October).

Tarpon Bay is fringed with red mangrove (*Rhizophora mangle*) except in the small developed concession area on the south side. It opens to San Carlos Bay at the northeast end and through a small cut on the northwest side. Salinities in Tarpon Bay have been measured as low as 10 ppt, but ranged from 25 to 35 ppt during the period of study reported here. Water depth averages 1 m at mean low tide. Tides are predominantly semi-diurnal (McNulty et al. 1972), with a mean diurnal range of 0.5–1.1 m. The seagrass community of Tarpon Bay is dominated by three species: turtle grass (*T. testudinum*, Hydrocharitaceae), manatee grass (*S. filiforme*, Potamogetonaceae), and shoal grass (*H. wrightii*). *Halophila engelmanni* (Hydrocharitaceae) is present, but very rare. It occurs both in association with *S. filiforme* and in water deeper than that well colonized by *S. filiforme*.

Sampling

Seagrass communities were sampled in Tarpon Bay from September 29 to November 19, 1998 (Figure 9.1) using a transect-based approach. The sampling followed the seagrass monitoring protocol adopted by the Southwest Florida Water Management District (SWFWMD) for its Surface Water Improvement and Management Program (Scheda Ecological Associates 1998) with additional data recorded as well. For each of the 23 transects, which varied in length from 9 to 200 m, meter-square quadrats were assessed at the shoreward and bay ends of seagrass cover and at a maximum of 50 m intervals in between. Transect placement was not randomized, but positioned to provide uniform coverage of the shoreline and shoal areas.

A total of 80 quadrats were assessed. In each quadrat, the percent cover of each of the three species of seagrass, turtle grass (*T. testudinum*), shoal grass (*H. wrightii*), and manatee grass

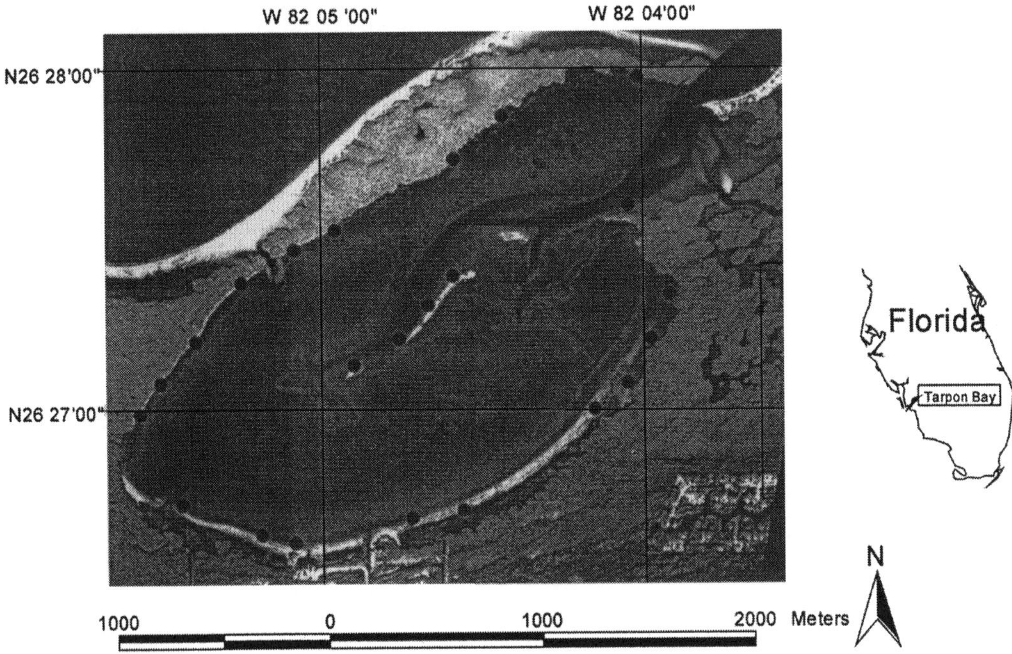

FIGURE 9.1 Aerial photograph of Tarpon Bay (December 1997), showing location of seagrass transects.

(*S. filiforme*) was rated using the Blaun-Blanquet criteria which assigns a numerical code to percent cover (+ = trace, 1 = 0–5%, 2 = 5–25%, 3 = 25–50%, 4 = 50–75%, and 5 = 75–100%). In the data reported here, the trace ratings were combined with category 1 and all open sediment quadrats were reported as 0. The percent cover of drift algae also was recorded for each quadrat.

Quadrats were defined by a perforated PVC frame. A 100 cm^2 smaller PVC frame was used to take three sub-samples, within the square meter frame, of short-shoot density by species. The shoreward and bayward ends of each transect were located at sub-meter resolution with a Trimble Pro XR™ global positioning system having a real time sub-meter resolution. At each quadrat, a series of environmental variables were measured:

- time of day (to determine tide stage)
- depth (± 1cm)
- transmission of photosynthetically active radiation (PAR) using a Li-COR™ photometer (μEin $m^{-2}sec^{-1}$)
- salinity (± 1.0 ppt) with a Vista A366ART™ refractometer
- temperature (± 0.1°C)
- dissolved oxygen (± 0.01mg/l)
- pH (Å 0.00) using a Yellow Springs Instruments 600XL™ sonde

Percent transmission of PAR was calculated as the light measured 5 cm off the bottom as a percent of the light 1 cm under the surface.

In addition to seagrass cover and shoot density estimated by species for each quadrat, the length of five blades of each species present was measured to the nearest mm. The epiphyte density on the blades in each quadrat was rated as clean, sparse, moderate, or heavy. Sediments were categorized as shelly sand, sand, muddy sand, mud, or shelly mud by visual inspection, according to the SWFWMD protocol. To evaluate colonization by seagrasses of sediment types in Tarpon Bay relative to sediment availability, sediment types were ranked according to relative availability (*s*).

Usage was measured as *r* – *s*, where *r* represented a ranking of occurrence on different sediment types for a species (rank preference index, Johnson 1980). Negative values denoted relative avoidance of a substrate, positive values denoted relative preference, and zero indicated a sediment type was used in accordance with a ranking of its availability. Secchi depth was measured at the bayward point of each transect when water was of sufficient depth.

To calculate above-ground standing stock biomass for each seagrass species in Tarpon Bay, blade length/mass relationships were developed using wet mass, dry mass, and ash-free dry mass. Sets of 100 blades of each species were collected from seagrass beds in Tarpon Bay, placed in a cooler, and returned to the laboratory for processing. Blades were collected by clipping at the base of the short shoot. The length of each blade was measured by species to the nearest mm and weighed to the nearest 10 mg fresh mass. Each blade was blotted once on each side prior to weighing, but no effort was made to remove epiphytes because of the cuticular damage that it causes. The blade was then cut into pieces, placed in a small aluminum pan, and oven dried at 50°C for 24 h. After drying, blades were weighed to the nearest 1 mg dry mass. Each blade was then placed in a crucible and ashed at 500°C for 4 h. The ash was then weighed to the nearest 1 mg. Ln-ln plots were developed for each of the three mass categories and coefficients calculated for conversion of mean blade length measurements to biomass (fresh, dry, and ash-free dry mass). Plots of length–mass relationships for each of the species are shown in Figure 9.3 (a,b,c).

Standing crop biomass estimates for each species in Tarpon Bay were calculated as follows:

1. short-shoot density, d, for each m^2 quadrat = mean of three 100 cm^2 quadrats · 100
2. mean number of blades per shoot, b, = mean of 100 shoot counts
3. mean blade length, l, for each m^2 quadrat = mean of 5 blade lengths collected from quadrat
4. biomass per blade, estimated as $\ln y = a + b \ln l$, where a and b represent the coefficient and slope of the regression equation, respectively
5. above-ground standing crop biomass for each quadrat, in g m^{-2} = d · b · y
6. mean biomass standing crop for each species m^{-2} for Tarpon Bay seagrass beds = mean of quadrat estimates (n = 80)

Additional estimates of standing crop biomass were made by completely removing the above sediment portion (blades) of seagrass plants in a defined area of bottom in Tarpon Bay. Plants were removed in November and December 1998 in the southwest sector of the Bay. Two samplers were used: a 0.0986 m^2 metal-flanged, open-bottom box (n = 6 samples), and a m^2 frame divided into a 4 × 4 grid, from which plants in five randomly selected squares, each 0.0625 m^2 in area, were collected. Grid squares were selected with use of a random number table. After removal from the sample area, the blades were separated by species, blade lengths were measured, and the total fresh mass of each was determined. The species lots of fresh material were then dried at 50°C for 24 h and weighed. Then they were ashed at 500° for 4 h.

RESULTS AND DISCUSSION

Of the three common seagrass species in Tarpon Bay, *T. testudinum* was most ubiquitous, occurring in 66% of the quadrats. *H. wrightii* was present in 37% of the quadrats and *S. filiforme* in 22%. Distributions of each species by cover class (percent areal coverage in a quadrat) are shown in Figure 9.2. For all species, the predominant areal coverage of a quadrat was less than 50% (cover classes 1, 2, and 3). Complete coverage of a quadrat by a species occurred in only a small number of cases. Drift algae were present in 35 (44%) of the quadrats. In the majority of cases where it occurred, its cover ranking was 1 (1–5% areal coverage). Species in the drift algae assemblage included *Acanthophora spicifera, Gracilaria caudata* (soft and tough forms), *G. tikvahiae, Solieria filiformis,* and *Caulerpa fastigiata* (determined by M. Brown and C. Dawes, University of South Florida).

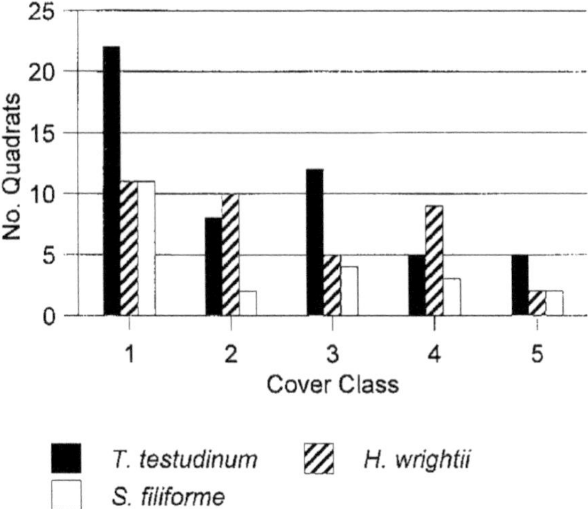

FIGURE 9.2 Frequency of occurrence of seagrass species in cover classes. Cover class 1 represents trace to 5% areal coverage of a m² quadrat; class 2 = 6–25%, class 3 = 26–50%, class 4 = 51–75%, and class 5 = 76–100%); n = 80 quadrats.

Although *T. testudinum* was the most widespread in its distribution, its short-shoot density (no. m^{-2}) was the least of the three species (Table 9.1). Short-shoot density was greatest for *H. wrightii*. When estimated from quadrat-based data (as the product of shoot density, no. blades per short shoot, and mass per blade, in g dry m^{-2}), above-ground standing crop biomass was greatest for *T. testudinum* and least for *H. wrightii* (Table 9.1). It is important to note that these standing crop estimates are based on quadrats in which coverage by a species was predominantly sparse (Figure 9.2). When restricted to cases in which coverage by a species was dense (cover class 5), biomass estimates approximately tripled for all three species.

Direct measurements of above-ground standing crop of seagrass blades (Table 9.2 and 9.3), made in locations where coverage of *T. testudinum* (Table 9.3) and *S. filiforme* (Table 9.2) was

TABLE 9.1
Short-shoot densities (mean no. m^{-2}) and above-ground standing crop biomass (g dry mass $m^{-2)}$ of seagrasses in Tarpon Bay.

	Thalassia testudinum	*Halodule wrightii*	*Syringodium filiforme*
mean no. m^{-2} [a]	155 (205,54)	521 (1067, 39)	162 (173,22)
mean no. m^{-2} [b]	199 (212,42)	564 (1101,36)	223 (165,16)
g dry mass m^{-2}	2.74 (3.30,40)	0.33 (0.50,34)	0.65 (0.80,16)

Note: Means of short-shoot densities are based on counts of three 100 cm² quadrats within a m² quadrat, for all quadrats in which cover of a species was ranked as 1 or greater.

[a] Includes zero values.

[b] Excludes cases where a species was absent from all three 100 cm² quadrats. Standing crop biomass is estimated from quadrat-based measurements as the product of shoot density, no. blades per short shoot, and mass per blade. Mass per blade is estimated from measurements of blade length (±1 mm, n = 5 per quadrat) and derived relationships between blade length: g dry mass (Figure 9.3 a,b,c). Standard deviations and sample size, respectively, are given in parentheses.

TABLE 9.2
Directly measured and estimated biomass of *Syringodium filiforme* in Tarpon Bay.

Subsample	g dry mass measured	g dry mass estimated	g AFDM measured	g AFDM estimated	blade length (mm)	blade density
1	5.14	3.02	2.43	1.98	286.1	66
2	6.55	11.49	3.32	7.30	382.2	167
3	4.12	5.65	1.75	3.71	293.9	119
4	6.23	7.16	2.64	4.77	257.0	182
5	5.66	8.18	2.71	5.40	277.2	187
mean	5.54					
(0.86)	7.10					
(2.80)	2.57					
(0.51)	4.63					
(1.77)	299.3	144.2				
no. m^{-2}	88.6	113.6	41.4	74.11	—	2307

Note: A m^2 quadrat, in which cover of *S. filiforme* was ranked as 5 (75–100%), was divided into 16 equal sub-sample sections for direct measurements. Sub-samples were used to measure total m^2 above sediment biomass and to estimate biomass based on blade counts per sub-section, a sub-sample of blade lengths, and derived blade length:mass regressions given in Figure 9.3c. Standard deviations indicated in parentheses.

TABLE 9.3
Comparison of standing crop biomass of above-sediment seagrass blades between Tarpon Bay and selected literature values, in grams dry mass meter^{-2}.

Location	*Thalassia testudinum*		*Halodule wrightii*		*Syringodium filiforme*		References
	Range	Median	Range	Median	Range	Median	
Tampa Bay[a]	25–180	103	38–50	44	50–170	110	Lewis and Phillips 1980
		49		19		21	Jensen and Gibson 1986
South Florida[a]	58–267	163	5–54	30	28–102	65	Zieman 1982
South Florida[b]	7–832	420	2–67	35	5–352	179	Zieman and Zieman 1989
Florida Bay[a]							
Control	21–90	56	0–5	0			Powell et al. 1989
Enriched	4–192	93	0–152	14			
Tarpon Bay[c]	15–119	67	2–28	15	66–105	86	This study

Note: Tarpon Bay values are based on direct measurements of standing crop in November–December 1998 from six 0.10 m^2 sampling units placed in a dense bed of *Thalassia testudinum* (for *T. testudinum* and *Halodule wrightii*), and on direct measurement of standing crop from five random 0.06 m^2 sections of a m^2 quadrat placed in a dense bed of *Syringodium filiforme*.

[a] Above sediment portion of plant.

[b] Calculated above sediment portion of plant using 13% for *Thalassia testudinum*, 22% for *Halodule wrightii*, 32% for *Syringodium filiforme*, after Zieman and Zieman (1989).

[c] Above sediment blades.

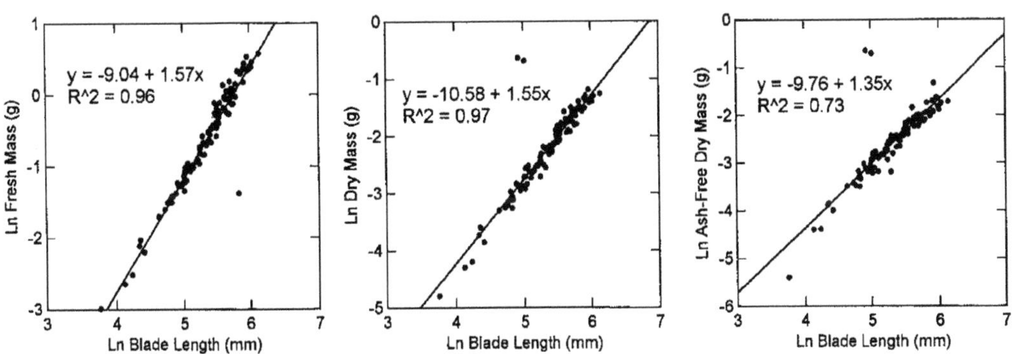

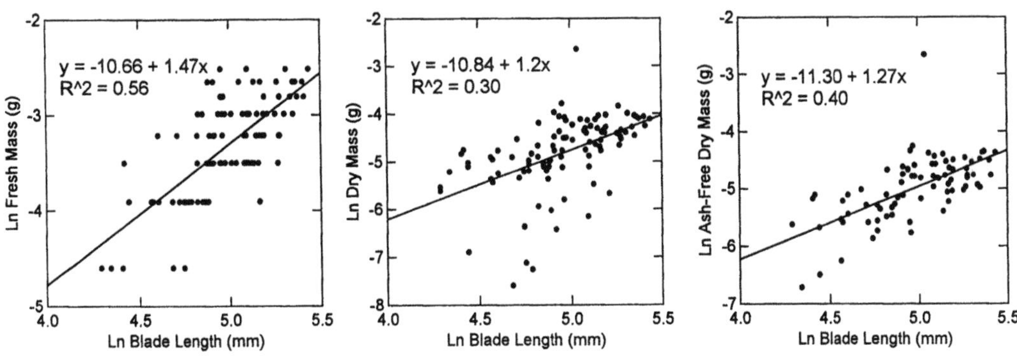

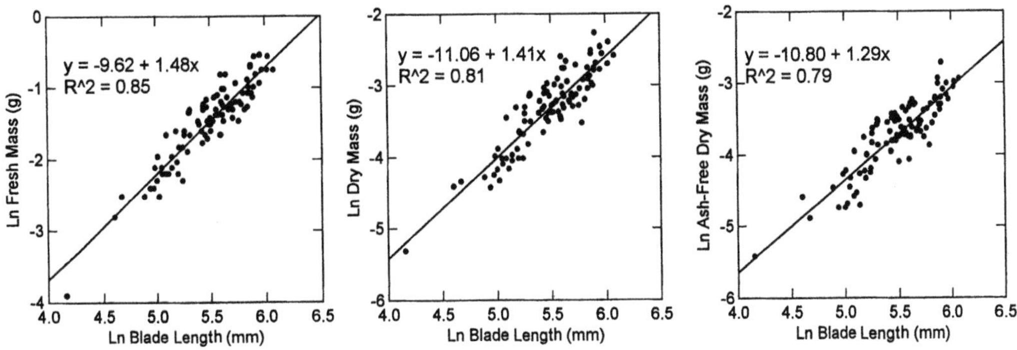

FIGURE 9.3 Blade length:mass relationships. A. *T. testudinum*; B. *H. wrightii*; C. *S. filiforme*; n = 100 for each species.

dense, were also higher than calculated values in areas of sparse coverage. In comparing directly measured and estimated biomass values from the same samples (Table 9.2), estimated values were greater than measured values in four out of five cases. Estimated values ranged from 59–220% of the directly measured values. Errors in estimation may result from error associated with use of averaged values for shoot density, blades per shoot, and blade lengths, as well as error associated with use of blade length:mass regressions.

The regression of blade mass against blade length in grams of fresh mass, dry mass, and ash-free dry mass for all three species are given in Figure 9.3(a,b,c). The regressions, fitted to a ln:ln plot following Ricker (1975), are all significant. Fit of the data to the regression line is strongest for *T. testudinum* and weakest for *H. wrightii*. The greater variability for *H. wrightii* likely reflects the greater variation in blade width (or diameter in the case of *S. filiforme*) that it exhibited. Because the surface:volume ratio is greater for *H. wrightii* than the other two species, encrusting organisms likely contribute to variation in its dry mass estimates. The higher correlation coefficient seen for ash-free dry mass relative to dry mass of *H. wrightii* (Figure 9.3b) reflects this. For each of the three seagrass species, the average epiphyte load was ranked as light.

Estimates of seagrass standing crop biomass from Tarpon Bay are compared to selected literature values in Table 9.3. The Tarpon Bay measures are all within the range of values reported from other studies, although *T. testudinum* and *H. wrightii* standing crops are at the low end of these reported ranges. This may be attributed in part to the fact that the data reported here are for winter populations. Zieman and Zieman (1989) reported that winter seagrass m^2 biomass was approximately half the summer standing crop. The wide ranges of literature values represent seasonal differences as well as spatial ones. Other explanations for the relatively low biomass include variation among study sites in the presence of encrusting organisms on seagrass blades as well as possible differences among studies in the portion of the plant measured. In this study, biomass estimates are for seagrass blades only. Inclusion of central meristem and surrounding old leaf sheaths at the base of each plant would significantly increase a biomass estimate.

The general pattern between mean depth at which each seagrass species occurred, irrespective of seagrass cover class or density, conformed to expectations (e.g., Ferguson et al. 1980; Zieman 1982) that *H. wrightii* was most frequently found at the shallowest sites, *T. testudinum* at intermediate depths, and *S. filiforme* at the deeper locations (Figure 9.4). This same relationship held when data were restricted to locations at which areal coverage by a species was greater than 50%. Depths were normalized to mean low tide by correcting each depth measurement on a tide chart for the

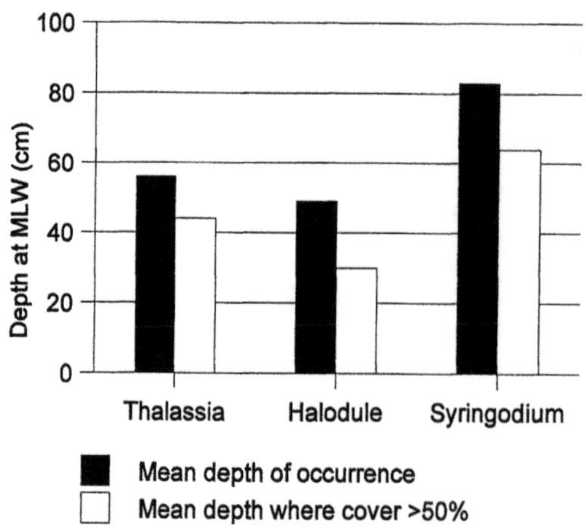

FIGURE 9.4 Zonation of seagrasses in Tarpon Bay with respect to water depth, standardized to mean low tide.

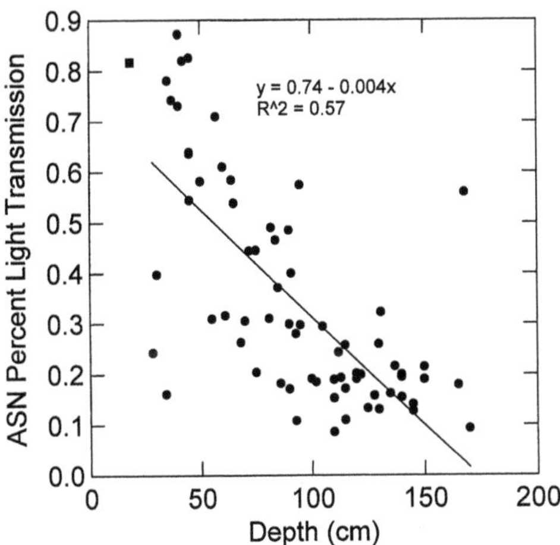

FIGURE 9.5 Relationship between arcsine (ASN) of percent light transmission and depth in Tarpon Bay. Percent light transmission represents solar irradiance 5 cm above the sediments, divided by irradiance measured 1 cm below the water surface.

sampling time to standardize for differences in tidal cycle at the time of sampling. The difference among seagrass species in depths at which each occurred was not significant in this study ($p > 0.05$), but a negative correlation was found between cover class and depth for *H. wrightii* ($r = -0.61$, $n = 38$). For *S. filiforme*, a slightly positive correlation was observed between shoot density and depth ($r = 0.28$, $n = 22$).

Because light transmission is a function of depth, there is an expected relationship between the two (Figure 9.5). The scatter of these data indicate that other factors such as water color and suspended load significantly modify light transmission, independent of depth. In addition, records of light transmission integrated over a longer time frame are needed in order to elucidate its influence on seagrass cover and growth. Tomasko and Hall-Ruark (1998) found 20% surface irradiance to be the lower limit for *T. testudinum* at salinities of less than 21 ppt, but point out that the relationship between light transmission and salinity make interpretation of the influence of either, measured separately on seagrass biomass, difficult to understand. In this study, PAR light transmission varied much more than salinity. Although light transmission varied with depth, its relationships with short-shoot density or with blade lengths were not significant ($p > 0.05$) for any of the three seagrass species. This is likely because of interactions with salinity and other parameters such as substrate characteristics, available nutrient levels, herbivory (e.g., grazing by manatee and the pinfish *Lagodon rhomboides*), and mechanical damage (e.g., watercraft prop scarring). However, as expected (e.g., Phillips 1960), *H. wrightii*, which tends to occur at the shallower sites (Figure 9.4) where the plants are often completely exposed at low tide, had the shortest mean blade length (mean = 16.5 cm, SD = 7.3, n = 172); *T. testudinum*, which occurs at intermediate depths, had an intermediate mean blade length (mean = 23.3 cm, SD = 10.0, n = 248); and *S. filiforme*, tending to grow in the deepest water, had the longest mean blade length (mean = 25.2 cm, SD = 10.3, n = 105).

The three seagrasses differed with respect to sediment colonization (Table 9.4). The majority of sediments (61% of 148 measurements) were comprised of muddy sand. Other sediment types, in order of frequency, were mud (22%), sand (9%), shelly sand (5%), and shelly mud (3%). All seagrasses showed a preference for sand, and all showed an apparent avoidance of shelly mud. *H. wrightii* and *S. filiforme* also avoided shelly sand, and *H. wrightii* exhibited a strong preference for mud. Sediments did not vary consistently with depth ($p > 0.05$, Spearman rank test).

TABLE 9.4
Colonization of sediment types relative to sediment availability of seagrasses in Tarpon Bay.

Sediment Type	Relative Availability (rank)	*Thalassia testudinum* usage	*Halodule wrightii* usage	*Syringodium filiforme* usage
Shelly Sand	4	0	−1.5	−1.0
Sand	3	+1.0	+1.5	+1.5
Muddy Sand	1	0	0	0
Mud	2	0	+2.5	0
Shelly Mud	5	−1.0	−2.5	−0.5

Note: Sediment types are ranked according to relative availability (s). Usage is measured as $r - s$, where r represents a ranking of occurrence on different sediment types for a species (rank preference index, Johnson 1980). Negative values denote relative avoidance of a substrate; positive values denote relative preference. Zero indicates a sediment type is used in accordance with a ranking of its availability.

CONCLUSIONS AND FUTURE RESEARCH

The resurvey of the seagrasses in Tarpon Bay on an annual/seasonal basis should provide the basis for the long-term evaluation of their condition relative to changing environmental conditions. The major objectives are to produce a very detailed map of the seagrass distributions by species that will allow subtle changes in distribution to be detected and to build an extensive database on standing crop biomass within each density isopleth for each species. It is only with this level of detail that a model can be derived that can serve as an early warning monitor of stresses on the seagrass communities of Tarpon Bay. To reach the desired level of detail, larger samples of blade length per species per quadrat need to be obtained, and blade width (or diameter) measurements need to be built into the length–mass regressions for each species. Future research will also focus on establishing a continuous recording water quality monitoring station in the bay, and determinations of the seasonal density of drift algae (e.g., *Acanthophora* and *Gracilaria*) in relation to the distributions of the seagrasses and the importance of periphytic algal cover on the seagrass blades. The latter will be accomplished with the use of scale plastic models of the three seagrasses common in Tarpon Bay through *in situ* measurements of photosynthesis and respiration of natural seagrasses compared to the plastic models in enclosed recirculating chambers. All the seagrass work is being linked to macroinvertebrate and spotted seatrout (*Cynoscion nebulosus*) surveys in the bay that are referenced to seagrass distributions. These integrated efforts at developing a data base are focused on establishing Tarpon Bay as a major reference site for the long-term evaluation of estuarine ecosystem condition and health for the region.

REFERENCES

Chamberlain, R. H. and P. H. Doering. 1998. Preliminary estimate of optimum freshwater flow to the Caloosahatchee Estuary: a resource-based approach. In *Proceedings, Charlotte Harbor Public Conference Technical Symposium,* S. F. Treat (ed.). Charlotte Harbor National Estuary Program Technical Report 98-02, pp. 121–130.

Doering, P. H. and R. H. Chamberlain. 1998. Water quality in the Caloosahatchee Estuary, San Carlos Bay, and Pine Island Sound, Florida. In *Proceedings, Charlotte Harbor Public Conference Technical Symposium,* S. F. Treat (ed.). Charlotte Harbor National Estuary Program Technical Report 98-02, pp. 229–240.

Ferguson, R. L., G. W. Thayer, and T. R. Rice. 1980. Marine primary producers. In *Functional Adaptations of Marine Organisms.* Academic Press, New York, Chap. 2.

Harris, B. H., K. D. Haddad, K. A. Steindinger, and J. A. Huff. 1983. Fisheries habitat loss in Charlotte Harbor and Lake Worth, Florida. Final Report, Florida Department Environmental Regulation Coastal Zone Management, Grant CM-29, 211 pp.

Jensen, P. R. and R. A. Gibson. 1986. Primary production in three subtropical seagrass communities: a comparison of four autotrophic components. *Florida Scientist* 49:131–134.

Johnson, D. H. 1980. The comparison of usage and availability measurements for evaluating resource preference. *Ecology* 61:65–71.

Lewis, R. R. and R. C. Phillips. 1980. Seagrass mapping project, Hillsborough County, Florida. Tampa Port Authority, 30 pp.

Livingston, R. J. 1987. Historic trends of human impacts on seagrass meadows in Florida. In *Proceedings of the Symposium on Subtropical-Tropical Seagrasses of the Southeastern United States,* M. J. Durako, R. C. Phillips, R. R. Lewis (eds.). Florida Marine Research Publication 42, pp. 139–152.

McNulty, J. K., W. N. Lindall, Jr., and J. R. Sykes. 1972. Cooperative Gulf of Mexico estuarine inventory and study, Florida: Phase I: area description. NOAA Technical Report, National Marine Fisheries Service Circulation 378, 126 pp.

Phillips, R. C. 1960. Observations on the ecology and distribution of the Florida seagrasses. Florida State Board Conservation Marine Laboratory, Professional Paper Series No. 2, 72 pp.

Powell, G. V. N., J. W. Fourqurean, W. J. Kenworthy, and J. C. Zieman. 1989. Experimental evidence for nutrient limitation of seagrass growth in a tropical estuary with restricted circulation. *Bulletin of Marine Science* 44:324–340.

Ricker, W. E. 1975. Computation and interpretation of biological statistics of fish populations. *Bulletin of the Fisheries Research Board of Canada* 191:382 pp.

Scheda Ecological Associates. 1998. Seagrass monitoring protocol, summer 1998 results. Prepared for Southwest Florida Water Management District.

Tomasko, D. A. and M. O. Hall-Ruark. 1998. Abundance and productivity of the seagrass *T. testudinum* along a gradient of freshwater influence in Charlotte Harbor, Florida. In *Proceedings, Charlotte Harbor Public Conference Technical Symposium,* S. F. Treat (ed.). Charlotte Harbor National Estuary Program Technical Report 98-02, pp. 111–120.

Zieman, J. C. 1982. The ecology of the seagrasses of South Florida: a community profile. United States Fish and Wildlife Service, Biological Service Program 82/25, 185 pp.

Zieman, J. C. and R. T. Zieman. 1989. The ecology of seagrass meadows of the west coast of Florida: a community profile. United States Fish and Wildlife Service Report 85(7.25), 155 pp.

10 Monitoring Submerged Aquatic Vegetation in Hillsborough Bay, Florida

Walter Avery

Abstract In 1976, the City of Tampa, Bay Study Group began monitoring Hillsborough Bay in anticipation of sewage pollution abatement. Included in the multidisciplinary monitoring plan were a drift macroalgae program and a submerged aquatic vegetation program, which commenced in 1983 and 1986, respectively.

Since the inception of the macroalgae program, the maximum average monthly macroalgal biomass of 164 $gdwtm^{-2}$ in 1987 declined to 2.87×10^{-3} $gdwtm^{-2}$ in 1998. *Ruppia maritima*, *Halodule wrightii*, and the rhizophytic alga *Caulerpa prolifera* were monitored during the submerged aquatic vegetation program. *R. maritima* presence was typically ephemeral with areal coverage fluctuating between 2 ha and 40 ha. In contrast, *H. wrightii* areal coverage steadily increased from about 0.2 ha in 1986 to nearly 57 ha in 1998. *C. prolifera* began to colonize areas of Hillsborough Bay in 1986, and areal coverage reached a maximum of nearly 220 ha in 1988 before declining to zero by 1997.

Monitoring techniques have been modified as a result of expanding seagrass coverage and recent technological innovations. Presently, the Bay Study Group uses a combination of high and low altitude photography, on-site groundtruth surveys, and a global positioning system to delineate seagrass meadows and follow changes in seagrass coverage. Additionally, the Bay Study Group developed a seagrass program under the auspices of the Tampa Bay National Estuary Program. This program is an interagency effort to monitor changes in distribution and coverage of the major seagrass species in Tampa Bay.

INTRODUCTION

Monitoring trends of seagrass areal coverage and identifying the factors affecting seagrass distribution have become extremely important as coastal ecosystems continue to be impacted by burgeoning population growth. Specifically, within Tampa Bay, seagrass coverage between 1876 and 1982 declined approximately 81% percent (Lewis et al. 1985). This decline was attributed to alteration of existing seagrass habitat (Lewis 1977), industrial discharges (Lewis and Phillips 1980), and increasing eutrophication as indicated by elevated phytoplankton biomass (Johansson et al. 1985). Johansson and Greening (Chapter 21 of this book) give a historical overview of anthropogenic impacts to Tampa Bay and their effects on the estuary's seagrass meadows.

Hillsborough Bay, located in the northeastern section of Tampa Bay, historically has been considered to have the poorest water quality in Tampa Bay (Simon 1974), and eutrophic conditions had become evident in Hillsborough Bay by the 1960s. The Federal Water Pollution Control Administration, in response to complaints of decay of extremely high drift macroalgae biomass, conducted a study and, in 1969, issued a report stating that point source nutrient loading to the bay needed to be addressed. The report cited discharges to the bay from fertilizer industries near the

0-8493-2045-3/99/$0.00+$.50
© 2000 by CRC Press LLC

perimeter of Hillsborough Bay and from the City of Tampa primary wastewater treatment plant. In response to this report, the City of Tampa built a state-of-the-art advanced wastewater treatment plant (currently known as the Howard F. Curren Advanced Wastewater Treatment Plant) aided by federal funding. By 1979, this plant effectively removed more than 90% of the biological oxygen demand, suspended solids loading, and nitrogen (Garrity et al. 1985).

The City of Tampa, Bay Study Group began monitoring Hillsborough Bay in 1976 to document the effects of anticipated pollution load reduction on the receiving waters of the bay. The Bay Study Group has had an ongoing multidisciplinary monitoring program that included, since 1983, monitoring of drift macroalgae assemblages in Hillsborough Bay (Kelly 1995). The macroalgae monitoring program commenced following an 18-month study of drift macroalgae in Hillsborough Bay by Mangrove Systems, Inc. (1985).

Hillsborough Bay (Figure 10.1) has nearly 2,000 ha of intertidal and shallow subtidal (< 2 m) flats which historically supported seagrass meadows. Phillips (1962), in a survey of Tampa Bay in the early 1960s, found *Halodule wrightii*, *Ruppia maritima*, and *Syringodium filiforme* in Hillsborough Bay. However, he described only *R. maritima* in the northern section of Hillsborough Bay, noting the poor water clarity in this upper portion of the bay. By the early 1980s, seagrass coverage in Hillsborough Bay had declined to some sparse *R. maritima* coverage (Lewis et al. 1985). As water quality began to improve during the early 1980s, some minor seagrass recolonization began to occur in eastern Middle Tampa Bay just south of Hillsborough Bay (Johansson and Lewis 1992). Presented with a unique opportunity to document potential seagrass recolonization in an area virtually devoid of seagrass coverage, the Bay Study Group initiated a seagrass monitoring program in 1986.

This chapter presents techniques to monitor trends in submerged aquatic vegetation and macroalgae in Hillsborough Bay. Although the submerged aquatic vegetation program is designed to gather detailed information on seagrass characteristics (short-shoot density, blade length, and epiphytes) and the factors that may affect seagrass distribution (hydrographic and photosynthetic active radiation data), only information of general areal coverage trends is discussed.

MATERIALS AND METHODS

Monthly, low-level overflights (< 300 m altitude) have been conducted by helicopter since January 1986. Initially, each overflight was flown to document the position and size of macroalgal assemblages with oblique to near vertical true color photographs. However, these flights allowed the documentation of other noteworthy features such as phytoplankton blooms, sediment suspensions in the water column, and the development of submerged aquatic vegetation areas.

MACROALGAE

The Bay Study Group has collected data on drift macroalgae biomass and species composition along fixed transects since 1983. However, the number, location, and length of transects were modified in 1985. Since 1986, five transects from approximately 250 to 950 m long have been sampled monthly (Figure 10.1) using a 2-m wide otter trawl for macroalgae collection. After each trawl, macroalgae was weighed, the percent species composition visually estimated, and a representative subsample taken for laboratory analysis. Laboratory analyses included species composition and the percent dry weight per species. Laboratory results were extrapolated to yield grams dry weight per square meter per transect for each species collected.

SEAGRASS MONITORING

In April 1986, the Bay Study Group began a survey for seagrass in Hillsborough Bay. First, low level photographs were examined for indications of submerged aquatic vegetation presence. Areas

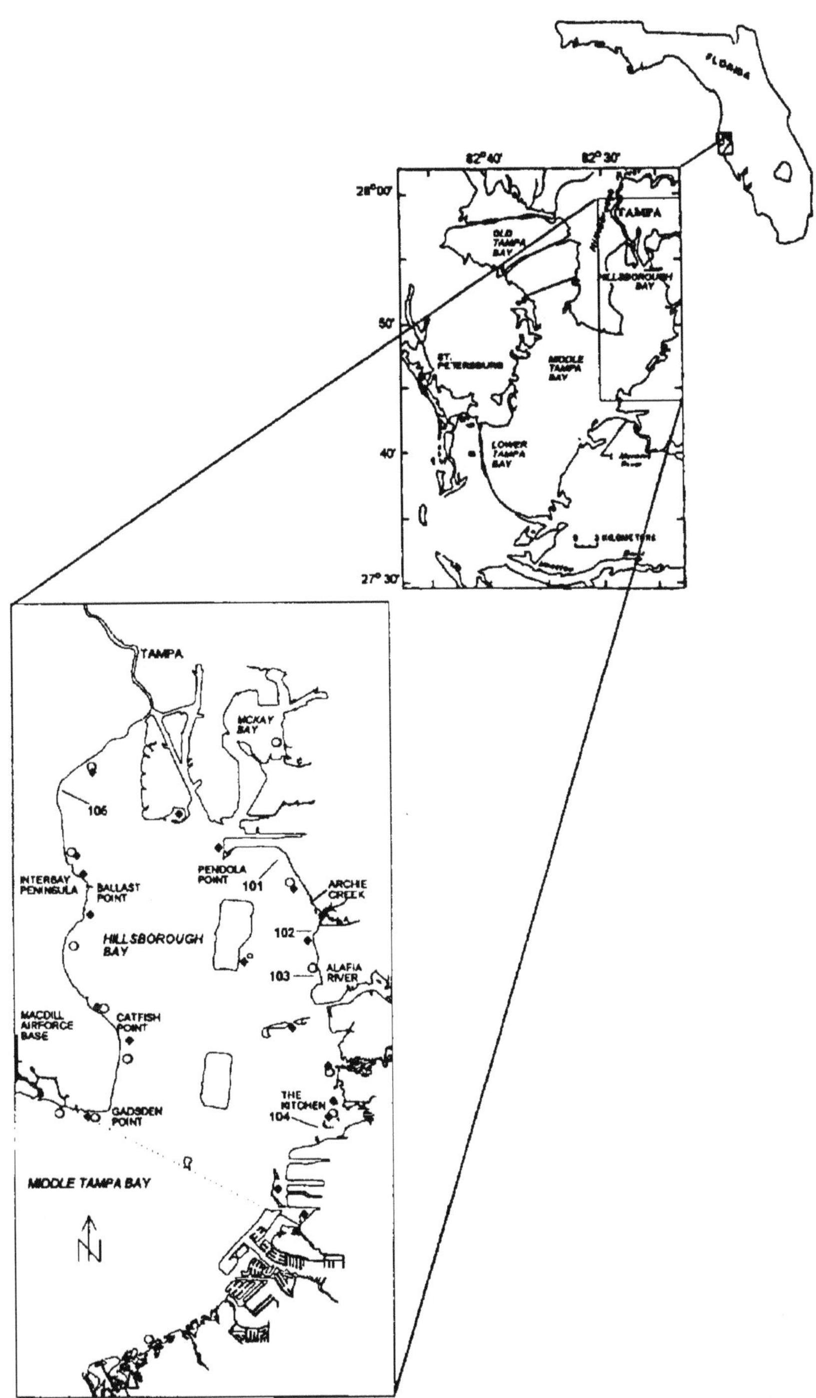

FIGURE 10.1 Location of the City of Tampa seagrass study sites (♦), seagrass transects (○), and five macroalgae transects: 101, 102, 103, 104, and 105. The dotted line indicates the southern boundary of Hillsborough Bay. (Map modified after Johansson, J. O. R. and R. R. Lewis. *Science of the Total Environment* Supplement. Elsevier Science Publishers, Amsterdam, 1199, 1992. With permission.)

that had submerged aquatic vegetation indicated in the photographs were subsequently visited during a low tide and groundtruthed. Additionally, areas that did not indicate seagrass coverage on the photographs were visited by divers or on foot during low tides. Thus, all intertidal and shallow subtidal (< 2 m in depth) areas that once had historical seagrass coverage or may potentially serve as seagrass areas were groundtruthed.

During the initial survey, two species of seagrass, *H. wrightii* and *R. maritima*, were found. However, due to the ephemeral nature of *R. maritima*, the focus of the program was on *H. wrightii* coverage. Each *H. wrightii* patch encountered during the survey was measured along the major and minor axis and the areal coverage calculated using

$$A = (ab/2)\pi$$

where A is the area, a the length of the major axis, and b the length of the minor axis.

In the early 1990s, several modifications were made to the seagrass monitoring program to allow for the increasing *H. wrightii* coverage through patch development and coalition. As the number of *H. wrightii* patches in Hillsborough Bay increased, field measurements of each patch became impractical. To expedite estimates of areal coverage, the low-level photographs were used for a patch count in selected areas. Subsequently, a subset of patches in each selected area was chosen for field measurement and the area of each patch measured along the major and minor axes, thus determining the area of an average-size patch. The areal coverage for a selected area was estimated using the number of patches counted on a photograph multiplied by the area of the average patch. However, some areas developed small meadows during this period and neither on-site patch measurements nor patch counts were suitable to estimate areal coverage.

To facilitate the calculation of areal coverage in developing meadows, the Bay Study Group employed a fixed-wing aircraft to obtain higher altitude (ca. 1500 m), near vertical, true-color photographs. In areas lacking sufficient fixed landmarks (piers, buildings, roads, etc.) 1-m diameter white disks were placed in the field at measured distances prior to each overflight, usually forming a rectangle within or around a meadow. During an overflight, several photographs were taken over each area. A scale for each photograph was calculated by measuring the distance between landmarks (including disks) on a photograph and determining a ratio between the distance separating landmarks in the field and the distance separating landmarks on a photograph. The ratio was called a scaling ratio. A grid consisting of 1×1 mm squares was placed over each photograph and the scaling ratio determined for each photograph was used to calculate the area represented by each square. The number of squares covering a seagrass signature in a photograph was tallied to reach an estimate of seagrass areal coverage. This procedure was done for each photograph taken of an area and the results averaged.

In 1998, the Bay Study Group acquired a Trimble® (Trimble Navigation Limited, Sunnyvale, CA) global positioning system (GPS) consisting of a Pro XR differential receiver interfaced with a TDC-1 datalogger to aid in seagrass mapping. To map areas of seagrass, the perimeter of seagrass coverage was followed with the GPS unit recording with sub-meter accuracy on 5 second intervals. Data were subsequently downloaded using Trimble Pathfinder Office, Version 2.10. The software plotted the seagrass perimeter data on a base map and the areal coverage calculated.

Seagrass Study Sites

Eighteen *H. wrightii* study sites (Figure 10.1) were monitored to follow seasonal trends in short-shoot density, blade length, and epiphyte cover; however, winter monitoring was not conducted due to the senescence of *H. wrightii*. In addition, mid-water column hydrographic data were recorded with a precalibrated YSI® 600XL probe interfaced with a 610DM datalogger to measure for dissolved oxygen, pH, salinity, and temperature. Further, a water column grab sample was taken for laboratory analysis of chlorophyll *a* and turbidity. Chlorophyll *a* was determined in an acetone extraction using a fluorometric method modified from Phinney and Yentsch (1985) by the Bay Study Group (personal communication from Roger Johansson, City of Tampa, Bay Study Group). Turbidity was measured on a Monitek® 20 nephelometer calibrated against a 5.0 primary standard.

Ancillary Submerged Aquatic Vegetation Monitoring

The Bay Study Group has also monitored trends in *R. maritima* and *Caulerpa prolifera* coverage. Initially, transects were established to follow coverage trends, however, both species are ephemeral and transect monitoring was ineffective. Generally, areal coverage for *R. maritima* and *C. prolifera* was determined using aerial photography combined with limited on-site measurements. Further, data on short-shoot density, blade length, and length of the flowering shoots was collected for *R. maritima* from sites that may vary annually given this species' ephemeral nature.

Submerged Aquatic Vegetation Transects

Thirteen transects, eleven in Hillsborough Bay and two in the upper portion of Middle Tampa Bay (Figure 10.1), were established in 1997 to determine submerged aquatic vegetation distribution with water depth. Transects were selected in areas that either contained seagrass coverage or may potentially have seagrass coverage in the future. Transects run from the shoreline to the 2-m depth contour, are 1-m wide, and range from 300–1300 m in length. Transects are marked at 100 m intervals with a PVC pipe driven into the sediment, leaving about 10 cm of pipe above the sediment surface.

Seagrass abundance was estimated along each transect using the Braun–Blanquet (1965) rating system that employs a numerical category from 0–5 to define a percent seagrass cover within a meter square. Each meter square was placed on 25 m centers throughout a transect except during the last 100 m section which contained seagrass, where the placement interval increased to 10 m. In addition, any drift or attached algae encountered was identified and similarly rated. Ancillary information collected at each meter square placement included: (1) depth; (2) sediment description (sand, mud, etc.); (3) a rating of epiphyte density; (4) type of dominant epiphyte (barnacle spat, diatoms, etc.); and (5) species of seagrass.

Water column hydrographic data, measurements of photosynthetic active radiation, and measurements of seagrass short-shoot density and blade length were collected at two or three sites along each transect, depending on seagrass presence. Also, a mid-depth water sample was collected for chlorophyll *a* and turbidity analyses at each site. If seagrass was present along the transect, hydrographic data, photosynthetic active radiation measurements, and a water sample were collected at the middle of the seagrass bed, the edge of the seagrass bed, and the two-meter contour. Further, seagrass short-shoot density was counted in a 100 or 625 cm^2 square and the blade length measured to the nearest centimeter for each seagrass species. If seagrass was not present along the transect, hydrographic data, photosynthetic active radiation measurements, and a water sample were collected at the middle of the transect and the two-meter depth contour. Finally a Secchi depth was determined near the two-meter depth contour or at a greater depth if the Secchi depth was greater than 2 m.

RESULTS AND DISCUSSION

Macroalgae

The Bay Study Group identified 18 species of macroalgae in Hillsborough Bay between 1986 and 1997 (Avery 1997). The most common species found during the study were *Gracilaria* spp. followed by *Spyridia filimentosa* and *Ulva lactuca*.

The average monthly macroalgal biomass per transect varied from a maximum of 164 $gdwtm^{-2}$ in 1987 to only trace amounts in 1998 (Figure 10.2). Generally, significant macroalgae biomass

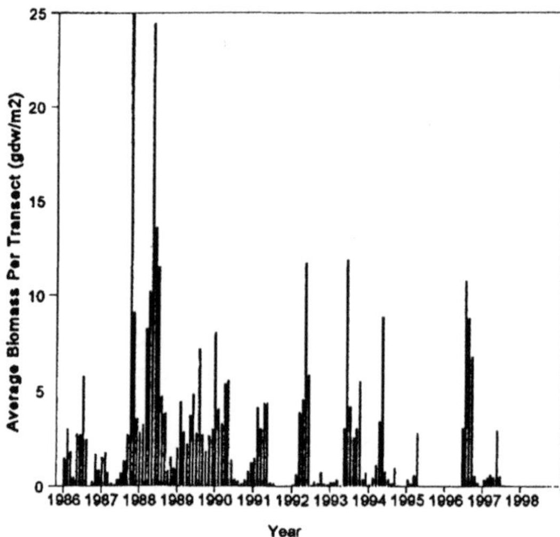

FIGURE 10.2 Summary of the monthly average macroalgae biomass per transect in Hillsborough Bay from 1986 to 1998. Biomass is measured in grams dry weight per square meter. The October 1987 peak is 164.5 gdwt/m^2.

was present in each monthly sample between 1986 and 1991. However, in the late summer of 1991, biomass declined to near zero. Between 1992 and 1997, episodes of high macroalgae biomass occurred; however, the bulk of the biomass was found along one transect in northeastern Hillsborough Bay and little or no macroalgae was found at the other four transects. In 1998, only sporadic macroalgal biomass was noted at all transects.

Nuisance macroalgae biomass was documented as early as 1944 in Hillsborough Bay (Florida State Board of Health 1965, cited by Florida Water Pollution Control Administration 1969). Subsequent reports by the Florida Water Pollution Control Administration in 1969 and Mangrove Systems, Inc. in 1985 indicated that high macroalgae biomass persisted in the late 1960s and the early 1980s. Results generated by the Bay Study Group show that as nutrient loading to the bay was reduced and eutrophic conditions ameliorated, drift macroalgae residence time and biomass were greatly reduced by 1991.

Halodule wrightii

In the initial submerged aquatic vegetation survey of Hillsborough Bay in 1986, the Bay Study Group measured and mapped nearly 400 patches of *H. wrightii* and the areal coverage for this species totaled about 2000 m^2 or 0.2 ha. The Bay Study Group mapped *H. wrightii* in each successive year except 1987, 1988, and 1990, and the areal coverage increased substantially each year through 1997. However, there was only a slight increase in *H. wrightii* coverage in 1998 as some areas had significant expansion while other areas declined in coverage. Coverage for 1998 was determined to be about 57 ha. Johansson and Greening (Chapter 21 of this book) discuss the potential reasons for the diminished increase in areal coverage in 1998. Results from the seagrass surveys are presented in Figure 10.3.

Ruppia maritima

R. maritima areal coverage fluctuated between 2 and 40 ha during the study. Prior to 1996, *R. maritima* areal coverage ranged between 2–3 ha; however, in 1996, areal coverage increased to about 40 ha. Although *R. maritima* was found in all sections of Hillsborough Bay, major meadows developed along the eastern shore between Pendola Point and Archie Creek and in the eastern area

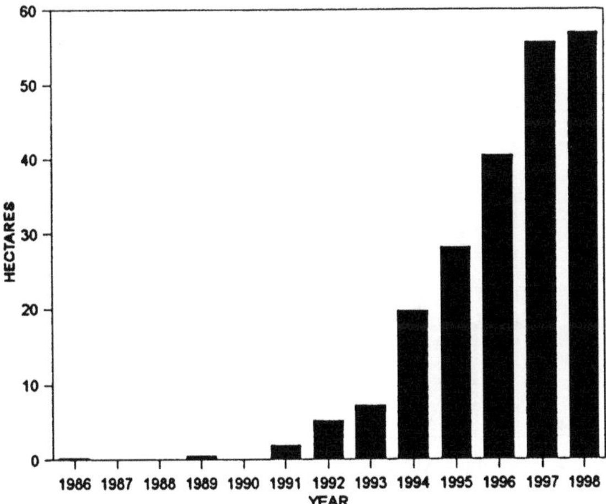

FIGURE 10.3 Summary of *Halodule wrightii* coverage in Hillsborough Bay from 1986 to 1998. There was no seagrass survey in 1987, 1988, and 1990.

of the Kitchen (Figure 10.1). In 1997, *R. maritima* areal coverage was reduced to 6 ha and further decreased to about 2 ha in 1998.

Caulerpa prolifera

In the spring of 1986, about 2000 m^2 of *C. prolifera* was found in shallow dredged areas near Gadsden Point and just north of Ballast Point in western Hillsborough Bay (Figure 10.1). By December 1986, *C. prolifera* areal coverage had expanded three orders of magnitude to about 200 ha. The *C. prolifera* meadow included most of the intertidal and shallow subtidal flats between Ballast Point and Gadsden Point to a depth of 2 and occasionally 3 meters.

The alga's areal extent remained relatively stable in western Hillsborough Bay until a "25-year" rainfall event occurred in September 1988. The rainfall event rapidly reduced the salinity along western Hillsborough Bay from about 25 to less than 5 because of increased freshwater flow from the Hillsborough River, the Alafia River, and the Tampa By-Pass Canal. Within 2 weeks, *C. prolifera* areal coverage was reduced more than 90%. *C. prolifera* coverage in western Hillsborough Bay continued to decrease through 1989 followed by sporadic occurrences of the alga through 1996. *C. prolifera* coverage for western Hillsborough Bay is presented in Figure 10.4.

In the fall of 1987, about 1000 m^2 of *C. prolifera* was documented in northeastern Hillsborough Bay. The areal coverage increased to 190 ha by 1990 before receding to about 25 ha in 1991. The reduction of areal extent in this area does not follow an episode of freshwater discharge to Hillsborough Bay. In fact, *C. prolifera* coverage increased in this area following the rainfall event in 1988. However, the 1988 salinity reduction was not as severe in this area of Hillsborough Bay compared with the western section of the bay, as the salinity minimum was measured at 15. *C. prolifera* coverage increased from 25 ha in 1991 to about 60 ha in 1993 before the meadow was reduced to about 5000 m^2 in 1994. No *C. prolifera* has been observed in eastern Hillsborough Bay since 1995. *C. prolifera* coverage in eastern Hillsborough Bay between 1986 and 1998 is presented in Figure 10.4.

Bay Study Group Transects

The first examination of the Hillsborough Bay submerged aquatic vegetation transects in 1997 established a baseline for future data compilation and comparison. The 1997 data indicated that *R. maritima* and *H. wrightii* were generally found in two distinct zones along the shoreline, with *R.*

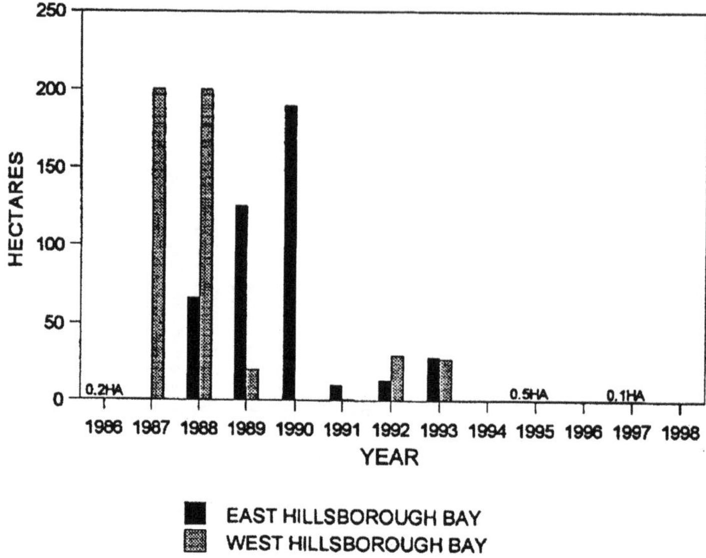

FIGURE 10.4 Summary of *Caulerpa prolifera* coverage in Hillsborough Bay from 1986 to 1998.

maritima found more shoreward. Usually, a minimum of overlap was noted between the species. The coverage of both species combined to form a narrow band of seagrass 20 to 400 m wide in many intertidal to shallow subtidal areas of Hillsborough Bay.

Interagency Seagrass Monitoring

In 1997, under the auspices of the Tampa Bay National Estuary Program, concerned private, local, county, and state agencies from around Tampa Bay convened to discuss the feasibility of establishing permanent seagrass monitoring transects using the Bay Study Group transect monitoring protocol as a model (Kurz et al., Chapter 12 of this book). In the succeeding months, these agencies worked together to establish an interagency monitoring program comprised of nearly 60 transects around the perimeter of Tampa Bay annually. About 30 transects traverse study sites monitored by the Southwest Florida Water Management District's Surface Water Improvement and Management Program and the Bay Study Group. An additional 30 transects were added in areas of interest or concern.

Each agency took responsibility for a certain number of transects within its area of interest. Prior to the initial monitoring effort in October 1998, all field personnel participated in "seagrass monitoring calibration classes" where monitoring methods were discussed and compared at several transects. These classes provided guidelines to facilitate the collection of consistent and comparable data collection among agencies.

CONCLUSION

The Bay Study Group found that combinations of monitoring techniques are necessary to closely monitor trends in attached and drift macrophytes. On-site monitoring yielded data in areas of sparse seagrass coverage that would not be documented in photographs taken from low or high altitude overflights. However, the overflights have proven useful in tracking areas of developing seagrass coverage and trends in *C. prolifera* growth and drift macroalgae assemblages. Additionally, global positioning systems have allowed seagrass meadows to be mapped with a great deal of accuracy. Future advances in monitoring submerged aquatic vegetation may include the use of high resolution satellite imagery to produce accurate submerged aquatic vegetation maps. Access to satellite imagery could decrease the reliance on aircraft needed for photographic overflights during the short-lived temporal windows of maximum seagrass biomass and acceptable atmospheric clarity. Finally, seagrass transect data could be integrated with satellite imagery to provide accurate, real-time information to agencies charged with coastline management.

ACKNOWLEDGMENTS

I thank Dr. Steve Bortone, Kerry Hennenfent, Roger Johansson, and Dr. Ray Kurz for their review and comments on this chapter.

REFERENCES

Avery, W. M. 1997. Macroalgae and seagrass in Hillsborough Bay. In *Proceedings Tampa Bay Area Scientific Information Symposium 3,* S. F. Treat (ed.). Tampa Bay Regional Planning Council, Clearwater, FL, pp. 151–165.

Braun-Blanquet, J. 1965. *Plant Sociology: The Study of Plant Communities.* Hafner, London.

Florida State Board of Health. 1965. A study of the causes of obnoxious odors, Hillsborough Bay, Hillsborough County, Florida. Florida State Board of Health, Bureau of Sanitary Engineering, Jacksonville, FL (article not read, cited by Florida Water Pollution Control Administration 1969).

Florida Water Pollution Control Administration. 1969. Problems and management of water quality in Hillsborough Bay, Florida. Hillsborough Bay Technical Assistance Project, Technical Programs, Southeast Region, Federal Water Pollution Control Administration, Tampa, FL.

Garrity, R. D., N. McCann, and J. D. Murdoch. 1985. A review of environmental impacts of municipal services in Tampa, Florida. In *Proceedings, Tampa Bay Area Scientific Information Symposium,* S. F. Treat, J. L. Simon, R. R. Lewis, and R. L. Whitman, Jr. (eds.). Bellwether Press, Edina, ME, pp. 526–550.

Kelly, B. O. 1995. Long term trends of macroalgae in Hillsborough Bay. *Florida Scientist* 58:179–192.

Kurz, R. C., D. A. Tomasko, D. Burdick, T. F. Ries, K. Patterson, R. Finck. 2000. Recent trends in seagrass distributions in southwest Florida coastal waters. In *Seagrasses: Monitoring, Ecology, Physiology, and Management,* S. A. Bortone (ed.). CRC Press, Boca Raton, FL, pp. 157–166.

Johansson, J. O. R. and R. R. Lewis. 1992. Recent improvements in water quality and biological indicators in Hillsborough Bay, a highly impacted subdivision of Tampa Bay, Florida, U.S.A. *Science of the Total Environment* Supplement 1992:1199–1215.

Johansson, J. O. R. and H. S. Greening. 2000. Seagrass restoration in Tampa Bay: a resource-based approach to estuarine management. In *Seagrasses: Monitoring, Ecology, Physiology, and Management,* S. A. Bortone (ed.). CRC Press, Boca Raton, FL, pp. 279–293.

Johansson, J. O. R., K. A. Steidenger, and D. C. Carpenter. 1985. Primary production in Tampa Bay, Florida: a review. In *Proceedings, Tampa Bay Area Scientific Information Symposium,* S. F. Treat, J. L. Simon, R. R. Lewis and R. L. Whitman, Jr. (eds.). Bellwether Press, Edina, ME, pp. 279–298.

Lewis, R. R. 1977. Impacts of dredging in the Tampa Bay estuary, 1876–1976. In *Proceedings of the Second Annual Conference of the Coastal Society,* E. L. Pruitt (ed.). The Coastal Society, Arlington, VA, pp. 31–55.

Lewis, R. R. and R. C. Phillips. 1980. Seagrass mapping project, Hillsborough County, Florida. Mangrove Systems, Inc., Tampa, FL.

Lewis, R. R., M. J. Durako, M. D. Moffler, and R. C. Phillips. 1985. Seagrass meadows in Tampa Bay: a review. In *Proceedings, Tampa Bay Area Scientific Information Symposium,* S. F. Treat, J. L. Simon, R. R. Lewis, and R. L. Whitman, Jr. (eds.). Bellwether Press, Edina, ME, pp. 210–246.

Mangrove Systems, Inc. 1985. Hillsborough Bay macroalgae study-final report, Tampa, FL.

Phillips, R. C. 1962. Distribution of seagrasses in Tampa Bay. Florida State Board of Conservation, Special Scientific Report No. 6, St. Petersburg, FL.

Phinney, D. A. and C. S. Yentsch. 1985. Novel phytoplankton chlorophyll technique: toward automated analysis. *Journal of Plankton Research.* 7:633–642.

Simon, J. L. 1974. Tampa Bay estuarine system — a synopsis. *Florida Scientist* 37:217–244.

11 Monitoring the Effects of Construction and Operation of a Marina on the Seagrass *Halophila decipiens* in Fort Lauderdale, Florida

Donald R. Deis

Abstract A monitoring program to determine the impact of marina development and operation on seagrasses was designed and performed for the City of Fort Lauderdale at the Birch/Las Olas Fort Lauderdale Municipal Marina. The marina site is located on the east side of the Atlantic Intracoastal Waterway (ICW) on the north and south sides of the Las Olas Boulevard bridge in Fort Lauderdale, Florida. Basically, the proposed marina development plan includes removing the existing marina and replacing it with piers and slips extending further waterward toward the ICW.

At least three previous seagrass surveys had been conducted at the marina site and indicated that *Halophila decipiens* occurs in varying densities within the site between years (i.e., over time). This species of *Halophila* has been described as an annual seagrass species regrowing each year from seed. Regrowth each year may be controlled by light (photoperiod); flowering is primarily controlled by temperature. This seasonal regrowth may explain some of the variability between past investigations at the site.

The intent of the seagrass investigation is to determine whether a statistically significant change in the distribution and abundance of *H. decipiens* occured as a direct result of demolition of the existing structures, construction of the new waterfront facility, and subsequent operations of the marina. The basis for this determination will be a statistical comparison to a nearby control area of approximately the same dimensions, depth, and bottom contours as the study site. At the same time, a pre- and post-construction water quality monitoring program was also designed and will be conducted to identify and characterize how selected water quality parameters might be influenced by demolition/construction/operation activities. The seagrass survey and water quality monitoring efforts will be coordinated to optimize the potential to correlate water quality trends with observed changes in seagrass abundance/diversity. Three seagrass investigations are proposed — one during the pre-construction period in the summer of 1997 and two over the 3 subsequent years. Water quality sampling began during the initial seagrass monitoring event. Quarterly water quality sampling will continue throughout the seagrass monitoring program.

INTRODUCTION

This chapter describes the design and initial survey of a monitoring program to document the impacts of marina construction and operation on the seagrass *Halophila decipiens*. The monitoring program is the result of the denial of a request by the City of Fort Lauderdale and the marina owner

0-8493-2045-3/99/$0.00+$.50
© 2000 by CRC Press LLC

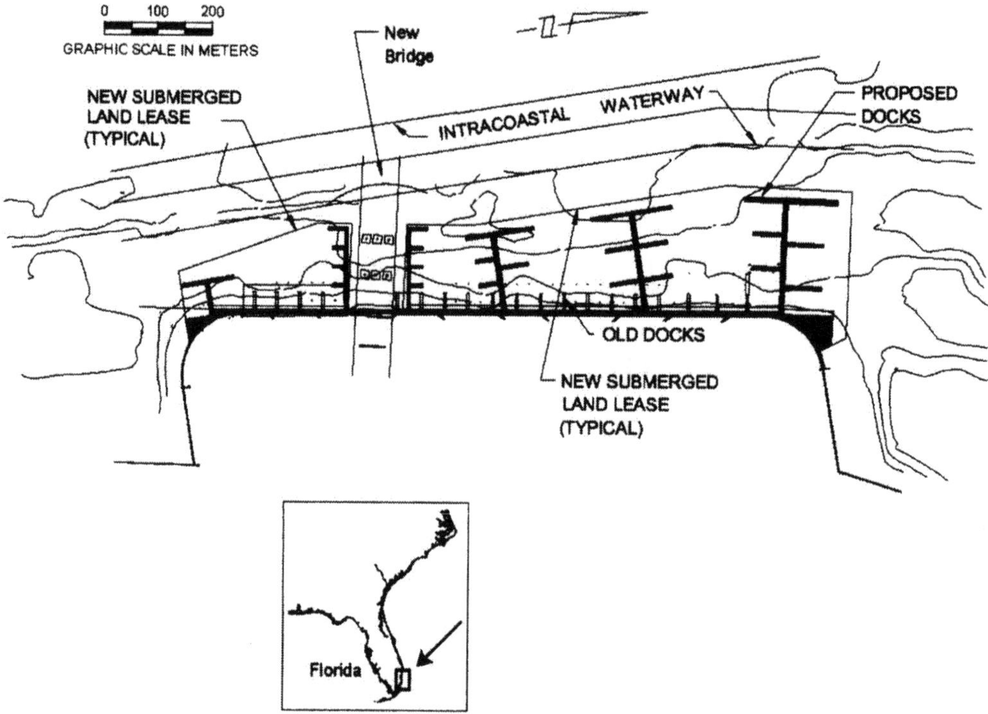

FIGURE 11.1 Location of the marina in Fort Lauderdale, Florida, and the old and proposed design.

and operator for a modification to an existing Florida Department of Environmental Protection (FDEP) permit. The original permit, issued in September 1994, contained conditions for elevating and grating docks at specific locations in the marina where FDEP found seagrasses and restricting use of these docks during the seagrass growing season (April 15 to November 1, according to the permit). The denial of the request for permit modification, issued in March 1996, was based upon the presence of seagrasses at specific locations where docks were proposed to be constructed in the marina design.

The marina site is on the east side of the Atlantic Intracoastal Waterway (ICW) on the north and south sides of the Las Olas Boulevard bridge in Fort Lauderdale (Figure 11.1). The proposed marina development plan is to remove the existing marina, which consists of slips and piers along the seawall, and replace it with piers and slips extending further waterward toward the ICW.

At least three seagrass surveys had been performed at the marina site prior to 1997. All of these studies were conducted during the growing season for *Halophila*. Staff from the FDEP conducted a survey in June 1994 that resulted in the seagrass distribution presented in the original permit drawings. The results from this survey indicated that the seagrass *Halophila* occurred in discrete locations within the proposed marina area (Figure 11.2).

Another survey was conducted for the City of Fort Lauderdale on June 23, 1995. The survey showed varying percent coverage of *Halophila* spp. at specific locations throughout the proposed marina area, with greatest percent coverage in shallower water north of the Las Olas Boulevard bridge (Figure 11.2).

A third seagrass survey of the proposed marina area was performed on September 13, 1996. The area was surveyed for seagrasses along east–west transects established at approximately 15-meter intervals. Results of that survey indicated that very sparse (less than 10 blades per m^2) amounts of *Halophila* cf. *johnsonii* were found at locations throughout the proposed marina area

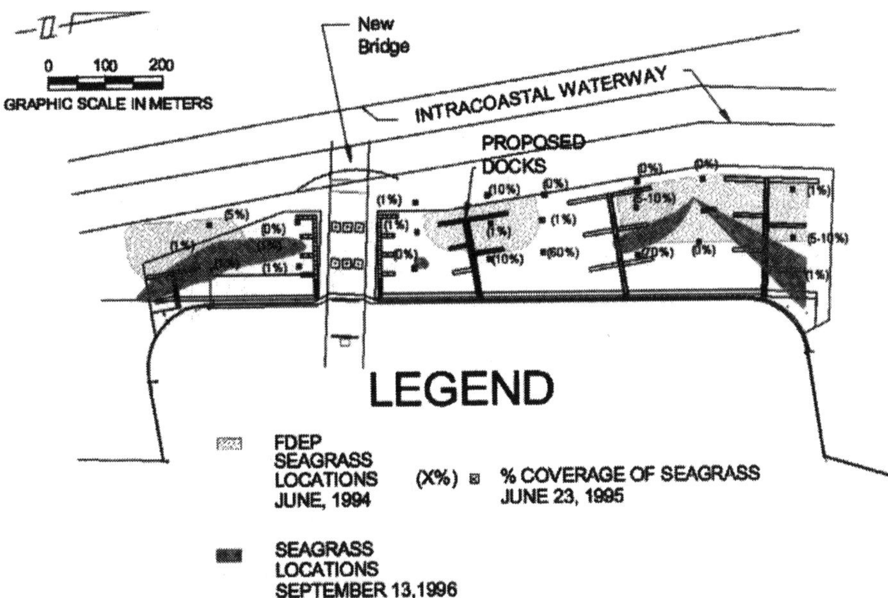

FIGURE 11.2 The results of past seagrass studies at the marina site in Fort Lauderdale, Florida.

(Figure 11.2). No seagrass beds were found; the seagrasses were observed as small clumps of blades and rhizomes on the bottom. The algae *Batophora oerstedii*, which resembles a small seagrass in morphology, was also found in several areas within the proposed marina site.

The City of Fort Lauderdale requested an administrative hearing in August 1996. To avoid a hearing, all parties met on October 30, 1996 to resolve the issues and begin developing a settlement agreement. The results of these past surveys were used to demonstrate to FDEP that seagrasses were not always found in the same locations within the proposed marina area from year to year and that seagrasses had been identified as growing nearly throughout the area of the proposed marina.

A settlement agreement was reached in April 1997. The development of a seagrass monitoring program was included as a settlement stipulation between the City of Fort Lauderdale and the FDEP.

PBS&J, Inc. was contracted to design a seagrass monitoring program, obtain approval of the design from FDEP, and perform the pre-construction survey for the program in accordance to the settlement stipulation. This report describes the seagrass monitoring program which was modified and approved by FDEP and the results of the pre-construction survey of the approved program.

In the past, the seagrasses in the genus *Halophila* have not received the same level of attention, both in research and environmental regulation, as the other seagrass species. This is mainly because these seagrasses are ephemeral, disappearing during the winter months. In addition, *Halophila decipiens* is known for its ability to grow in low light conditions. It has been found growing in deep water over 30 meters throughout the Caribbean and Gulf of Mexico and in dark waters (Busby and Virnstein 1993). In the Fort Lauderdale area, the estuarine waterways are dark with tannins. The seagrass is found growing in water depths of 1.8 to 3.6 meters. The bottom at these depths is not visible from the surface. Any studies of the seagrasses must be done with scuba diving equipment.

From past investigations two species of seagrasses in the genus *Halophila*, *H. decipiens* and *H. johnsonii*, could be found at the marina site and in the surrounding area. The two species differ in distributional range, morphology, and location of flowering parts, among other attributes. Durako and Wettstein (1994) provide a tabular comparison of the two species (Table 11.1). *H. johnsonii* has been recently listed as a threatened species because of its limitation in range (southeastern

TABLE 11.1
A comparison of *Halophila decipiens* and *H. johnsonii* (from Durako and Wettstein 1994).

Halophila decipiens	*Halophila johnsonii*
leaves rounded with sawtooth edges	leaves linear with smooth edges
minute, prickle hairs on leaf surface	no hairs on leaf surface
leaf veins make 60-degree angles	leaf veins make 45-degree angles
monoecious flowering (both male and female flowers occur on the same plant)	probably dioecious flowering (different male and female plants and the female plant has not yet been identified)
tolerates a fairly narrow range of salinity (more towards seawater) and temperatures	tolerates a broad range of salinity (brackish to seawater) and temperatures
wide pantropical range	narrow range in southeastern Florida
grows in waters with diminished sunlight and depths over 30 meters	grows in shallow water in intense sunlight and in deeper channels

Florida). Little is known about the seed production and method of distribution of *H. johnsonii*. Growth and seed development of *H. decipiens* are detailed in the Discussion section.

In addition to the above mentioned seagrass monitoring program, the settlement agreement contains limitations on the percentage occupancy at slips within the marina during seagrass growing season, a water quality monitoring program, and two mitigation measures directed at improving water quality and seagrass habitat within the marina area. One measure was the installation of sediment traps on three stormwater outfalls in marina area. This measure could potentially improve both water and sediment quality around the outfalls. The other measure was the installation of sewage pumpout facilities at the docks in the marina potentially reducing human-waste-related pollutants into the water column.

Both the seagrass and a water quality monitoring program were included in the settlement agreement to compare the conditions in the marina before and after project construction and after a period of marina operation. The results of these studies will be used to demonstrate the effectiveness of the settlement agreement's mitigation measures in protecting and improving habitat and water quality within the project area.

SEAGRASS MONITORING PROGRAM DESCRIPTION

TECHNICAL APPROACH

The intent of the seagrass investigation is to determine whether a statistically significant change in the distribution and/or abundance of seagrasses, particularly *Halophila* and/or *H. johnsonii*, occurred as a direct result of demolition of the existing structures, construction of the new waterfront facility, and/or subsequent operations of the marina. The basis for this determination will be a statistical comparison to a nearby control area of approximately the same dimensions, depth, and bottom contours as the study site. At the same time, a pre- and post-construction water quality monitoring program was also designed and will be conducted by the City of Fort Lauderdale staff to identify and characterize how selected water quality parameters might be influenced by demolition/construction/operation activities. The seagrass survey and water quality monitoring efforts will be coordinated to optimize the potential to correlate water quality trends with observed changes in seagrass abundance/diversity.

SAMPLING DESIGN

The seagrass monitoring program proposed to and approved by FDEP included a pre-construction sampling and two post-construction sampling events. The pre-construction monitoring was

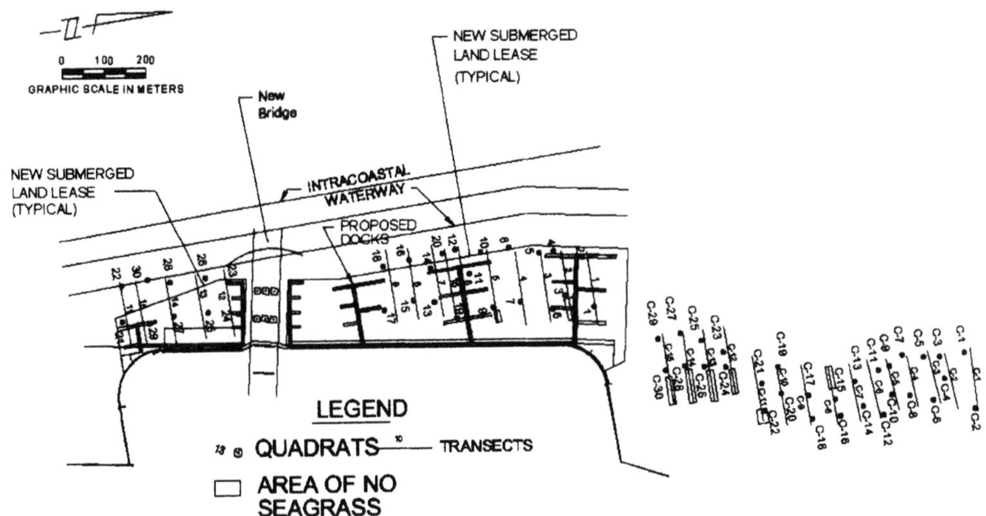

FIGURE 11.3 The location of sampling transects and quadrats, and the areas along transects containing no seagrasses at the marina and control sites.

conducted in early September 1997 followed by two post-construction sampling events in early September 1999 and September 2000 or later. The purpose of the September 1999 sampling event was to evaluate the adverse impacts, if any, associated with the demolition and construction activities (results not available as of publication date). Sampling during September 2000, or later, will be performed to evaluate adverse impacts, if any, associated with facility operations. Facility operations include the use of the facility for mooring boats in accordance to the settlement stipulations and the operation of the other mitigation measures, i.e., the sediment traps at the stormwater outfalls and the sewage pumpout facilities.

The seagrass monitoring program included sampling seagrasses in a control area for comparison to the project area. The function of the control in an experimental program is to physically or statistically control the variables that may occur between the sites so that any potential differences between the tested variable, in this case, the relative density of seagrass, can be compared. The agreement stipulation requested a control site in the vicinity of the project site with similar bottom contours, sediments, and water quality characteristics. A site immediately north of the project site was selected as the control site (Figure 11.3). This area contains similar bottom depths as the project site (2.4 to 3.6 meters National Geodetic Vertical Datum (NGVD)). Bottom sediment was variable at the project site (sandy to silty) and varied within the same range at the control site. Bottom sediment quality was qualitatively observed during the sampling program. The sampling locations are in close proximity and should be within similar water masses so that water quality should not vary between the sites. Water quality, however, is being sampled within both areas throughout the program by the City of Fort Lauderdale staff. The results of the water quality survey can be used to statistically control any differences between the two sites.

In comments, FDEP requested that the control site have a comparable "shoreline development level" as the project site. The selected control site is immediately waterward of marinas located along a bulkheaded shoreline and a portion of the control site is influenced by a large stormwater drainage outfall.

Data Collection/Analysis

The seagrass monitoring program consists of both transect sampling and quadrat sampling to assess the effects of demolition, construction and facility operation on seagrass distribution and abundance.

Fifteen east–west transects of approximately 2-m width were selected and permanently located with differential global positioning system (DGPS) points and fixed with physical markers for sampling within the study site and the similarly sized control area located north of the marina site (Figure 11.3). In addition, approximately 60 quadrats of 1 m^2 were randomly selected at distances along the transects within the study site and control area (approximately 30 quadrats in each area) to assess seagrass density. The 1 m^2 quadrat was divided into 100, 100 cm^2 cells; 10 cells from each quadrat were selected, using random number tables, for petiolate counts. After construction of the facility, the identical transects will be sampled in both areas to assess potential changes. Similarly, a randomly selected (new) set of approximately 60 quadrats will be sampled to evaluate seagrass density.

The number and location of seagrass beds within each transect were enumerated and evaluated by scientists/divers using an integrated video mapping system (IVMS). This system integrates positioning data, video data, and photographic data in an on-line computer database to allow investigators to concentrate on making real-time observations (in this case of seagrass bed size and density). The real-time positioning data, DGPS, are also visible and recorded on the video record as well as directly linked and plotted into geographical information system (GIS) map format. Video data are recorded on videotape and 24-bit images of selected views are digitized and stored onto optical disks.

Positioning data were stored in the computer database, continually correlated to all other data inputs, and stored on videotapes. This facilitated computer-assisted review of the videotapes during post-collection processing. Notes and other observational data were logged in the database and correlated with specific frames or sections of video data. In the field, the video was attached to the helmet of the scientific diver. The diver was linked to shore or a surface vessel for constant communication.

The assessment of demolition/construction/operation influences on seagrass was conducted by comparison of before/after changes between the site study and control area. The difference between the number of seagrass beds before and after construction (for each transect) was calculated. The same variables were calculated for each transect in the control area. These two sets of differences were statistically compared using a two-sample t-test and the nonparametric Wilcoxon rank-sum test. These tests determine where changes at the study site are significantly greater than changes that occurred without the effects of the waterfront demolition, construction, and marina operations at the control area.

Estimates of the mean density of seagrass pre- and post-construction were calculated as the sample mean density of the randomly sampled quadrats. The standard errors of these estimates were also calculated based on simple random sampling estimation methods.

A preliminary power analysis was conducted based on transect data gathered prior to demolition. This power analysis was used to assess the probability of detecting a change in seagrass coverage based on various levels of transect sampling effort and also on the hypothesized construction effects on seagrass coverage (i.e., there is a greater probability of detecting a more significant change in seagrass distribution attributable to waterfront activities than a lesser-sized change).

RESULTS AND DISCUSSION

TRANSECT AND QUADRAT LOCATIONS

The pre-construction seagrass survey was done on September 2–4, 1997. Fifteen transects were located in an east–west orientation within the project area and a control area north of the marina site (Figure 11.3). The transects were initially distributed equally spaced at each location. Since dredging to locate pipelines and transmission cables was occurring north of the bridge at the time of the survey, transects anticipated to be located in that area were shifted north (Transect 10) and south (Transect 11) of the bridge. Thirty seagrass quadrats were located at a random distance (0 to 10 m, using a random number table) within 10 m from the east and west ends of the transects in both the project area and the control area.

Seagrass Transects

The seagrass *Halophila decipiens* was observed along the full length of most transects within the project area. Seagrasses were not found along a portion of the eastern end of Transects 2 and 5 (Figure 11.3). The seagrass density was variable both along individual transects and throughout the project area. In general, the greatest density was found at the north end of the project area, along Transect 1, and south of the Las Olas Boulevard Bridge. The area within the center of the project, north of the Las Olas Boulevard Bridge from Transects 2 through 10, contained soft, silty sediments. The density of *H. decipiens* was lower in these areas of silty sediment.

Transects C-1 through C-7 within the control area were located in a dense *H. decipiens* bed which, in general, had its greatest density on the shallow, eastern end of the transect, and was less dense on the deeper, western end. A large stormwater discharge was located along the seawall in the southern end of the control area. The bottom sediment in the area of influence of the discharge changed from a sandy bottom to a soft, silty bottom that, in areas, did not support seagrass growth (Figure 11.3). Transects C-11 through C-15 appeared to be most affected by the silty sediment created by the outfall. No seagrasses were found along the eastern ends of these transects (Figure 11.3). Firmer sandy sediments were found along Transect C-15 and resulted in increased *H. decipiens* density in that area along the western end of this transect.

Seagrass Quadrats

The density of *H. decipiens* varied from 10 to 1330 petiolates/m^2 in the project area and 0 to 1520 petiolates/m^2 at the control area. To preliminarily interpret this data, some characteristics of the species *H. decipiens* must be mentioned. The overall knowledge about the ecology of the species within the genus *Halophila*, particularly within Florida, is limited. Some species of seagrasses growing in the coastal areas of Florida, particularly north of Biscayne Bay, disperse asexually by plant fragments demonstrating only minimal seed production and dispersion by seed. *H. decipiens* has not been observed to disperse by plant fragments; it appears as though dispersion is by seed production only (Durako and Wettstein 1994). *H. decipiens* is monoecious and has high seed production (Eisman 1980; McMillan 1982, 1988a; Durako and Wettstein 1994). Like the other species of seagrasses, this species is bed forming with the leaves growing in petiolates (two leaves to a petiolate) along a rhizome (lateral root). Beds of this species experience die-back in the winter; in severe winters, possibly only the seeds remain in the sediment. Because of this trait, *H. decipiens* has been described as an annual species (Kenworthy 1993). Unlike some seagrass species which have floating fruits and/or seeds, *H. decipiens* produces some fruits and seeds that float briefly; however, most of the fruits and seeds are either already buried in the sediment or do not float, forming a seed reserve for next season's growth (McMillan 1988b). This type of fruit and seed distribution and lateral growth along a rhizome leads to a clumping distribution when growth is sparse or patchy. The creation of a seed reserve or bank, however, ensures that the dense beds will regrow in the next year if no perturbation greatly disturbs the bottom sediment.

This information would indicate that, in areas of dense growth of *H. decipiens*, a seed reserve will build and, if no perturbation occurs in the area, a dense growth should be anticipated in the future. This information also indicates that, in areas of sparse *H. decipiens* growth, the expansion of the seagrass both in area and density may be slow (i.e., multiple years) because the individual plants need to grow from seeds dropped from the plants of the previous year.

This pre-construction survey indicates that the project and control sites apparently have similar conditions providing for the growth of *H. decipiens*. The docks have not yet been built within the marina and, of course, no structures exist at the control site. Water quality data is, at present, insufficient to indicate that different water quality exists at either site or that a specific water quality parameter at either site may impact seagrass growth. Two environmental conditions, water depth and sediment quality, vary within each site and appear, at present, to be influential on the density of *H. decipiens* within the study area. Water depth along the transects in the project area varies from about 2.4 meters NGVD on the east end to 3.6 meters NGVD on the west end on the slope into the ICW. The same depth variation exists at the control site except at Transect 15 where the west end is located on a shallow mound (see Figure 11.3). The general trend at both sites is that *H. decipiens* is found in greater density on the eastern (shallow) end of the transect than on the western (deeper) end. The highest density of *H. decipiens* along the transects was found in water depths of 2.4 to 3.0 meters NGVD. This indicates that the quantity and quality of light available for seagrass growth reaching the bottom is similar in both the project and control site. This inference could be supported and strengthened by light measurements.

Sediment quality greatly affected the density of *H. decipiens* at the sites. The highest density was found on firm shelly, sandy sediment. The soft, silty sediment that was found near the stormwater outfalls and seemed to be created by the outfalls supported a greatly reduced density of *H. decipiens* or no seagrasses. The construction of the sediment traps within the marina areas will reduce the amount of fine sediment entering that area. The potential result will be improved sediment quality within the marina area over time and expansion of more dense *H. decipiens*.

Power Analysis

Several power analysis scenarios were performed on the current sampling design and possible future modifications. A power analysis was used to assess the probability of detecting a change in seagrass coverage based on various levels of transect and quadrat sampling effort and hypothesized effects of demolition/construction and operation on seagrass coverage. The current design uses fixed transects and random 1 m^2 quadrats from the east (shallow) end and west (deep) end of each transect. As was stated in the results and discussion section, the number of petiolates/m^2 of seagrass varies greatly among the quadrats at both sites. The current program will have sufficient power to detect potential decreases in seagrasses from project construction and operation, i.e., mainly decreases resulting from decreased light penetration to the bottom (shading from docks and boats). Increases in density from changes in bottom sediment composition, such as those that may occur from the installation of baffle boxes at the storm water outfalls into the marina, may be more difficult to detect because of the small changes in density that may occur within a 2- or 3-year period.

CONCLUSIONS

The seagrass *H. decipiens* was found in variable density nearly throughout the project area and control area. This fact alone continues to demonstrate that the specific conditions of the original permit would not have been effective in protecting seagrass growth in the project area after the construction of the marina. *H. decipiens* is found in the highest density within the survey area in approximately 2.4 to 3.0 meters NGVD water depth on firm sandy sediment. With the absence of docks and boats in the project area, the limiting factors to the seagrass achieving maximum density within the survey area appear to be water depth and sediment type. The optimal water depth for seagrass growth is obviously related to adequate quantity and quality of light penetrating the water column to the seagrasses on the bottom. Measures that are intended to improve water quality within the marina area have been included in the project design (i.e., sediment traps at the stormwater outfalls and sewage pumpout facilities at the docks). However, it is not anticipated that these measures alone will cause a significant difference in reducing light attenuation in the water column within the survey area because the water quality within the area is the result of many sources and factors. It can be anticipated that the placement of the docks and boats over the water column will certainly limit light penetration to the bottom and eliminate seagrass growth in an area beneath them.

The installation of the sediment traps at the stormwater outfalls into the marina does have the potential to improve sediment quality in the marina area. Stormwater outfalls appear to be the source of the fine sediment found within both the project and the control areas where seagrasses are either excluded or found in low density. Improvements in sediment quality within the marina area may take many years and, because of the limitations on expansion of *H. decipiens* (both in area and density), significant increases in seagrass density may take years.

REFERENCES

Busby, D. S. and R. W. Virnstein. 1993. Executive summary. In *Proceedings and Conclusions of Workshops on Submerged Aquatic Vegetation Initiative and Photosynthetically Active Radiation,* L. J. Morris and D. A. Tomasko (eds.). Special Publication SJ93-SP13, St. Johns River Water Management District, Palatka, FL, pp. iii–viii.

Durako, M. and F. Wettstein. 1994. Johnson's seagrass: the Rodney Dangerfield of seagrasses (gets no respect, survives harsh conditions, little or no sex). *Palmetto* Winter:3–5.

Eisman, N. J. 1980. An illustrated guide to the sea grasses of the Indian River Region of Florida. Technical Report No. 31, Harbor Branch Foundation, Inc., Fort Pierce, FL.

Kenworthy, W. J. 1993. Defining the ecological light compensation point of seagrasses in the Indian River Lagoon. In *Proceedings and Conclusions of Workshops on Submerged Aquatic Vegetation Initiative and Photosynthetically Active Radiation,* L. J. Morris and D. A. Tomasko (eds.). Special Publication SJ93-SP13, St. Johns River Water Management District, Palatka, FL, pp. 195–210.

McMillan, C. 1982. Reproductive physiology of tropical seagrasses. *Aquatic Botany* 14:245–258.

McMillan, C. 1988a. Seed germination and seedling development of *Halophila decipiens* Ostenfield (Hydrocharitaceae) from Panama. *Aquatic Botany* 31:169–176.

McMillan, C. 1988b. The seed reserve of *Halophila decipiens* Ostenfield (Hydrocharitaceae) in Panama. *Aquatic Botany* 31:177–182.

12 Recent Trends in Seagrass Distributions in Southwest Florida Coastal Waters

Raymond C. Kurz, David A. Tomasko, Diana Burdick, Thomas F. Ries, Keith Patterson, and Robert Finck

Abstract In southwest Florida, significant declines in historical seagrass populations have been observed. Specifically, as much as 80% of the seagrass beds in Tampa Bay have been lost since the late 1800s as a result of dredging, pollution, and reduced water clarity. Historical seagrass losses have also been documented in other coastal areas such as Charlotte Harbor and Sarasota Bay. This study was initiated to assess temporal and spatial changes in seagrass coverage in these estuaries and to monitor the effects of improvements in water quality on seagrass bed expansion.

Using 1:24,000 scale aerial color photography, seagrass distributions were photointerpreted and transferred to a geographical information system (GIS) database for analysis. Based on trend analysis of the data, Tampa Bay and Sarasota Bay have experienced consistent, measurable gains in seagrass coverage since 1988. Seagrass coverage in Tampa Bay has increased an average of 2% per year between 1988 and 1996. A remarkable 7% increase in seagrass coverage was observed between 1988 and 1994 in Sarasota Bay followed by an 11% increase between 1994 and 1996. In Charlotte Harbor, a small decline in seagrass coverage occurred between 1982 and 1992; however, a 2% average annual increase has been observed between 1992 and 1996.

Observed increases in seagrass coverage in Tampa Bay and Sarasota Bay are believed to be directly linked to improving water quality and light penetration resulting from reductions in point-source pollutant loads. The timing and duration of riverine inflow, nutrient loading, and changes in water color can affect light penetration in Charlotte Harbor. The relationships between these factors and seagrass productivity appear to be more complex than in Tampa Bay and Sarasota Bay. Further data collection will be necessary to associate water quality trends with seagrass community dynamics in this estuary.

INTRODUCTION

Seagrasses are a vital component of marine ecosystems and are present in most shallow coastal waters throughout the world. By binding sediments and baffling waves and strong currents, seagrasses can prevent coastal erosion (Almasi et al. 1987) and provide important nursery and foraging habitat for a variety of economically important fish and shellfish species such as spotted seatrout and shrimp. Through decomposition, seagrasses provide detritus and organic matter which are important for the cycling of nutrients in coastal estuaries (Dunton 1990; Hemminga et al. 1991).

In Florida, seagrasses have been identified as a valuable habitat, both economically and ecologically. Monitoring seagrass distribution and health has become a useful resource management tool for several important estuaries including Tampa Bay, Sarasota Bay, and the Indian River

0-8493-2045-3/99/$0.00+$.50
© 2000 by CRC Press LLC

Lagoon. Seagrass distributions can be affected by a number of factors including light availability (i.e., photosynthetically active radiation, or PAR; Hall et al. 1990), water depth (Duarte 1991), pollution (Cambridge and McComb 1984), sediment characteristics, temperature (Anderson 1969; Barber and Behrens 1985), salinity (Dunton 1990), epiphyte colonization (Humm 1964), and morphological and physiological adaptations to a combination of these various physical factors.

Five species of seagrass are common in southwest Florida including *Thalassia testudinum* (turtle grass), *Syringodium filiforme* (manatee grass), *Halodule wrightii* (shoal grass), *Ruppia maritima* (widgeon grass), and *Halophila engelmannii* (star grass). Historic losses of seagrasses have been estimated for Tampa Bay (Lewis et al. 1985), Sarasota Bay (Tomasko et al. 1996), and Charlotte Harbor (Harris et al. 1983). The greatest losses (approximately 80%) occurred in Tampa Bay between the late 1800s and the early 1980s as a result of extensive dredging, pollution, and reduced water clarity. Seagrass losses in Sarasota Bay and Charlotte Harbor have occurred for many of the same reasons as in Tampa Bay, albeit to a lesser degree (approximately 30% loss in coverages).

In 1988, the Southwest Florida Water Management District (SWFWMD) initiated a biennial mapping project to assess trends in seagrass distributions in the Tampa Bay estuary. In 1992, Charlotte Harbor was added, followed by Sarasota Bay and Lemon Bay in 1994. Historical photographs were obtained for Charlotte Harbor (1982, 1988), Lemon Bay (1988), and Sarasota Bay (1988) and later interpreted to determine trends in seagrass distributions. Concurrently, field monitoring studies have been performed to assess the importance of various water quality and light parameters on seagrass growth and productivity (Tampa Bay National Estuary Program 1994; Tomasko et al. 1996; Tomasko and Hall, in press; SWFWMD 1995; SWFWMD 1998). The primary objectives of this study were to: (1) identify distributions of seagrasses within the political boundaries of the SWFWMD with respect to location, depth (where possible), and density; and (2) assess recent trends and causes of changes in seagrass coverage. Data generated from this study will be used by local government agencies, including the Tampa Bay, Sarasota Bay, and Charlotte Harbor National Estuary Programs, to assess the effectiveness of pollutant load reduction strategies on water quality.

METHODS

Seagrass distributions were mapped through a series of steps from 1:24,000 true color, aerial photography for the Tampa Bay, Sarasota Bay, and Charlotte Harbor estuaries (Figure 12.1). Aerial photography was typically obtained biennially during late fall/early winter months (October–January) to capitalize on times of maximum water clarity. The photos were then analyzed for distinct ecological signatures (seagrass beds, submerged shallow platforms, hard bottom, and algae) using zoom transfer methodology registered to 7½ minute USGS quadrangles. Seagrass signatures were identified and classified according to the Florida Department of Transportation (FDOT) Florida Land Use/Cover Classification System (FLUCCS) with a minimum mapping unit of 0.5 acres. Individual polygons were delineated on mylar overlays and digitally transferred to an ARC/INFO database for analysis. To ensure accuracy, representative areas were examined in the field to verify both presence and extent of seagrass coverage. A statistically based method was derived which required verifying field conditions at a number of different locations throughout each estuary. The results indicated that the maps have an accuracy of approximately 95.0 percent. This level of accuracy was achieved consistently during the study since the maps are only indicating presence, absence, and a coarse scale of seagrass density. Two forms of seagrass bed density categories were mapped: patchy (> 25% of unvegetated bottom visible within a bed) and continuous (< 25% of unvegetated bottom visible within a bed).

Mapping boundaries were based on various segmentation schemes developed for each estuary. The segmentation scheme was developed to assist local resource managers in developing various pollutant load reduction goals and water quality targets based on biological and physicochemical

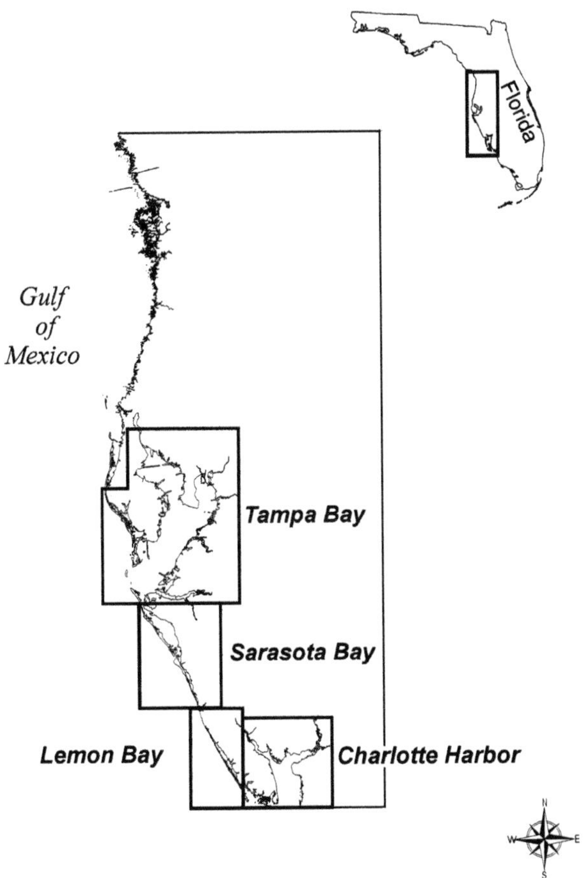

FIGURE 12.1 Location of seagrass mapping areas in southwest Florida.

characteristics of a particular geographic region of each estuary. During the first trend analysis (1988 to 1990), problems arose from the GIS-generated trend analysis since a new seagrass basemap was created for the 1990 coverage. This resulted in the potential for false change detection since some of the changes detected by the spatial analysis software were primarily caused by the enumeration of "slivers" created by the reinterpretation and digitizing of seagrass polygons and did not reflect the actual changes in the resource. As a result, photography was compared to the previous set of seagrass overlays to determine which polygons had changed between successive mapping periods. Only those areas showing either an increase or decrease in seagrass coverage or a change in seagrass bed density from the previous years' map were delineated.

RESULTS AND DISCUSSION

Tampa Bay

Trend analyses of the data collected since 1988 indicated both increasing and receding seagrass coverages in various segments of the bay over time (Table 12.1). For comparison, seagrass acreage is also shown using data generated by the Florida Department of Natural Resources and the U.S. Fish and Wildlife Service (USFWS) for the 1950s and from the Florida Marine Research Institute (FMRI) for the 1982 time period. Generally, total seagrass acreage for the entire bay has increased consistently since 1982 (Table 12.1, Figure 12.2). The greatest increase occurred between 1988

TABLE 12.1
Seagrass coverage (in hectares) by year in four southwest Florida estuaries determined through photointerpretation and geographical information system (GIS) mapping.

	1950	1982	1988	1990	1992	1994	1996
			Tampa Bay				
Segment			448.9	462.3	463.4	467.7	476.8
Boca Ciega	4,281.1	2,334.5	2,467.3	2,664.5	2,721.6	2,815.0	3,027.5
Hillsborough Bay	1,109.8	0.0	6.2	93.0	46.1	63.5	81.5
Lower Tampa Bay	2,470.5	2,029.5	1,916.0	2,192.0	2,234.3	2,192.9	2,273.3
Manatee River	51.0	53.0	110.3	116.7	116.7	117.7	117.9
Middle Tampa Bay	3,843.3	1,635.4	2,022.2	2,257.7	2,138.5	2,270.9	2,156.2
Old Tampa Bay	4,391.9	2,404.5	2,119.2	2,338.9	2,468.8	2,525.0	2,431.7
Terra Ceia	297.0	303.9	356.3	367.0	368.3	366.9	364.8
Total	**17,233.5**	**9,562.7**	**9,446.6**	**10,492.3**	**10,557.6**	**10,819.6**	**10,929.6**
			Charlotte Harbor				
Segment							
East		1,548.0	1,371.6		1,361.1	1,416.0	1,371.1
Middle		70.3	50.3		49.8	60.1	76.3
Myakka		237.5	201.7		130.4	188.7	209.0
Peace		377.5	157.7		166.1	195.5	225.1
Placida		948.0	1,407.8		1,376.1	1,336.5	1,450.1
South		3,512.9	3,683.9		3,635.8	3,633.2	3,625.5
West		672.2	584.6		494.9	675.1	794.3
Total		**7,366.5**	**7,457.5**		**7,214.3**	**7,505.2**	**7,778.5**
			Sarasota Bay				
Segment							
Manatee Co.		2,212.6		2,355.0	2,540.2		
Sarasota Co.		772.2		850.0	1,042.9		
Little Sarasota Bay		215.4		239.0	290.0		
Roberts Bay		133.9		139.7	145.2		
Blackburn Bay		165.8		165.8	162.4		
Total		**3,500.0**		**3,749.6**	**4,180.6**		
			Lemon Bay				
		1,055.0		**1,072.7**	**1,054.0**		

and 1990 with approximately 11.1% or 1046 ha (2585 ac) of regrowth. Between 1990 and 1992, seagrass coverage increased by 65 ha (161 ac) or 0.6%. Between 1992 and 1994, seagrass coverage increased by 262 ha (648 ac) or about 2.5%. Between 1994 and 1996 an additional 110 ha (272 ac) of seagrass was observed, representing a 1.0% increase. Although the greatest seagrass regrowth occurred between 1988 and 1990, the net result has been a steady increase in seagrass coverage for Tampa Bay at an average rate of approximately 2% per year. Since seagrass mapping began in 1988, seagrass coverage has increased a total of about 14% or 1373 ha (3393 ac).

Using large-scale bathymetric data provided by the FMRI, seagrass distributions were analyzed with respect to water depth. Approximately 90% of all seagrass beds occur between depths of 1 to 2 m. The percentage of seagrass occurring in this zone has been relatively constant since 1988 (Figure 12.3). However, marked increases (2–50%) have occurred in the percentage of seagrasses occupying depths between 2 to 6 m, especially between the years 1988 and 1990. This phenomenon

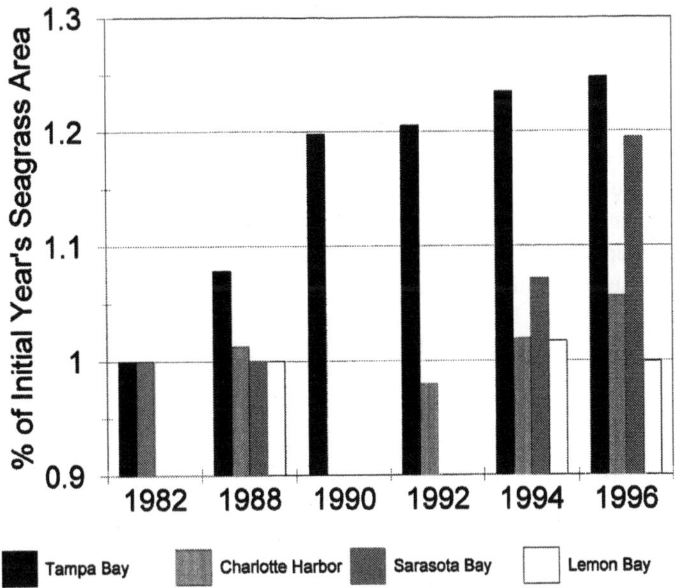

FIGURE 12.2 Changes in areal coverage of seagrass over time in Tampa Bay, Charlotte Harbor, Sarasota Bay, and Lemon Bay.

is probably a result of increasing water clarity (and greater light availability in deeper waters) which has been improving steadily in all bay segments since the early 1980s.

Significant water quality improvements have resulted since the City of Tampa's wastewater facility on Hillsborough Bay implemented advanced wastewater treatment technologies in the late 1970s. These improvements have led to reductions in pollutant loads, notably nitrogen, which can stimulate phytoplankton growth in the water column. The resulting reduction in light penetration

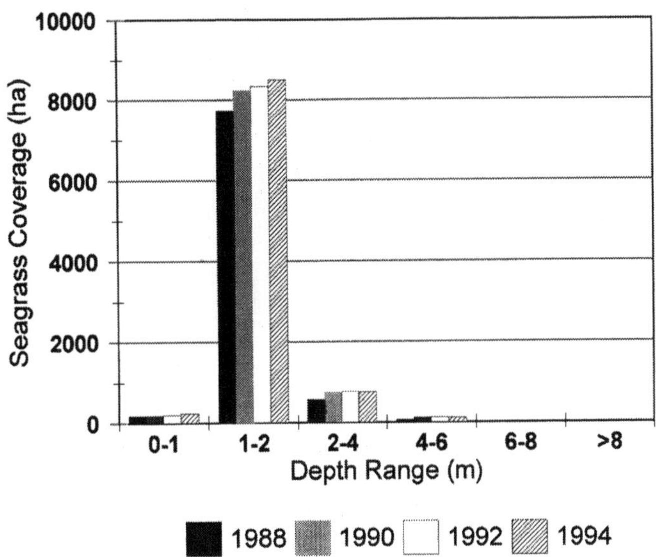

FIGURE 12.3 Seagrass coverage at various depths in Tampa Bay.

or PAR can lower productivity and growth of seagrasses. Light penetration, as measured by Secchi disk depth, has improved dramatically in Hillsborough Bay and Old Tampa Bay (Figure 12.4). As a result, the greatest proportional increases in seagrass growth have been observed in these bay segments. In fact, in the late 1970s no seagrass was ever documented in Hillsborough Bay compared to more than 81 ha (200 ac) of seagrass beds in 1996.

Improvements in controlling nonpoint pollution, including urban stormwater runoff, may have helped further reduce nutrient and total suspended solid concentrations in the estuary. During the first few years of mapping (1988–1992), the Tampa Bay area experienced unusually low rainfall conditions. This, coupled with a number of regional stormwater improvement projects, may have led to decreased pollutant loading to the bay and the much greater rate of seagrass regrowth observed bay-wide between 1988 and 1992 vs. the period between 1992 and 1996. However, as rainfall conditions returned to more normal trends between 1992 and 1996, seagrass coverage only increased

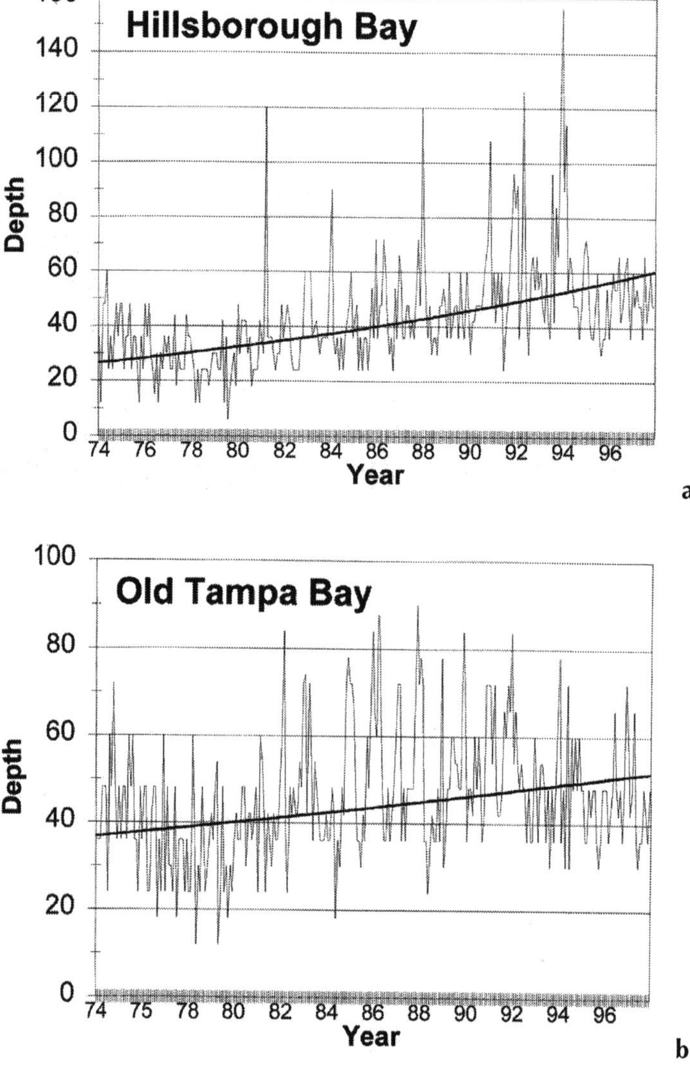

FIGURE 12.4 Secchi depth measurements over time in the Hillsborough Bay and Old Tampa Bay segments of Tampa Bay.

by a rate of approximately 1 to 2% per year. At the same time, water clarity (measured by Secchi disk) became more constant, particularly in Hillsborough Bay.

Despite increasing trends in seagrass acreage, trends in seagrass bed density declined between 1988 and 1994. The majority of thickening (patchy beds coalescing to become continuous beds) and thinning (continuous beds fragmenting to become patchy beds) occurred within the 1 to 2 m depth zone. Most of the thinning has occurred between 1992 to 1994, whereas the majority of thickening occurred between 1988 and 1990. Between 1988 and 1990, thickening of beds occurred in all segments of the bay but was most evident in middle Tampa Bay in the nearshore areas between Cockroach Bay and Bishops Harbor. Recent (1992–1994) thinning of seagrass beds was most evident in an area extending from the west end of the Howard Frankland Causeway south to Bayboro Harbor in St. Petersburg and also along the southern tip of the Interbay Peninsula in Tampa. The cause of this extensive thinning phenomenon has not yet been determined.

As noted above, seagrass populations in many areas of Tampa Bay appear to be extremely dynamic and fluctuate biennially due to a variety of factors. The amount of light available to seagrass blades is one of the primary factors which limit productivity and growth. PAR is affected by a variety of factors including chlorophyll *a* concentrations (a measure of algal growth in the water column which can be stimulated by nutrient loading), concentrations of nutrients in both the water column and sediments, temperature, and color (often a result of tannins in runoff from forested areas in the contributing watershed).

As water clarity continues to improve in Tampa Bay, seagrasses should respond positively and expand in their distribution. This has been observed by increases in seagrass acreage at deeper depths of the bay. Conversely, seagrass losses may be caused by reductions in water clarity, dessication during extreme low tides, suboptimal temperatures during the growing season, or physical disturbances like propeller scarring or wind-driven sediment deposition. Excess nutrients can also cause excessive epiphyte growth on seagrass blades which can drastically reduce light penetration to the blade and cause plant die-back. Occasional stochastic events such as major storms or hurricanes or anthropogenic impacts from dredging also appear to result in seagrass losses in Tampa Bay.

Sarasota Bay

Changes in seagrass coverage in Sarasota Bay have been dramatic since 1988 (Table 12.1). The Bay was divided into five segments named, from north to south, Manatee County, Sarasota County, Little Sarasota Bay, Roberts Bay, and Blackburn Bay. Seagrass regrowth has been greatest in the Manatee and Sarasota County segments that encompass most of Big Sarasota Bay. These areas generally exhibited poor water quality/clarity during the early 1980s due to significant wastewater plant discharges to Big Sarasota Bay. Between 1989 and 1990, wastewater treatment plants in the City of Sarasota and Manatee County reduced nitrogen loads to Sarasota Bay by as much as 25% through advanced treatment processes and reduced discharges (Sarasota Bay National Estuary Program 1995). Consequently, water transparency improved in this region of the bay from a mean Secchi depth of 1.1 m to 1.5 m (Figure 12.5). More importantly, the percentage of Secchi depth measurements which exceeded 1.5 m and 2.0 m depths increased by approximately 7% and 21%, respectively (Figure 12.6).

As a result, seagrass coverage in the Manatee County segment increased by 6.4% or 142 ha (352 ac) between 1988 and 1994. Between 1994 and 1996, a 7.8% or 185 ha (458 ac) increase in seagrass coverage was observed. In the Sarasota County segment, the rate of seagrass regrowth was much greater with an increase of 10.1% or 78 ha (192 ac) between 1988 and 1994 and a 22.7% or 193 ha (477 ac) increase between 1994 and 1996. Most of these increases occurred along the deep (> 1.0 m) edges of existing seagrass (primarily *Halodule*) beds in the mid-western sections of Big Sarasota Bay.

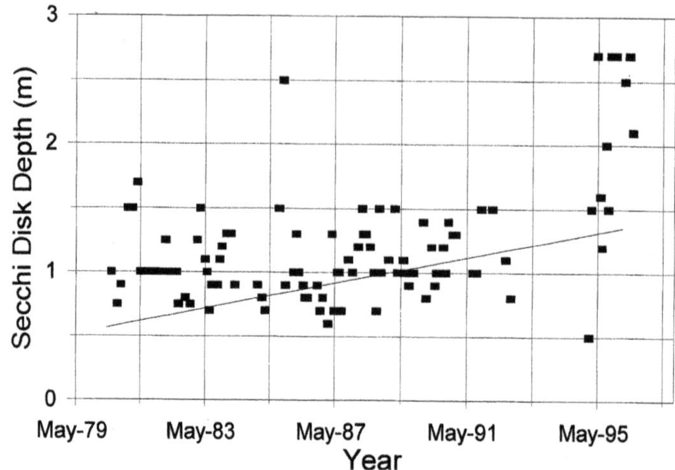

FIGURE 12.5 Secchi depth measurements vs. time in the Sarasota County segment of Sarasota Bay.

Despite the lack of significant water quality improvements in Little Sarasota Bay, rates of seagrass regrowth were similar to those for Big Sarasota Bay (11% or 22 ha (54 ac) between 1988 to 1994, 21.3% or 51 ha (126 ac) between 1994 and 1996). However, extensive, unvegetated, shallow, subtidal platforms with the potential to support seagrass populations still exist in Little Sarasota Bay. Further improvements in water quality in this segment could lead to more significant seagrass regrowth.

Changes in seagrass acreage in the southernmost portions of Sarasota Bay (Roberts Bay, Blackburn Bay) have not been as dramatic nor have any recent improvements in water quality been noted (D. Tomasko, personal communication). In fact, seagrass acreage in Roberts Bay increased by only approximately 4.3% or 14 acres between 1988 to 1994 and then again at 3.9% or 14 acres

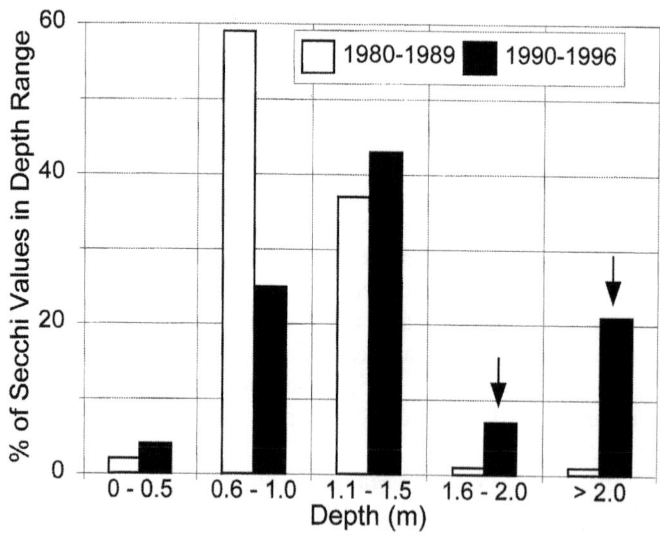

FIGURE 12.6 Percentage of Secchi depth measurements occurring within various depth ranges in the Sarasota County segment of Sarasota Bay. Open bars represent measurements taken between 1980 and 1989 (before nitrogen load reductions); dark bars represent measurements taken between 1990 and 1996 (after nitrogen load reductions).

between 1994 and 1996. Seagrass coverage in Blackburn Bay did not change at all between 1988 and 1994 and actually declined between 1994 and 1996 by 2.1% or 9 acres.

Lemon Bay

Seagrass trends for the entire intercoastal waterway and lower reaches of Lemon Bay were analyzed as a single waterbody segment. Between 1988 and 1994 an increase of 18 ha (44 ac) or 1.7% of seagrass was observed followed by a nearly identical decline of 19 ha (46 ac) or –1.8% between 1994 and 1996 (Table 12.1). Most of these changes occurred in the same general geographic location at the mouth of a large creek system. Unfortunately, little data exist for water quality or light penetration in this area. However, data collected by Tomasko and Bristol (unpublished) suggest that water transparency can be predicted by phytoplankton concentrations. This relationship was stronger than previously found in the lower segments of Sarasota Bay (Tomasko, personal communication). Further years of data collection will be necessary to associate water quality trends with seagrass community dynamics in Lemon Bay.

Charlotte Harbor

Charlotte Harbor was divided into seven segments including the Myakka River, Peace River, East Harbor, West Harbor, Middle Harbor, South Harbor, and Placida. Greatest rates of change (increases and decreases) in seagrass coverage have occurred in the Myakka River and West Harbor segments (Table 12.1). These areas of the harbor can be subject to highly variable changes in salinity and color as a result of variable river flow from both the Myakka and Peace Rivers. The East, Middle, and South segments have all experienced less than 1% gain or loss during any of the biennial mapping events. The Peace River segment had an increase of approximately 5.4% between 1988 and 1992 followed by smaller gains of 0.2% during both the 1992 to 1994 and 1994 to 1996 mapping periods. The Placida segment had losses of 2.3% between 1988 and 1992 and 2.9% between 1992 and 1994. However, between 1994 and 1996, an 8.5% gain was observed.

Harbor-wide, seagrasses increased by approximately 1.2% or 91 ha (225 ac) between 1982 and 1988 (Figure 12.2). Between 1988 and 1992, a decline of approximately 3.3% or 243 ha (601 ac) was observed. However, between 1992 and 1994, a 4.0% or 291 ha (719 ac) increase was observed followed by an additional 3.6% increase or 274 ha (676 ac) between 1994 and 1996. This has resulted in a harbor-wide increase in seagrass of 412 ha (1018 ac) or 5.6% between 1982 and 1996.

Tomasko and Hall (in press) suggest that the timing and duration of riverine inflow, nutrient loading, and changes in water color can affect light penetration in Charlotte Harbor. However, the relationships between these factors and seagrass productivity appear to be more complex than in Tampa Bay and Sarasota Bay. Further data collection will be necessary to predict the effects of water quality trends on seagrass community dynamics in Charlotte Harbor.

Mapping of seagrasses has proven to be a valuable tool in assessing trends in anthropogenically altered estuarine systems in Florida. Additional data that could be extremely useful in assessing the relationships between water quality/clarity and seagrass growth include obtaining detailed bathymetric data and light attenuation coefficients. By analyzing trends in seagrass coverages with depth, relationships between water quality parameters and seagrass productivity can be more accurately predicted, allowing greater precision in setting restoration goals for seagrass communities.

ACKNOWLEDGMENTS

We thank the Pinellas-Anclote, Manasota, Alafia River, Hillsborough River, Northwest Hillsborough, and Peace River Basin Boards of the Southwest Florida Water Management District for their continued funding support of this study.

REFERENCES

Almasi, M. N., C. M. Hoskin, J. K. Reed, and J. Milo. 1987. Effects of natural and artificial *Thalassia* on rates of sedimentation. *Journal of Sedimentary Petrology* 57:901–906.

Anderson, R. R. 1969. Temperature and rooted aquatic plants. *Chesapeake Science* 10:157–164.

Barber, B. J. and P. J. Behrens. 1985. Effects of elevated temperature on seasonal *in situ* leaf productivity of *Thalassia testudinum* and *Syringodium filiforme*. *Aquatic Botany* 22:61–70.

Cambridge, M. L. and A. J. McComb. 1984. The loss of seagrasses in Cockburn Sound, Western Australia. I. The time course and magnitude of seagrass decline in relation to industrial development. *Aquatic Botany* 20:229–243.

Duarte, C. M. 1991. Seagrass depth limits. *Aquatic Botany* 40:363–377.

Dunton, K. H. 1990. Production ecology of *Ruppia maritima* L. s.l. and *Halodule wrightii* Aschers. in two subtropical estuaries. *Journal of Experimental Marine Biology and Ecology* 143:147–164.

Hall, M. O., D. A. Tomasko, and F. X. Courtney. 1990. Responses of *Thalassia testudinum* to *in situ* light reduction. In *Results and Recommendations of a Workshop Convened to Examine the Capability of Water Quality Criteria, Standards and Monitoring Programs to Protect Seagrasses from Deteriorating Water Transparency,* W. J. Kenworthy and D. E. Haunert (eds.). South Florida Water Management District, West Palm Beach, FL, pp. 77–86.

Harris, B. A., K. A. Haddad, K. A. Steidinger, and J. A. Huff. 1983. Assessment of fisheries habitat: Charlotte Harbor and Lake Worth. Florida Department of Natural Resources, St. Petersburg, FL.

Hemminga, M. A., P. G. Harrison, and F. van Lent. 1991. The balance of nutrient losses and gains in seagrass meadows. *Marine Ecology Progress Series* 71:85–96.

Humm, H. J. 1964. Epiphytes of the sea grass *Thalassia testudinum* in Florida. *Bulletin of Marine Science of the Gulf and Caribbean* 14:306–341.

Lewis, R. R., III, M. J. Durako, M. D. Moffler, and R. C. Phillips. 1985. Seagrass meadows of Tampa Bay: A review. In *Proceedings, Tampa Bay Area Scientific Information Symposium, May 1982,* S. F. Treat, J. L. Simon, R. R. Lewis, III, and R. L. Whitman, Jr. (eds.). Minneapolis Burgess Publishing Company, Minneapolis, MN, pp. 210–246.

Sarasota Bay National Estuary Program. 1995. Comprehensive Conservation and Management Plan, Sarasota, FL.

Southwest Florida Water Management District. 1995. Light requirements of *Thalassia testudinum* in Tampa Bay, Florida. Technical report prepared by Mote Marine Laboratory for the Surface Water Improvement and Management Section, Tampa, FL.

Southwest Florida Water Management District. 1998. Seagrass Monitoring Protocol, Summer 1998 Results. Technical report prepared by Scheda Ecological Associates, Inc. for the Surface Water Improvement and Management Section, Tampa, FL.

Tampa Bay National Estuary Program. 1994. Habitat protection and restoration targets for Tampa Bay. Technical publication #07-93, St. Petersburg, FL.

Tomasko, D. A., and M. O. Hall. In Press. Productivity and biomass of *Thalassia testudinum* along a gradient of freshwater influence in Charlotte Harbor, Florida (USA). *Estuaries*.

Tomasko, D. A., C. J. Dawes, and M. O. Hall. 1996. The effects of anthropogenic nutrient enrichment on turtle grass (*Thalassia testudinum*) in Sarasota Bay, Florida. *Estuaries* 19:448–456.

13 Monitoring Seagrass Changes in Indian River Lagoon, Florida Using Fixed Transects

Lori J. Morris, Robert W. Virnstein, Janice D. Miller, and Lauren M. Hall

Abstract Long-term monitoring of fixed transects is an important management tool for determining the health of seagrass ecosystems. In the Indian River Lagoon, Florida, semiannual monitoring (summer-winter) of 76 fixed transects was established in 1994 to quantify changes in the seagrass. The transects are perpendicular to shore along the 250-km-long axis of the lagoon. Transects are sampled using non-destructive techniques to measure percent cover and canopy height of each seagrass species every 10 m along the line until the deep edge of the bed is reached. Shoot counts are also made at the center and deep edge of the beds at each transect location.

Halodule wrightii was found to be the dominant species on a lagoon-wide scale. It was present in every transect but one and comprised, on average, 36% cover. Other species found included: *Syringodium filiforme* (9.5% cover), *Ruppia maritima* (5.2% cover), *Thalassia testudinum* (2% cover), *Halophila johnsonii* (2% cover), *Halophila decipiens* (0.6% cover), and *Halophila engelmannii* (0.5% cover). Percent cover was significantly lower in the winter compared to the summer for *Halodule wrightii, Halophila decipiens*, and *Halophila engelmannii.* Canopy height was significantly lower in the winter for *Halodule wrightii, S. filiforme, T. testudinum*, and *Halophila decipiens. T. testudinum* and *Halophila johnsonii* reached their Atlantic coastal northern limit in the mid-lagoon region.

During the period 1994 to 1998, seagrass beds in the lagoon increased in abundance in most areas; at 65 of the 76 transect sites, seagrass beds increased in width. On average for all sites, the distance from shore to the deep edge of the seagrass beds increased 26% during this period. At some specific locations the increase in transect length was quite dramatic. In the Cocoa Beach segment of the lagoon, for example, average transect length increased 207%. Year-to-year, lagoon-wide variability in seagrass abundance was high, especially that of the less abundant species such as *R. maritima, Halophila engelmannii, Halophila decipiens*, and *Halophila johnsonii* where increases in cover exceeded 500% in some cases. Intermittent droughts and hurricanes also had significant impacts on annual seagrass abundance. This study illustrates how long-term, repeated survey measurements are necessary to develop an understanding of the natural variability of seagrass systems, and ultimately, how to separate this natural variability from that induced by anthropogenic causes.

INTRODUCTION

The Indian River Lagoon stretches 250 km along the east coast of Florida from Ponce de Leon Inlet south to Jupiter Inlet (Figure 13.1). The Indian River Lagoon estuarine system is made up of three interconnected lagoons, Mosquito Lagoon, Banana River, and the Indian River Lagoon; five inlets to the ocean; and more than ten tributaries and major canals contributing fresh water. Warm-

0-8493-2045-3/99/$0.00+$.50
© 2000 by CRC Press LLC

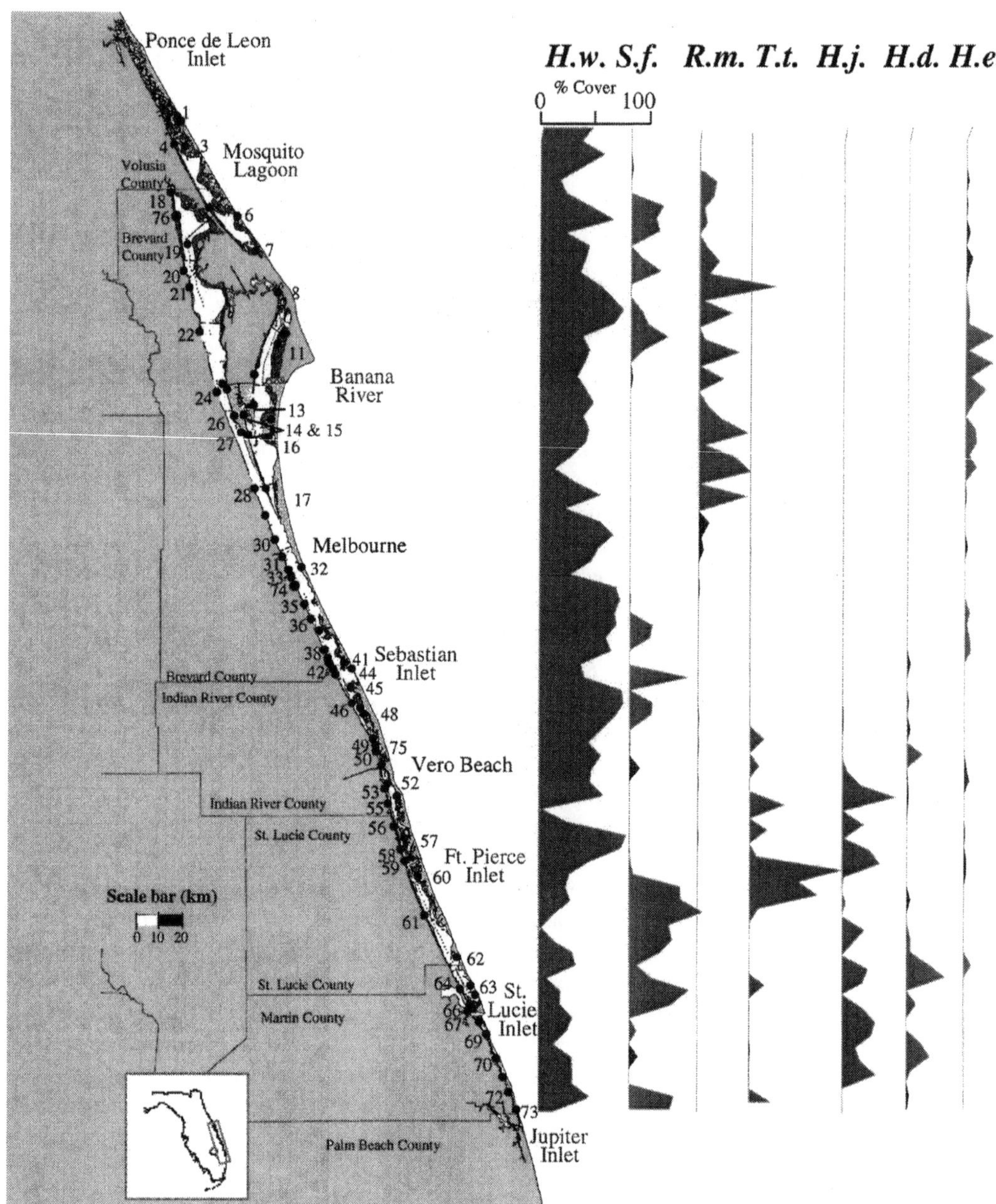

FIGURE 13.1 Seagrass transects and average percent cover of each species in the Indian River Lagoon. Transects (established in 1994) are numbered 1–73, north to south. Transects 74, 75, and 76 were added after 1994. Patterns of species distribution and abundance are graphically presented for all seven seagrass species found in the lagoon. The vertical axes of the percent cover graphs match up horizontally with the transect site location on the map to the left. *Halodule wrightii* (H.w.), *Syringodium filiforme* (S.f.), *Ruppia maritima* (R.m.), *Thalassia testudinum* (T.t.), *Halophila johnsonii* (H.j.), *Halophila decipiens* (H.d.), and *Halophila engelmannii* (H.e.).

temperate and subtropical biogeographic regions meet in the Indian River Lagoon. Thus, the flora and fauna consist of both temperate and subtropical species, making it one of the most diverse estuaries in North America (Gilmore 1985).

One of the goals of the St. Johns River Water Management District program for the Indian River Lagoon is to protect and restore the integrity and functionality of seagrass habitats. The potential goal is to extend and maintain seagrass beds in water depths up to 1.7 m (Virnstein and Morris 2000), the average maximum depth of seagrass occurrence in "healthy" areas of the lagoon.

Fixed seagrass transect monitoring is used to detect changes within seagrass beds. The lagoon's configuration lends itself well for this type of monitoring because of its linear nature with extensive shoals on both shores extending for most of the lagoon's length, generally in a north-to-south direction. Although spatial inference is limited by monitoring only a fixed line, the power of spatial inference within an area can be increased by having many transects in that area. Sampling fixed transects allows reliable detection of small-scale seagrass changes in depth distribution, abundance, and species composition over time. Fixed transects offer a precise record of what is present in a given location at a particular time, and thus provide the capability to detect change. Because each transect extends to the edge of the seagrass bed, expansions or contractions of the bed also can be detected. Another important use of the fixed transects is to aid ground-truthing efforts needed for mapping seagrass from aerial photography.

The objective of this chapter is to present overall seasonal and interannual trends from 5 years of sampling fixed seagrass transects. This chapter is not intended to be a complete analysis of all data collected, but rather a summary of lagoon-wide trends with examples of patterns at various levels of detail. Changes in abundance and species composition are shown lagoon-wide, within a segment, and at an individual transect site.

METHODS

Establishment of Transects

Transect locations were selected to represent all segments of the lagoon. A segment includes any area delimited by water body boundaries, causeways, or an area where seagrass abundance or maximum depth distribution appears similar throughout, based on mapping. In areas of transition, e.g., near inlets, tributaries, or major canals, transects were located closer together. Currently, there are 76 transects throughout the lagoon; new transects have been and will be added in areas that need to be more intensely monitored (Figure 13.1). Fixed transects are located near Indian River Lagoon water quality monitoring network stations in order to relate seagrass to water quality (Sigua et al. 1996). The average distance from the transects to the closest water quality monitoring network site is 1 km.

Each transect consists of a marked line roughly perpendicular to shore extending from shore out to the deep edge of the grass bed. The line is marked with a series of stakes driven into the sediment at not more than 50-m measured intervals. Transect lengths vary from 15 m to over 500 m.

Transect Monitoring

Transects were established and first monitored lagoon-wide in the summer of 1994. All transects are monitored twice a year, summer and winter, corresponding roughly to times of maximum and minimum seagrass abundance, respectively. The intent of the design of this methodology is to sample (1) repeatedly at the same location along the same line; (2) quantitatively; (3) non-destructively; and (4) rapidly.

To survey the transect, a measured line is tautly strung between the stakes. At 10-m intervals, seagrass parameters are measured within a 1-m^2 quadrat, centered on the line. The quadrat is divided by strings into 100 squares, each 10×10 cm, to simplify quantitative estimates of cover. Monitoring includes parameters related to abundance, coverage, species composition, and epiphytes (for details, see Virnstein and Morris 1996). The variables measured and discussed in this chapter are:

- Seagrass percent cover — for each species. Each seagrass species is counted as being either present or absent within each of the 10 × 10-cm cells. This procedure allows for direct count of percent cover.
- Canopy height — in centimeters, for each species present. Blades are vertically "combed" with fingers along a meter stick.
- Shoot density — in ten randomly selected cells of the 100, 10 × 10-cm cells. Counts of individual shoots of each species present are made at mid-bed and near the deep edge of the transect.

Other parameters measured, but not discussed in this chapter, are epiphyte load, drift algae abundance, light attenuation coefficient, water depth, and visual estimates of cover. The transect is videotaped each summer for a visual archive.

Statistical Analysis

Statistical analyses were performed with SAS statistical software (SAS Institute 1988). Statistical significance among means was tested with analysis of variance (ANOVA). The Tukey–Kramer test (T–K) was used when ANOVA indicated the existence of significant difference. Prior to performing ANOVA between years (1994–1998) and seasons (summer/winter), data were transformed (Sokal and Rohlf 1981) with arcsine (for percent cover) or logarithmic transformations (for canopy height). The minimal level of significance for any analysis was $p \leq 0.05$.

The data can be analyzed at various geographic and temporal scales. Comparisons are presented at three geographic scales (lagoon-wide, segments, and site-specific) and two temporal scales [between years and seasonal (summer and winter)]. A complete data analysis of all data is not presented; rather, overall patterns and examples of geographical and temporal patterns are presented.

RESULTS

Lagoon-Wide Patterns

On a large, lagoon-wide scale, the data from all 76 sites are combined and averaged to describe trends, by species, by years, and by season. The seven seagrass species present in the Indian River Lagoon are not uniformly distributed (Figure 13.1). *Halodule wrightii* (average 36% cover over all transects) is the most abundant and common species found — all but one transect have *Halodule wrightii* present. *Syringodium filiforme* (9.5% coverage) is the second most abundant species followed by *Ruppia maritima* (5.2% coverage), *Thalassia testudinum* (2% coverage), and the three *Halophila* spp. [*H. johnsonii* (2% coverage), *H. decipiens* (0.6% coverage), and *H. engelmannii* (0.5% coverage)]. *Halodule wrightii, S. filiforme, R. maritima,* and *Halophila engelmannii* occur lagoon-wide, but *T. testudinum*, *Halophila decipiens*, and *Halophila johnsonii* only occur in the southern half of the lagoon, south of the Sebastian Inlet.

There were significant annual differences in percent cover for *R. maritima*, *Halophila decipiens*, and *Halophila johnsonii.* There were also significant seasonal (summer/winter) differences for *Halodule wrightii* ($p < 0.001$ for percent cover and canopy height), *S. filiforme* ($p < 0.001$ for canopy height), *T. testudinum* ($p = 0.01$ for canopy height), *Halophila engelmannii* ($p < 0.001$ for percent cover), and *Halophila decipiens* ($p < 0.001$ for percent cover and canopy height) (Figure 13.2).

Because the transects were run from the shore to the deep edge of the seagrass bed, the total length of the transect (distance from shore, in meters) is an effective measure of bed width. Transect length in the Indian River Lagoon ranges from 15 to 500 m with an average of 156 m. Using the

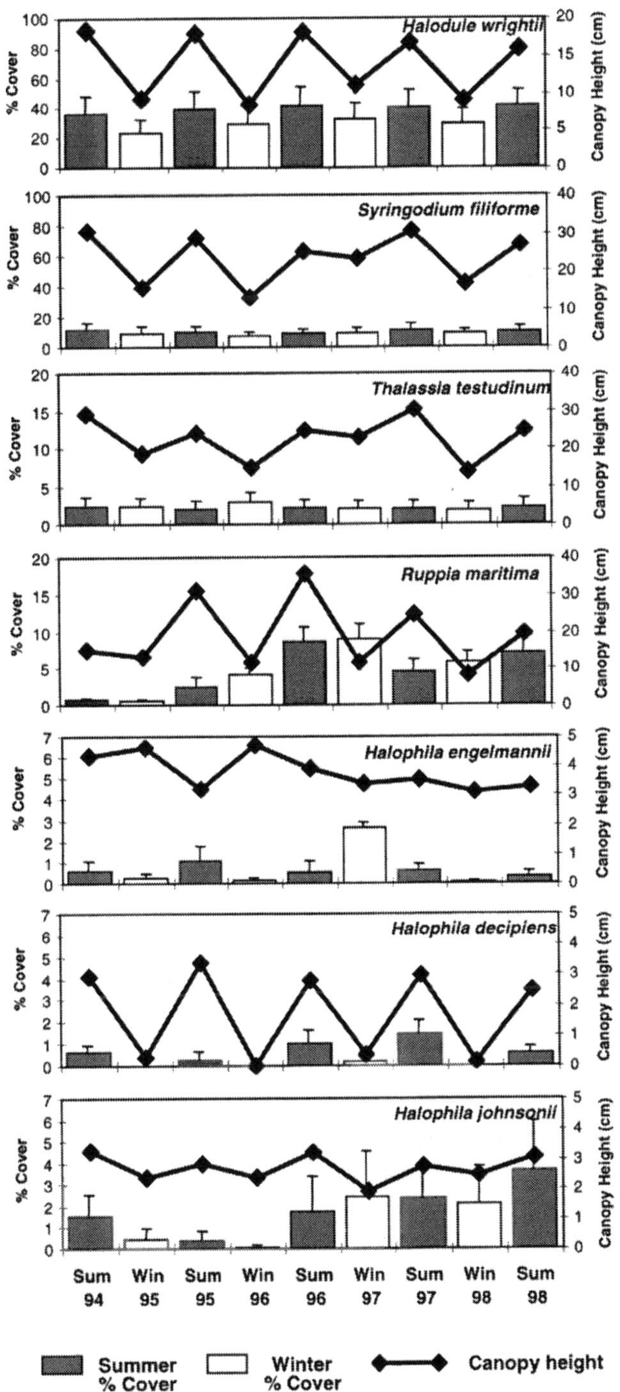

FIGURE 13.2 Annual and seasonal (summer/winter) patterns of percent cover and canopy height of each seagrass species in the Indian River Lagoon (from 1994 to 1998). Annual differences in percent cover were significant for *Ruppia maritima, Halophila decipiens,* and *Halophila johnsonii.* Seasonal (summer/winter) differences in percent cover were significant for *Halodule wrightii, Halophila engelmannii,* and *Halophila decipiens.* Seasonal canopy height differences were significant for *Halodule wrightii, Syringodium filiforme, Thalassia testudinum,* and *Halophila decipiens.* Note that the scales vary for both percent cover and canopy height.

TABLE 13.1
Change in total transect length from 1994 to 1998. Lagoon-wide mean (m) and (percent) changes.

	Year			
Change	**1994–1995**	**1995–1996**	**1996–1997**	**1997–1998**
Gain	240 (3.1%)	1465 (19.2%)	1125 (10.8%)	1280 (11.3%)
Loss	–330 (–4.2%)	–160 (–2.1%)	–430 (–4.1%)	–275 (–2.4%)
Net Change	–90 (–1.1%)	+1305 (+17.1%)	+695 (+6.7%)	+1005 (+8.9%)

original length measured in summer 1994 as the starting point, percent change was calculated as the difference in transect length from 1994 to 1998 (Figure 13.3) for a net increase in average transect length of 26% (Table 13.1). The mean transect length was 140.4 m in 1994 and 176.6 m in 1998.

Segment Patterns

By dividing the lagoon into segments, smaller-scale/regional changes are easier to detect. For example, the lagoon can be divided into 14 geographic segments. Some segments are delineated by natural boundaries (Mosquito Lagoon and Banana River) or inlets (Sebastian, Fort Pierce, St. Lucie, and Jupiter). Other segments have man-made boundaries, e.g., causeways and canals (Titusville, Cocoa Beach, Melbourne, Grant, Wabasso, Vero Beach, and Stuart) (Figure 13.3). There was a significant difference ($p = 0.005$) between the years in average transect length lagoon-wide, but if analyzed by segment, half the segments (Mosquito Lagoon, Northern Banana River, Grant, Sebastian, Fort Pierce, Stuart, and St. Lucie) show no significant difference between the years (Figure 13.3).

Site-Specific Patterns

At the individual transect scale, not only can the total length of the transect be compared among seasons and years, but changes in species distribution and percent cover within the bed can also be compared (Figure 13.4). For example, at the South Canal site (site #53), *Halodule wrightii* was always present and spread, extending the bed by 30 m over the years. In contrast, the federally threatened species *Halophila johnsonii* fluctuated in abundance and distribution along the transect and disappeared in 1998 (Figure 13.4).

DISCUSSION

The high fluctuation in the data reflects the variability and dynamics of the seagrass beds in the Indian River Lagoon. The interannual variability would not be known without the quantity and frequency of transect monitoring. The high temporal and spatial variability found in the seagrass beds within the Indian River Lagoon is probably due to:

1. overall length of the lagoon (250 km)
2. dividing zone between warm-temperate and tropical climates

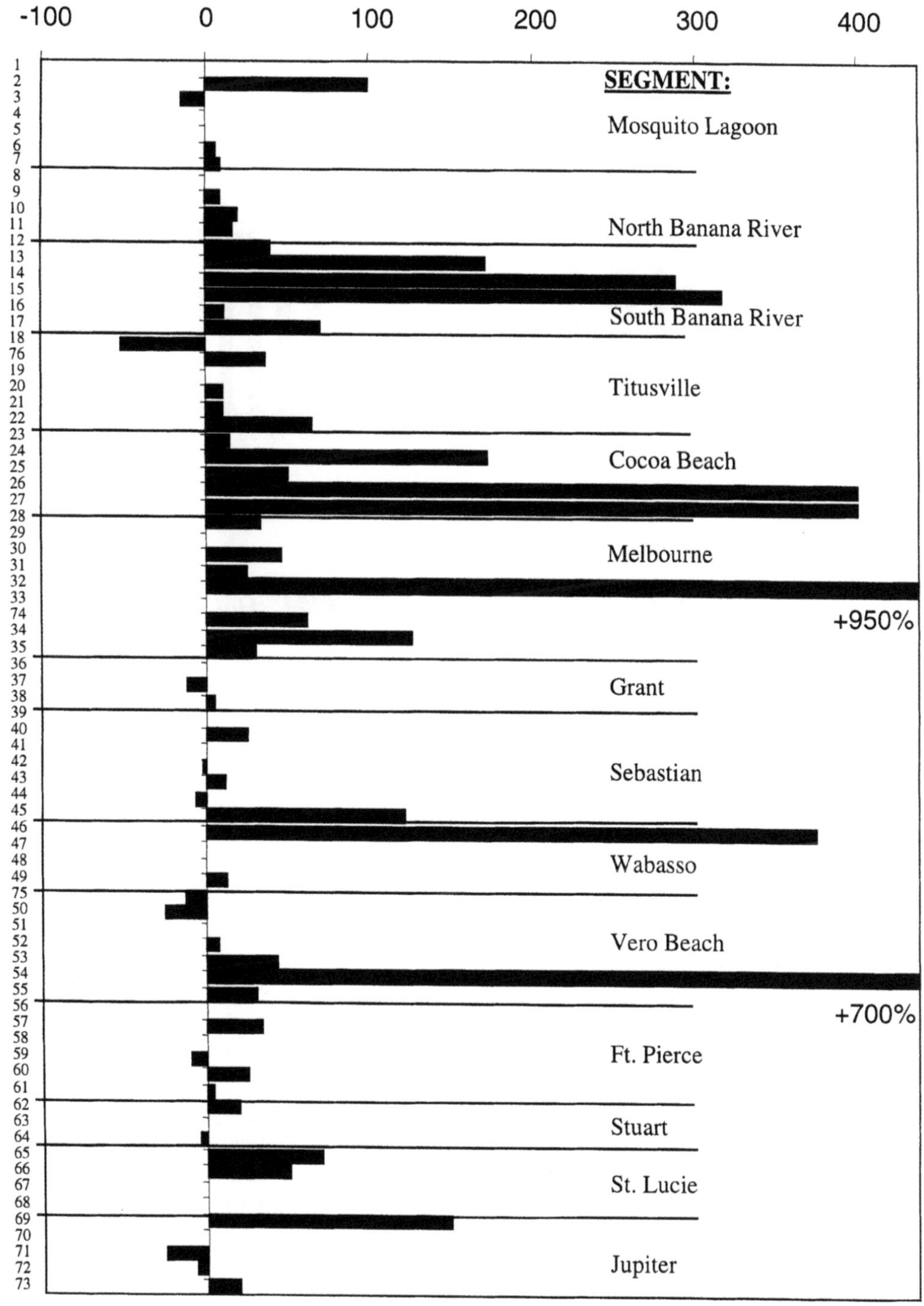

FIGURE 13.3 The percent change in transect length for all sites in the Indian River Lagoon from 1994 to 1998. Sites are in geographic order (north to south). The lagoon is divided into 14 geographic segments. See Figure 13.1 for specific transect location.

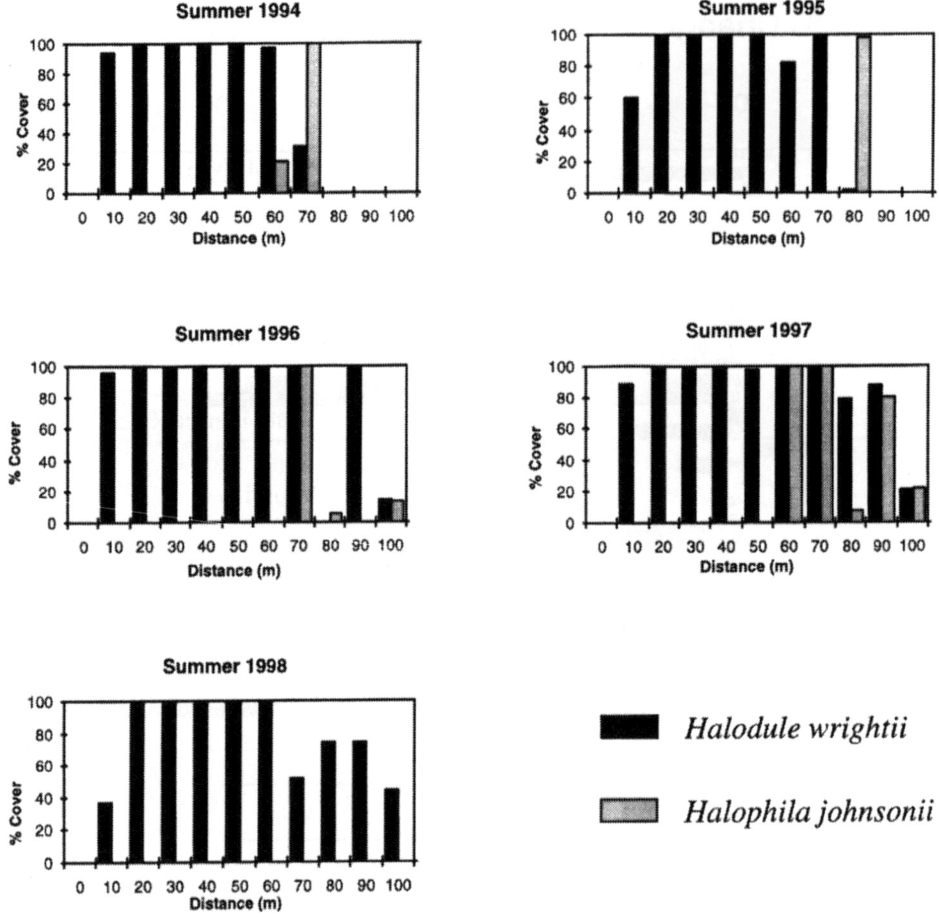

FIGURE 13.4 Site-specific patterns in species distribution and percent cover of *Halodule wrightii* and *Halophila johnsonii* at an individual site (South Canal #53, see Figures 13.1 and 13.2). The dynamic changes can be seen from summer to summer. At 60 m, for example, *Halophila johnsonii* disappeared in 1995, reappeared in 1997, and disappeared again in 1998.

3. highly developed areas vs. pristine, protected areas
4. five inlets to the ocean, four from Sebastian Inlet south, producing some well-flushed, tidal areas, while other areas are poorly flushed and non-tidal
5. at least ten major tributaries and canals contributing freshwater and stormwater runoff.

When the lagoon is divided into segments, the variability within a segment becomes less than the variability among segments. However, even within an individual transect, spatial and temporal variability can be high. Pulich et al. (1997) also found very high variability between individual sites in Corpus Christi Bay. They warn resource managers of the need to examine seagrass responses on a case-by-case basis to identify environmental stressors causing change.

Transects have proved to be extremely useful in monitoring the recovery of an area following a storm or season. For example, Hurricane Bonnie affected most of the lagoon in summer 1995. Mean transect length decreased (Table 13.1). In contrast, there was a major drought in spring and summer 1998, resulting in increased water clarity, and seagrass beds expanded. Most of this "expansion" occurred in areas that historically supported seagrass but had declined over the years (Virnstein and Morris 2000).

On a smaller scale, species composition of the grassbeds or shifts in species may indicate disturbances. For example, the increase of *R. maritima* in the summer of 1996 (Figure 13.2) was primarily seen in the Banana River as a species shift from *S. filiforme*. This shift coincided with increased rainfall and decreased water salinity (Provancha and Scheidt 2000). Fragmentation and loss of climax species, e.g. *Thalassia testudinum*, may not represent normal succession in a grassbed and may be a precursor of complete loss of the seagrass bed (Pulich et al. 1997).

Recommendations

Presently, the plans are to continue monitoring the fixed seagrass transects during the summer and winter; however, the following changes are recommended:

1. Replicating samples taken within transects. At present, sites are visited only once per season. This would be changed to either two or three times per season.
2. Increasing the number of locations where shoot counts are taken on each transect. Hanisak (1999) demonstrated that above- and below-ground, and total biomass are highly correlated with more rapidly obtained measurements (shoot density, percent cover, and canopy height) (R^2 = 0.78, 0.67, and 0.73, respectively).
3. Refining the use of visual estimates for epiphytes (Miller-Myers and Virnstein 2000) and macroalgae abundance. Kirkman (1996) refers to the use of a photo "folio" for seagrass biomass. This would allow more quantitative information to be obtained over larger areas.

ACKNOWLEDGMENTS

We gratefully acknowledge the federal and state agencies that have relentlessly monitored these transects, both summer and winter, since their inception. We thank the staff from the following agencies: National Park Service at Canaveral National Seashore; Dynamac Corporation; Brevard County, Surface Water Improvement Department; Florida Department of Environmental Protection, Aquatic Preserve offices in Fellsmere and Port St. Lucie; U.S. Fish and Wildlife Service, South Florida Ecosystem Program; South Florida Water Management District; and Loxahatchee River District. This study was funded primarily by the St. Johns River Water Management District.

REFERENCES

Gilmore, R. G. 1985. The productive web of life in the estuary. In *The Indian River Lagoon: Proceedings of the Indian River Resources Symposium,* D. D. Barile (ed.). Sea Grant Project No. I.R. 84-28, Marine Resources Council of East Florida, Florida Institute of Technology, Melbourne, FL.

Hanisak, M. D. 1999. A site-specific photosynthetically active radiation and submerged aquatic vegetation study in the Indian River Lagoon. Final report to St. Johns River Water Management District, Palatka, FL.

Kirkman, H. 1996. Baseline and monitoring methods for seagrass meadows. *Journal of Environmental Management* 47:191–210.

Miller-Myers, R. and R. W. Virnstein. 2000. Development and use of an epiphyte photo-index (EPI) for assessing epiphyte loadings on the seagrass *Halodule wrightii*. In *Seagrasses: Monitoring, Ecology, Physiology, and Management,* S. A. Bortone (ed.). CRC Press, Boca Raton, FL, pp. 115–123.

Provancha, J. A. and D. M. Scheidt. 2000. Long-term trends in seagrass beds in the Mosquito Lagoon and Northern Banana River, Florida. In *Seagrasses: Monitoring, Ecology, Physiology, and Management,* S. A. Bortone (ed.). CRC Press, Boca Raton, FL, pp. 177–193.

Pulich, W., Jr., C. Blair, and W. A. White. 1997. Current status and historical trends of seagrasses in the Corpus Christi Bay National Estuary Program study area. Corpus Christi Bay National Estuary Program Publication No. 20, Corpus Christi, TX.

SAS Institute Inc. 1988. SAS/STAT User's Guide, Release 6.03 Edition, Cary, NC.

Sigua, G. C., J. S. Steward, and W. A. Tweedale. 1996. Indian River Lagoon water quality monitoring network: proposed modifications. Technical Memorandum No. 12, St. Johns River Water Management District, Palatka, FL, 70 pp.

Sokal, R. R. and F. J. Rohlf. 1981. *Biometry.* W. H. Freeman & Co., San Francisco, 858 pp.

Steward, J. S., R. W. Virnstein, D. E. Haunert, and F. Lund. 1994. Surface Water Improvement and Management (SWIM) Plan for the Indian River Lagoon, Florida. St. Johns River Water Management District and South Florida Water Management District, Palatka and West Palm Beach, FL, 120 pp.

Virnstein, R. W. and L. J. Morris. 1996. Seagrass preservation and restoration: a diagnostic plan for the Indian River Lagoon. Technical Memorandum No. 14, St. Johns River Water Management District, Palatka, FL, 18 pp. plus appendices.

Virnstein, R. W. and L. J. Morris. 2000. Setting seagrass targets for the Indian River Lagoon, Florida. In *Seagrasses: Monitoring, Ecology, Physiology, and Management,* S. A. Bortone (ed.). CRC Press, Boca Raton, FL, pp. 211–218.

14 Long-Term Trends in Seagrass Beds in the Mosquito Lagoon and Northern Banana River, Florida

Jane A. Provancha and Douglas M. Scheidt

Abstract A long-term study of seagrass beds in lagoonal waters of Kennedy Space Center, Florida was conducted from 1983 to 1996 and included 8150 samples collected along 37 shallow water transects. Species composition and percent cover were determined at 5 m intervals along the transects using a canopy-coverage technique originally developed for terrestrial systems (Daubenmire 1968).

Four seagrass species were found as well as one attached algae. The overall frequency of occurrence for each species indicated the following dominance: *Halodule wrightii* (71.9%), *Ruppia maritima* (23.7%), *Syringodium filiforme* (9.4%), *Halophila engelmannii* (2.3%), and *Caulerpa prolifera* (5.4%). *Halodule wrightii* and *R. maritima* were represented on most transects. *S. filiforme* was never encountered on 14 of 37 transects and, when it occurred, the most frequently recorded percent coverage was < 5%. Temporal trends in percent cover for *Halodule wrightii* indicate a significant long-term decline. *R. maritima* maintained an average occurrence of 26% and cover of 6% from 1983 to 1989. These averages dropped to 11% occurrence and 1.2% percent cover for 1990–1994. A marked increase occurred in 1995–1996 when occurrence was 49% and percent cover was 19%. The increase in *R. maritima* and declines in *S. filiforme* and *Halodule wrightii* appear to be linked to recent declines in salinity. An increase in the number of bare plots was also observed over the study. *C. prolifera* was observed in remarkably high coverages from 1986 to 1987, but rapidly declined by 1989. These data provide benchmarks that will be useful to researchers and managers in comparing trends observed elsewhere in the lagoon and determining if these are site-specific or regional trends.

INTRODUCTION

Much public and scientific attention has focused on seagrasses due to the documentation of human influence on the worldwide decline of seagrass and the increased awareness of the importance of seagrass as habitat (Phillips 1960; Orth and Moore 1983). In addition, numerous recreational and commercial fish found offshore spawn and grow in seagrass beds. Seagrasses and submerged aquatic vegetation (SAV) are currently considered the ecological foundations of the Indian River Lagoon (IRL) system (Kenworthy and Haunert 1991).

The causes of decline of SAV in various estuaries have been attributed to increases in the runoff associated with agricultural herbicides, suspended nutrients, sediments, and toxic discharges. Research in various locations has determined that decreasing light intensity (analogous to decreasing water clarity) causes a major effect on production, biomass, and morphology (Kenworthy and Haunert 1991; Short 1991).

Seagrass beds are found in varying sizes along the IRL shoreline. There are seven species with distributions that vary along the north–south axis of the IRL. All seven species occur in the southern third (Dawes et al. 1995). Three of the seven (*Thalassia testudinum* and *Halophila johnsonii*, and *Halophila dicipiens*) are not found in the northern IRL where *Halodule wrightii*, *Syringodium filiforme*, *Ruppia maritima*, and *Halophila engelmannii* do occur. Primary production and habitat/species interactions research has been predominantly conducted in the southern part of the lagoon (Kenworthy et al. 1988; Dawes et al. 1995; Virnstein 1995).

The seagrass beds in Mosquito Lagoon provide direct forage for marine turtles (*Chelonia mydas*) and manatees (*Trichechus manatus*). The Banana River portion of the study area is remarkably devoid of marine turtles but provides habitat for large numbers of manatees (Provancha and Provancha 1988; Provancha and Hall 1991.) Several studies have begun to explore the relationships between this large herbivore and its seagrass forage (Lipkin 1975; Ogden 1976; Packard 1984; Ledder 1986; Lefebvre et. al. 1988).

The Kennedy Space Center's (KSC) Biomedical Office began funding baseline ecological studies in the 1970s in preparation for the space transportation system (Space Shuttle) operations. In 1983, Brevard County and the Space Center began a cooperative project to set up transects in various seagrass beds that would provide ground truth sites to coordinate with aerial photography. The objective was to create a baseline dataset from each transect to provide descriptive information regarding species composition, percent cover, and frequency of occurrence. Collected over a long term, these data provide time series information to allow for assessment of trends in seagrasses in northern IRL. These data provide a barometer at the local level, allowing for detection of anthropogenic vs. natural changes, and could be used in regional comparisons.

Long-term transects, documenting seagrass densities and composition on a lagoon-wide basis, have only recently (1994) been established (Virnstein and Morris 1996). Our study spans 14 years and is the first published, long-term study of permanent plots of seagrasses in the northern IRL.

Study Area

The study area is part of the IRL complex and is located on the east coast of central Florida between latitude 28°53′00″ N and 28°24′30″ N (Figure 14.1). Lands and lagoon waters of KSC comprise 57,000 ha in Brevard and Volusia counties and Merritt Island National Wildlife Refuge. The waters included in this study are the southern Mosquito Lagoon and Banana River. KSC is on northern Merritt Island, a barrier island complex separated from Cape Canaveral by the northern Banana River. It occurs within a biogeographical transition zone, having faunal and floral assemblages derived from both temperate Carolinian and subtropical Caribbean provinces (Ehrhart 1976; Gilmore et al. 1981; Provancha et al. 1986). These waters are shallow, aeolian lagoons with depths averaging 1.5 m and maximums of 9 m (in dredged basins). The salinity ranges from 10 to 42 ppt. Detailed descriptions of the study area can be found in Snelson (1976) and Down (1978).

Development on KSC is limited to more upland sections, and the direct or indirect runoff impacts to the surrounding waters appear to be minimal. Security requirements and the critical manatee habitat designation for most of the KSC section of the Banana River result in very low use of this area by the boating public. The waters south of the NASA causeway are restricted to non-motorized boating activities. The waters north of the NASA causeway are closed to the public, including boating. The southern section of our Banana River study area has a few, small direct freshwater runoff/input areas and one larger creek. The northern Banana River study site receives only local runoff. The Mosquito Lagoon has very minor boating and fishing restrictions. The Space Shuttle and Titan rocket launch complexes are located near the shoreline of the northern Banana River. The southern portion of Banana River and Mosquito Lagoon are within 5–10 km of these complexes. However, none of the regions receives direct impacts from launch operations.

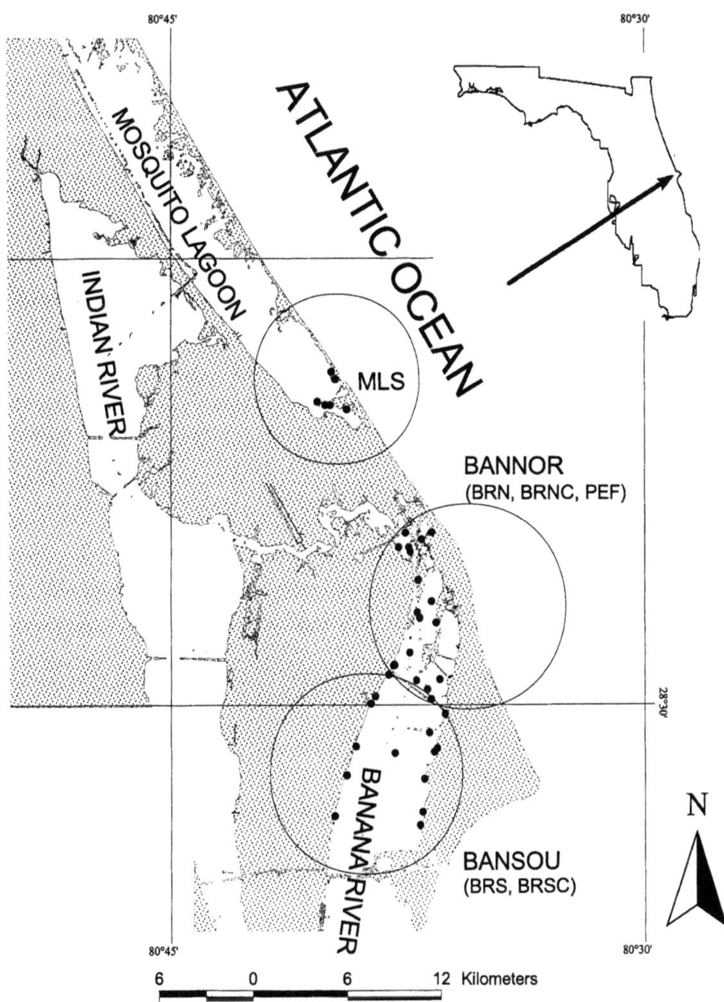

FIGURE 14.1 Map of study area located at the Kennedy Space Center, Florida. The dots indicate the location of individual transects and the circles show the location of region and area groupings. Regions are designated as Banana River North (BANNOR), Banana River South, (BANSOU) and Mosquito Lagoon South (MLS). Areas are sub-groups of regions. BRN, BRNC, and PEF are located north of the NASA causeway; BRS and BRSC are found south of the NASA causeway; and MLS (also classified as a region) is found in southern Mosquito Lagoon.

MATERIAL AND METHODS

Seagrass Transects

In 1983, 22 transects were established with 15 more added in 1986, and all were sampled annually. Transects were delimited with permanent markers at the origin and terminal positions of each by using PVC poles driven into the sediment. During each sampling event, a nylon cord or tape measure (marked off in 5 m increments) was temporarily connected between the two PVC poles to delineate the line against which all subsequent measures were taken. Transects that followed a general north–south orientation were measured on the eastern side of the line while those running in an east–west fashion were measured on the northern side. Transect lengths ranged from 50 to 200 m with most starting from land and extending waterward. Most seagrass beds on KSC extend considerable distances from shore, consequently our "short" transects generally ended within a bed, instead of at the waterward edge of a bed. Our intent was to evaluate conditions in mid-bed areas as well as nearshore where land-based run-off might first enter the waters and affect beds. This design admittedly did not allow for detection of changes caused by light limitations at the deeper waterward edges of beds.

The 15 transects created in 1986 were established after biologists noted a sharp increase in the alga, *C. prolifera,* in 1985. These 50 m transects often started in the middle of a grassbed where approximately half of the transect was dominated by *C. prolifera* and the other half was covered with some other seagrass or SAV species. All of the *C. prolifera* transects were identified by a "C" prefix next to each numerical designation. Eight of the transects were located south of the NASA causeway and the remainder were north of the causeway (Figure 14.1).

Percent cover and species composition were determined at 5 m intervals in plots along all transects using a canopy-coverage technique of vegetation analysis originally developed for terrestrial systems (Daubenmire 1968). This method employs the use of a frame with inside dimensions of 20 × 50 cm and appropriately marked off so that coverage estimates can be quickly made using a series of six unequal coverage classes. The class limits are: 0.1–5%, 5–25%, 25–50%, 50–75%, 75–95%, and 95–100%. These classes then yield mid-point density categories (2.5%, 15%, 37.5%, 62.5%, 85%, and 97.5%) for analyses. This system results in greater agreement of observers, as the range of percentage points within a class allows for each observer's deviance from the "correct" cover percentage.

All seagrass species and *C. prolifera* were evaluated, and attempts were made to identify and quantify other algal species and detritus. The identification of detritus and algae was inconsistent across years, limiting data analyses for this report to seagrass species and *C. prolifera.*

Depth was recorded at each sample plot along a transect using a pole marked in centimeters. Salinity and water temperature were recorded when each transect was sampled. Long-term water monitoring stations provided quarterly water quality data as well. The National Atmospheric Deposition Station located at the Space Center provided rain volume data.

Analyses

Analyses for this report are confined to samples collected from 1983 to 1996 during the summer months (June–September) when SAV is generally at its maximum abundance. For most analyses, 1984 was not included due to the low number of samples taken that year.

The terms *occurrence* and *cover* will be used throughout the results section to represent frequency of occurrence and percent cover, respectively. The cover relates to the Daubenmire cover values and is always reported as a percentage. Occurrence indicates the number of times a species is present divided by the number of plot samples and is reported as a percentage. Frequency of occurrence of a particular species was calculated for each year, as well as for the entire study period. Percent cover was calculated by summing all mid-point values including zeroes and then dividing by the number of plots sampled. Cover was calculated for years, regions, and area, and overall for the entire study.

Data for this report were evaluated for year, species, location, and depth. Location included three hierarchical levels. The first was the entire study site where transect plots were pooled. The next level separated the study site into three major regions: Mosquito Lagoon South (MLS); Banana River North (BANNOR), which is from the NASA causeway north; and Banana River South (BANSOU), extending from the NASA causeway south to State Route 528 causeway. The third level separated the regions into six subsets or areas. This was done by grouping transects within a region using the following criteria. Those that originate from shore, those that originate away from shore (designated with a C suffix), or those that share a close geographic relationship (e.g., PEF, the Pepper Flats portion of the BANNOR). The resulting areas are BRN, BRNC, and PEF, which are located north of the NASA causeway; BRS and BRSC, found south of the NASA causeway; and MLS (also classified as a region), found in southern Mosquito Lagoon (Figure 14.1).

Importance values were calculated for each SAV species by areas as well as for region. The values were calculated by summing the relative percent cover and relative frequency of each species within an area ([x/sum of occurrence + x/sum of cover] × 100 = importance value). This was accomplished using the entire data set, pooled over time. The importance value assists in ranking or scoring relative differences in species among areas by weighting them according to two parameters instead of one.

Most of the species data required nonparametric tests. Statistical analyses included linear regression, Spearman rank test correlations, and Kruskal–Wallis. These were performed using software from Jandel, Sigmastat Version 2.0, and Microsoft, Excel Version 7.0.

RESULTS

A total of 8150 samples fit the criteria, e.g., season, for analysis in this report. The SAV species encountered in the KSC lagoons included *H. wrightii*, *S. filiforme*, *R. maritima*, and *Halophila engelmannii* and the alga *C. prolifera*. The overall frequency of occurrence for each species was *Halodule wrightii* (71.9%), *R. maritima* (23.7%), *S. filiforme* (9.4 %), *C. prolifera* (5.4%), and *Halophila* (2.3%). The overall percent coverage for each species was *Halodule wrightii* (35.6%), *R. maritima* (6.5%), *S. filiforme* (1.7%), *C. prolifera* (2.60%) and *Halophila engelmannii* (0.61%). The overall occurrence of bare plots was 6.2% with 1996 having the highest occurrence of 10%.

Individual Species Observations

Halodule wrightii

H. wrightii was found at most sample depths, but the most frequent depth averaged 52 cm, with 68% of the *H. wrightii* samples occurring between 35 and 65 cm depths (Figure 14.2). On an annual basis the occurrence of *H. wrightii* ranged from 70 and 85% until 1996 when there was a notable drop in occurrence to less than 30% (Figure 14.3).

Temporal variation in cover for *H. wrightii* indicated the highest coverages occurred from 1983 to 1991 (Figure 14.4). From 1991 to 1996, values were lower with a significant drop in 1995 and 1996. The multi-annual trend-line showed a decline with $r^2 = 0.683$. The years 1983, 1995, and 1996 were significantly different (Kruskal–Wallis, $p < 0.05$) from all other years.

Ruppia maritima

R. maritima was the second most common species in terms of overall frequency of occurrence, (23.7%) over the 14-year period. It was found over a broad range of depths (20–70 cm), but had highest coverage in shallower waters. The mean depth of occurrence was 42.7 cm (Figure 14.2). The annual occurrence for *R. maritima* was relatively high from 1984 to 1988 and declined from 1989 to 1994 (Figure 14.3). A large increase in occurrence in 1995 was followed by a twofold increase in 1996. The annual cover remained below 10% except for 1996 when it reached 37% (a ninefold increase over the previous year). *R. maritima* cover generally declined over the period 1989–1994 followed by considerable increases in 1995. Highest values occurred in 1986, 1988, and 1996 (Figure 14.4). The regression line for occurrence of *R. maritima* yielded an $r^2 = 0.0573$. The percent cover for *R. maritima* showed the years 1995 and 1996 were significantly different from all other years (Kruskal–Wallis, $p < 0.05$).

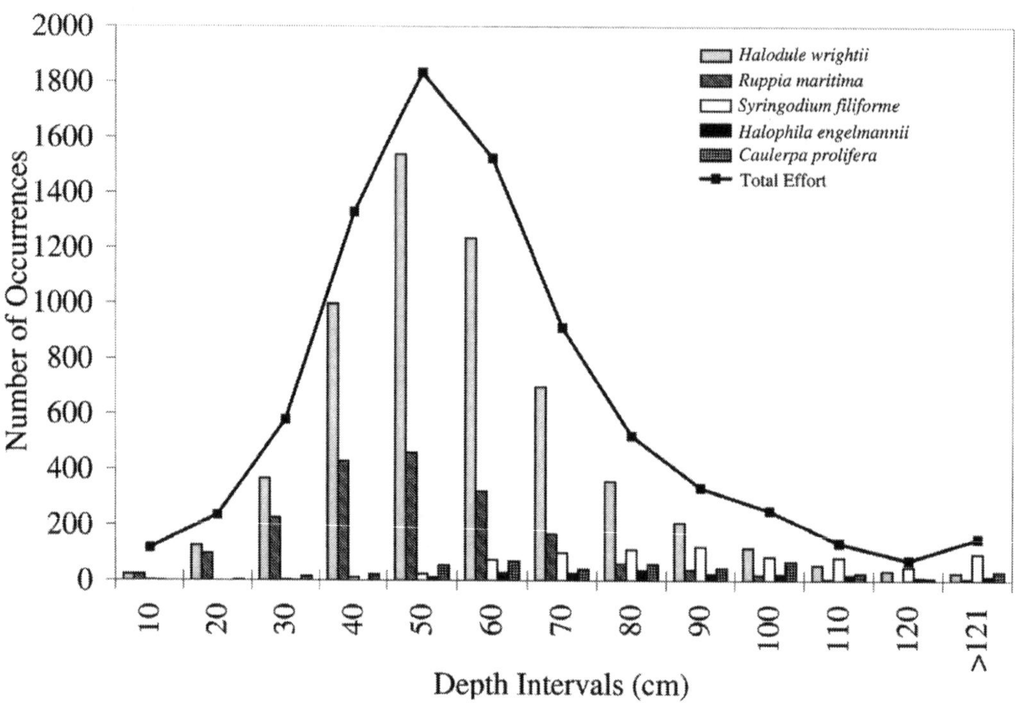

FIGURE 14.2 Sampling effort in relation to water depth and frequency of occurrence for the five SAV species (all years pooled). The line represents sampling effort at each depth category. Labeled categories constituted 10 cm depth increments (i.e., 0–10, 11–20, etc.).

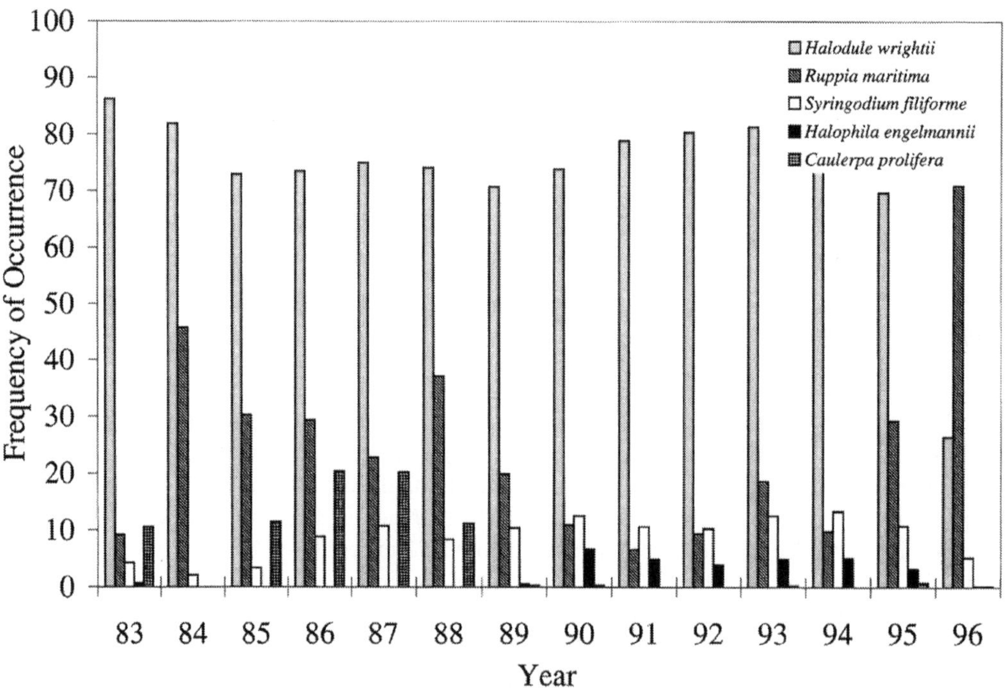

FIGURE 14.3 The temporal distribution of the frequency of occurrence of SAV species at the Kennedy Space Center from 1983 to 1996.

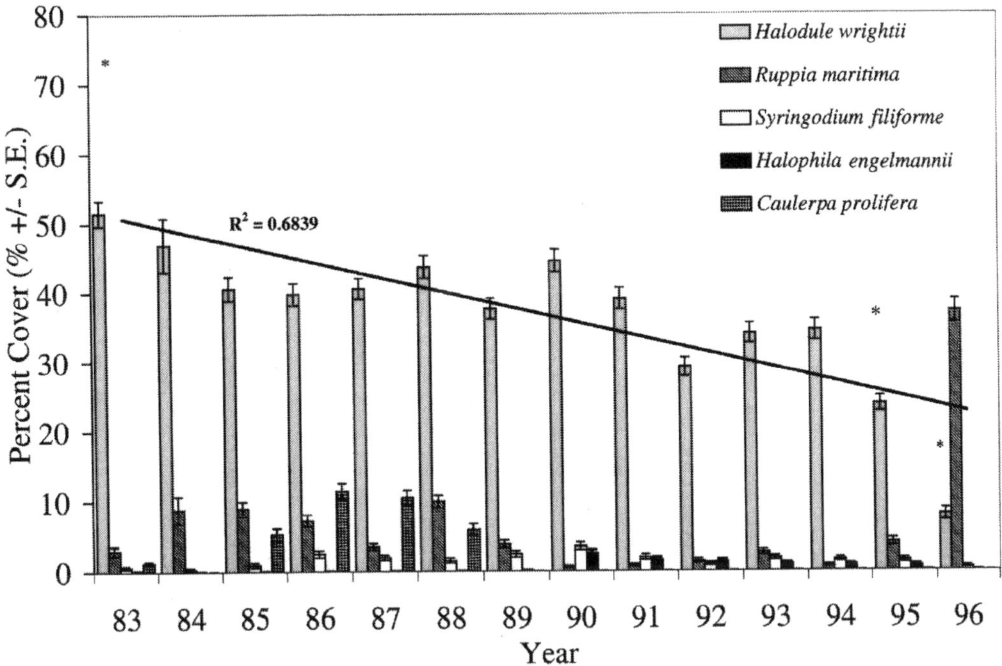

FIGURE 14.4 The temporal distribution of the mean percent cover of SAV species. Asterisk (*) denotes years when the values for *Halodule wrightii* were significantly different from all others.

Syringodium filiforme

S. filiforme was encountered on 23 of the 37 transects. The overall occurrence for *S. filiforme* was only 9.4% and the overall cover was 1.7%. *S. filiforme* was found at a range of depths, from 5–198 cm, but the most frequent depth averaged 90.3 cm (Figure 14.2). Approximately 45% of the plots had water depths greater than 50 cm, where 92.5% of the observations of *S. filiforme* occurred.

S. filiforme cover showed highest values in 1989 and 1990 (Figure 14.4). The frequency of occurrence for *S. filiforme* was relatively stable, even though several individual transects showed high year-to-year variability (Figure 14.3). On three transects, one in each water body, *S. filiforme* declined and then was absent from 1991 to the present.

Caulerpa prolifera

C. prolifera was found at least once on 21 of 37 (56.7%) transects. However, most of the observations of *C. prolifera* occurred over a very short period from 1986 to 1988, and it was rarely observed after 1989. The shoreward edges of *C. prolifera* beds were generally found at the 50–60 cm depths where the average coverage of *Halodule wrightii* started to decline. In the winter of 1984, patches of *C. prolifera* were noted growing in waters as shallow as 15 cm in the Indian River. However, *C. prolifera* extended well past the range of *S. filiforme* showing up in significant amounts at depths exceeding 3 m in 1983 (Provancha and Willard 1984). The mean depth of occurrence for *C. prolifera* on our transects was 76.5 cm with a range of 10–200 cm (Figure 14.2). The highest occurrences for *C. prolifera* were in the 90–120 cm depth range.

The overall occurrence for *C. prolifera* was 5.4% with a mean cover of 2.6%. However, during the 1986–1988 period, the yearly occurrence ranged from 11 to 20% and the cover ranged from 5.9 to 11.5% (Figure 14.3 and 14.4).

Halophila engelmannii

Of the four seagrass species found in our study area, this was the least frequently encountered, with an occurrence of 2.3% and overall cover of 0.61%. *H. engelmannii* was relatively ephemeral in space and time (Figures 14.3 and 14.4). It was observed on only one transect in 1983 and not seen anywhere else until 1989. Then it was observed through 1995. This species was found at least once on 21 of 37 transects (56%) but was not found on any MLS transects.

H. engelmannii was observed at a wide range of depths (10–200 cm) but most commonly encountered between 70–110 cm (Figure 14.2). The highest coverage and frequency of occurrences were seen at depths greater than the mean depth of occurrence (83 cm). The highest mean cover for this species was recorded in 1990 and 1991.

Bare Plots

Overall, the frequency of occurrence of bare plots increased in 1989–1990 (Table 14.1). Bare plots occurred at low levels from 1983-88, increased markedly in 1989–1992, dropped in 1993, and steadily increased afterwards with a record high in 1996. The trends for percent cover followed the same pattern as observed for occurrence.

Spatial Details

As stated earlier, the study site was partitioned either by three regions: BANNOR, BANSOU, and MLS, or six areas: BRN, BRNC, BRS, BRSC, PEF, and MLS (Figure 14.1). Figures 14.5 and 14.6 show the occurrence and cover, respectively, for each seagrass species for each of the six areas. *Halodule wrightii* was the most frequently occurring seagrass in all areas. *R. maritima* was the second most common occurring species in four of the six areas. In BRN and BRNC, *S. filiforme* and *C. prolifera* had higher occurrence values than *R. maritima* (Figure 14.5). The occurrence of *Halodule wrightii* in MLS and PEF was the highest, while at BRS and BRNC, frequencies were 10 to 20% lower than the overall mean (71.9%). For *R. maritima*, the highest frequencies were in BRS (48%), PEF (29%), and BRSC (27%), while the other areas ranged from 0 to 9%. For *S. filiforme*, BRNC yielded a 45% occurrence, and all the other areas ranged from 1 to 13%. The relatively deeper waters of the BRNC transects appear to be associated with this and the *Halophila engelmannii* distribution. *Halophila engelmannii* occurred most often in BRNC (14%) and BRN (4%), but was absent in MLS. *C. prolifera* had highest occurrences (1–17%) at BRN and did not occur on transects in MLS.

For *Halodule wrightii,* coverage was highest above 45% in MLS and PEF (Figure 14.6) while BRNC had the lowest cover at 8.7%. BRS, PEF, BRN and BRSC, generally shallow areas, had the

TABLE 14.1
The annual frequency of occurrence of bare plots across the entire study area (overall) and by region. All values are expressed as percent.

YEAR	83	85	86	87	88	89	90	91	92	93	94	95	96
OVERALL	7.8	11.6	8.1	8.7	8.7	17.4	15.8	11.2	14.2	7.9	13.3	13.5	17.5
BANNOR	5.2	6.6	5.0	5.6	5.6	15.8	14.7	7.2	12.3	3.9	13.4	15.6	6.5
BANSOU	9.6	5.3	3.3	7.7	8.3	20.7	18.1	12.4	11.4	5.7	15.5	8.8	35.2
MLS	10.6	27.5	21.3	17.0	16.3	16.3	14.9	18.4	22.0	20.6	9.9	14.9	19.5

Note: Regions are designated as Banana River North (BANNOR), Banana River South (BANSOU), and Mosquito Lagoon South (MLS).

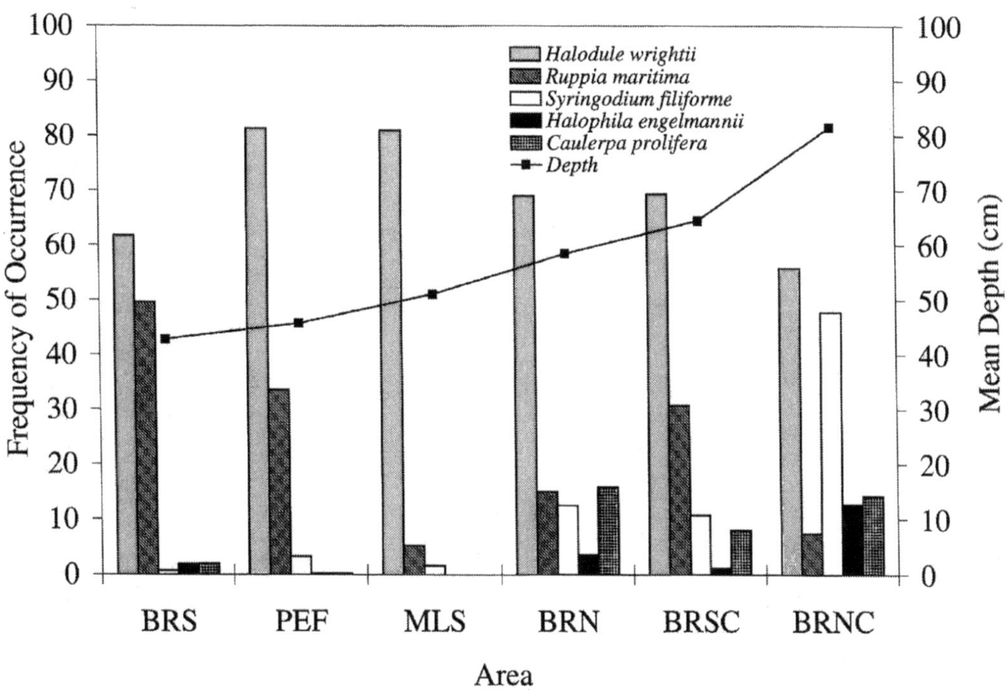

FIGURE 14.5 Frequency of occurrence by depth for each species of seagrass within each of the six areas from data pooled from 1983 to 1996. The line indicates mean water depth for each area.

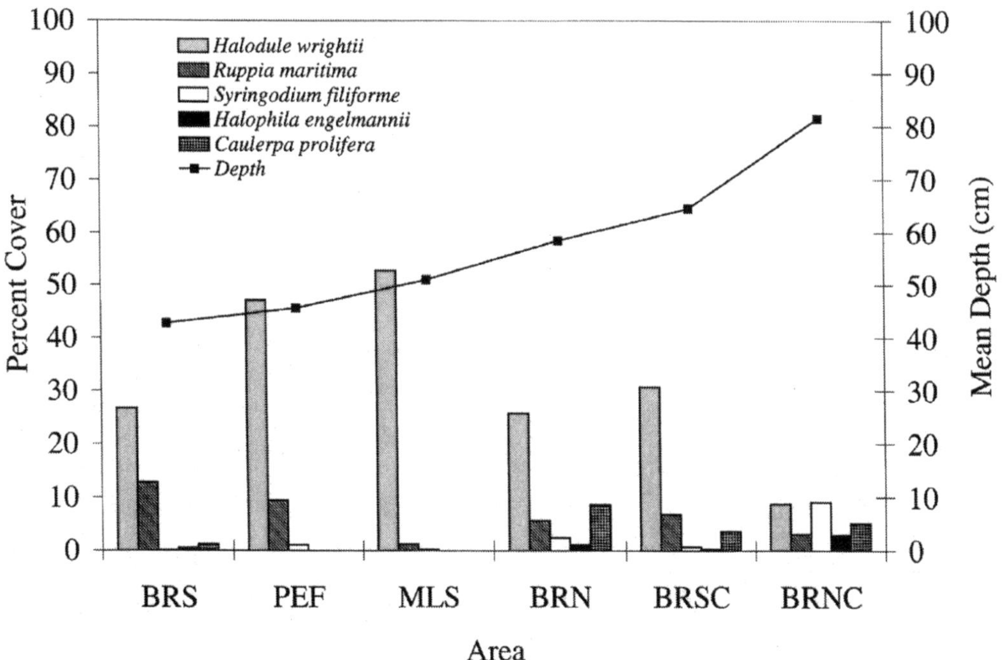

FIGURE 14.6 Percent cover by species by area. Data were pooled from 1983 to 1996. The line indicates mean water depth for each area.

TABLE 14.2
Importance value rankings calculated for each species found within the Kennedy Space Center study (1983–1996).

Location		Species				
Region	Area	*Halodule wrightii*	*Ruppia maritima*	*Syringodium filiforme*	*Halophila engelmannii*	*Caulerpa prolifera*
BANNOR		1	2	3	5	4
BANSOU		1	2	4	5	3
MLS		1	2	3		
	BRS	1	2	5	4	3
	PEF	1	2	3	4	5
	MLS	1	2	3		
	BRN	1	3	4	5	2
	BRSC	1	2	4	5	3
	BRNC	1	5	2	4	3

Note: The values were calculated by summing the relative percent cover and relative frequency of each species within an area ([x/sum of occurrence + x/sum of cover] × 100 = importance value). Regions are designated as Banana River North (BANNOR), Banana River South (BANSOU), and Mosquito Lagoon South (MLS). Areas are sub-groups of regions. BRN, BRNC, and PEF are located north of the NASA causeway; BRS and BRSC are found south of the NASA causeway; and MLS (also classified as a region) is found in southern Mosquito Lagoon.

high cover values of *R. maritima* (Figure 14.6). MLS, which is shallow, but is typically more saline and occasionally hypersaline, had extremely low *R. maritima* coverages.

Spatial trends for the cover and occurrence for *S. filiforme* were similar. The highest occurrences and coverages for *S. filiforme* were found at BRNC and were markedly greater than the overall mean occurrence (9.4%) and coverage (1.7%). Other areas had covers ranging only from 0.6 to 1.08% with BRS representing the lowest coverages of *S. filiforme*. These high values for *S. filiforme* were found in the area with the greatest depths, BRNC (Figures 14.5 and 14.6).

Importance Values

Halodule wrightii had the highest importance value ranking in all regions (Table 14.2). The second ranked species was *R. maritima*; *S. filiforme* ranked third in BANNOR and MLS; and *C. prolifera* was third in BANSOU. When ranking species by area, *H. wrightii* ranked first (Table 14.2). *R. maritima* was second in all areas except for BRN and BRNC where *C. prolifera* and *S. filiforme* had higher importance value rankings than *R. maritima*.

Temporal Variation by Region

Mosquito Lagoon South (MLS)

The seagrass species observed on the MLS transects were *Halodule wrightii*, *S. filiforme* and *R. maritima*. A considerable decline in *H. wrightii* coverage was noted in 1986, 1995, and 1996, while other species increased during these years (Figure 14.7). The regression line for *H. wrightii* yielded no significant trend with an $r^2 = 0.01$. Annual comparisons of *H. wrightii* yielded three groups of years significantly different from all others: 1985–1987, 1989–1994, and 1995–1996 (Kruskal–Wallis test, $p < 0.05$).

For *R. maritima* coverage, the years 1995 and 1996 were significantly different from all others; these were years with dramatic increases in *R. maritima*. Annual comparisons of *S. filiforme* showed

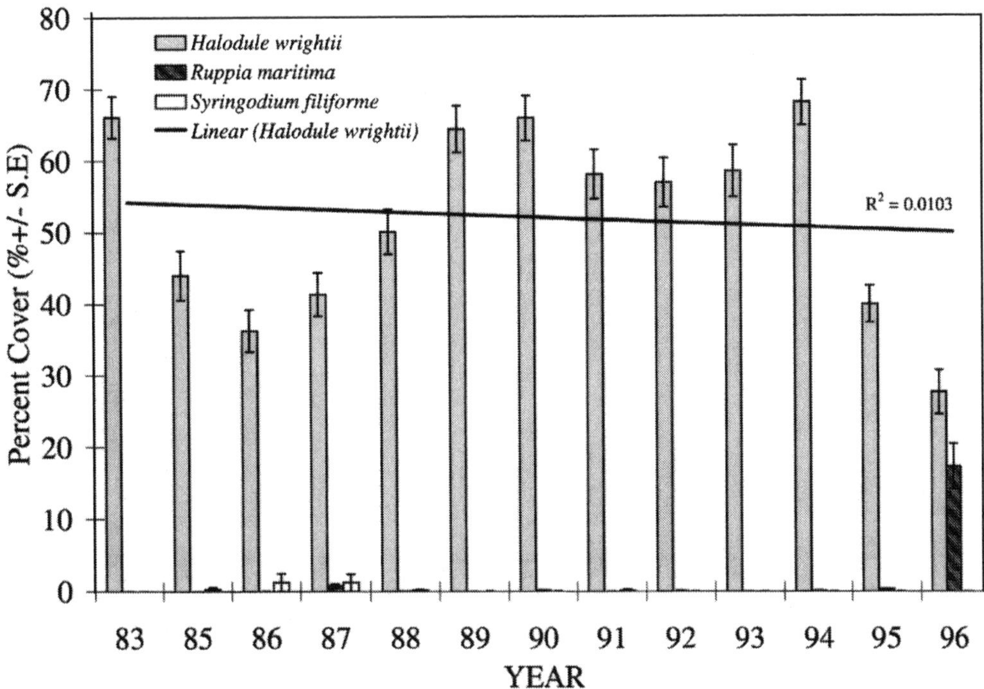

FIGURE 14.7 Temporal trend in percent cover of SAV species in the Mosquito Lagoon transects. The trendline assesses the values for *Halodule wrightii.*

1986 to be significantly different from the years 1992–1996, and 1987 to be different from 1989–1996 (Figure 14.7). Of the three regions, MLS had the highest occurrence of bare plots ranging from 9–27% with most values in the 15–20% range (Table 14.1).

Banana River North (BANNOR)

The decline in *Halodule wrightii* coverage over time in BANNOR is evident with a strong negative correlation $r^2 = 0.708$ (Figure 14.8). Similar to MLS, *H. wrightii* cover in BANNOR in 1995 and 1996 was significantly lower than all other years. *R. maritima* coverage was significantly higher during this time ($p < 0.05$).

S. filiforme showed significant differences in coverage by year, with 1983–1985 different than 1990–1995. Values for the year 1996 differed significantly from values for 1990–1995. For *caulerpa*, 1983–1988 were significantly different from 1989–1996. The occurrence of bare plots ranged from 5–16%, with a marked increase in 1989, 1990, 1992, 1994, and 1995. There was a marked decrease of bare plots in 1996 (Table 14.1).

Banana River South (BANSOU)

A strong negative trend existed between time and cover for *Halodule wrightii* ($r^2 = 0.804$). Also 1996 was significantly different (Kruskal–Wallis, $p < 0.05$) from all other years (Figure 14.9). For *R. maritima*, 1995–1996 were significantly different from all other years (Kruskal–Wallis, $p < 0.05$). Also, the early 1980s were significantly different from the early 1990s. The data for *R. maritima* appear to be somewhat cyclic for the period, with a low about every 3–4 years. For *caulerpa*, 1986–1987 were significantly different from all other years (Kruskal–Wallis, $p < 0.05$). For most years, bare plots in BRS represented 5–20% of all plots sampled with the exception of 1996 when

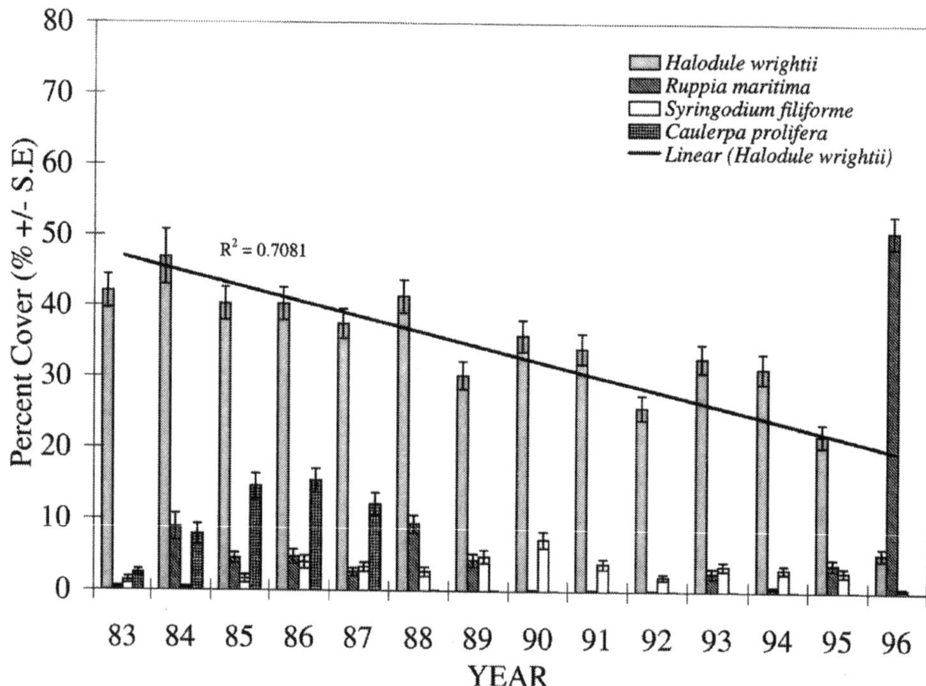

FIGURE 14.8 Temporal trend in percent cover of SAV species in the Banana River North transects. The trendline assesses the values for *Halodule wrightii.*

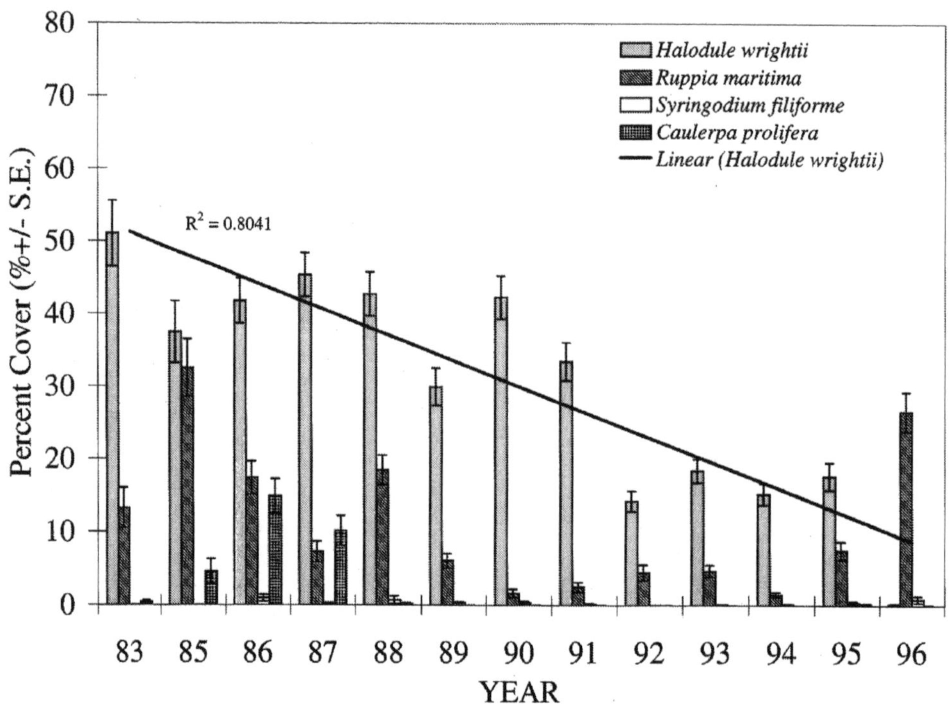

FIGURE 14.9 Temporal trend in percent cover of SAV species in the Banana River South transects. The trendline assesses the values for *Halodule wrightii.*

there was a high of 36%. The years 1989, 1990, and 1996 showed the highest occurrence of bareground in BRS (Table 14.1).

Physical Parameters

Mailander's (1990) review of the local air temperatures from 1887 to 1987 show January and February as months with lowest temperatures (mean minimums and maximums of 9.6°C and 22.3°C, respectively). During our study, the years 1986 and 1989 experienced colder winters than normal. The warmest months are typically July and August with mean minimums of 21.9°C and maximums of 33.3°C.

Mailander (1990) also reviewed local precipitation data from a 78-year period and described rainfall as bimodal with a wet season from May to October and relative dry period through the rest of the year. Over 50% of the precipitation occurred during the wet season. Annual rainfall amounts varied widely from about 76 to 203 cm. The 15-year average annual rainfall for KSC was 132 cm (Drese 1997, Figure 14.10). For the purpose of this study, rainfall was calculated as the amount that had fallen from the end of one sampling season to the next (summer to summer). By using this "rain year" data we believe it would more realistically reflect the impact that rainfall has on a growing season for the seagrass. The highest rainfall occurred in 1995 and 1996, while 1989 and 1990 had the lowest, with most years falling below the annual calendar year average of 132 cm (Figure 14.10). These data only reflect the total annual volume and do not address interseasonal and seasonal trends. Although a total volume may be high, it may result from a large event early or late in the year but is beyond the scope of this study. Mailander (1990) reported that several years in a row often have significantly less rainfall than the average, but that there is no detectable recurrence interval for this locality based on the 78-year study.

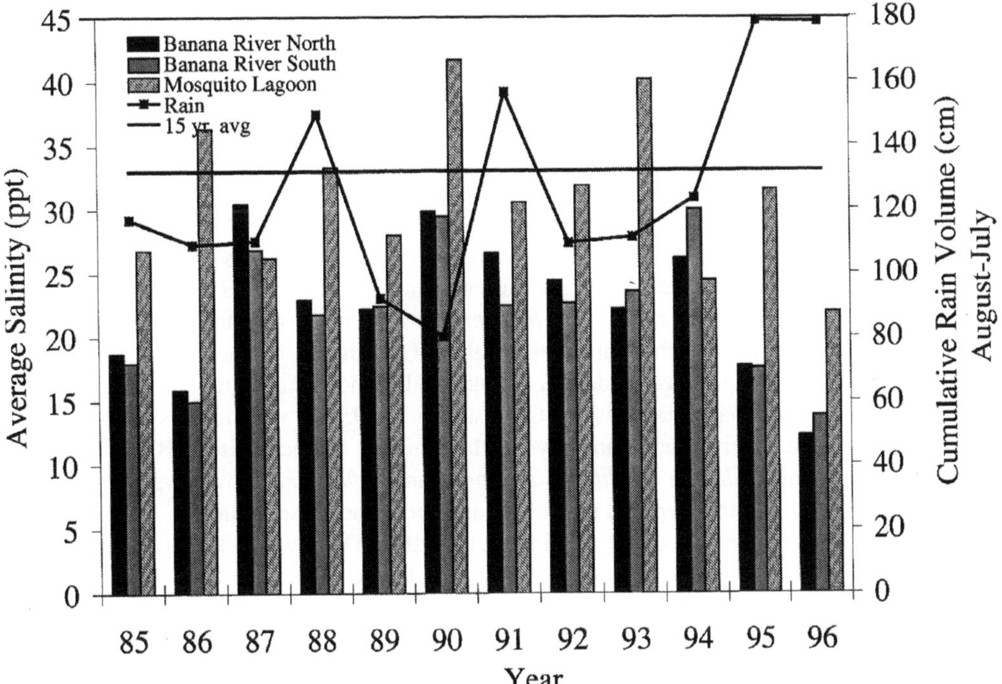

FIGURE 14.10 Salinity and rainfall values for the Kennedy Space Center (KSC) area. Annual total rainfall volume (summer to summer) compared to annual mean salinity recorded in each region. The horizontal line is the 15-year mean for KSC rainfall.

Salinity was significantly negatively correlated (Spearman rank test, $p < 0.001$) with *R. maritima* and *C. prolifera*. Salinity was positively correlated with *Halodule wrightii*, *S. filiforme*, and *Halophila engelmannii*. Rain volume was negatively correlated with salinity (Spearman rank test, $p < 0.001$) in all areas, with the Mosquito Lagoon having the highest correlation coefficient (–0.578). Rain volume was positively correlated with percent cover of *R. maritima* and negatively correlated with *Halodule wrightii*, *S. filiforme*, and *C. prolifera* (Spearman rank test).

DISCUSSION

Seagrasses in our study area were distributed across waters with depths ranging from less than 1 cm to over 1.3 m, with each species demonstrating a distinct zonation pattern. SAV species diversity, based on occurrence alone, was highest at the middle depth ranges of 60–90 cm. As depth increased, *S. filiforme* became the dominant seagrass with *C. prolifera* and *Halophila engelmannii* also occasionally present. *Halodule wrightii* had the broadest depth range with the most frequent occurrences and highest coverage in shallow to medium depths. Zieman (1982) described *Halodule wrightii* as a broadly euryhaline seagrass, capable of growing in the shallowest water and able to avoid desiccation during periods of exposure.

R. maritima, considered by some to be a brackish water species rather than a true seagrass, is reported to favor water conditions near shore where salinity can become reduced due to upland drainage (Phillips 1960). This generalization held true many times during our sampling, but we also confirmed Thompson's (1978) observations that *R. maritima* is occasionally observed in deeper and saline waters.

General observations during other mapping projects (Provancha, unpublished 1998) showed *S. filiforme* was generally detected first at the 50–60 cm depth class and extended out to depths of about 2 m. In a few areas, for short periods, this species was observed growing in shallows near the shoreline where exposure was frequent due to wind-driven events.

Irvine et al. (1979), while studying manatees in central Brevard County, reported that *C. prolifera* was widespread in the Banana River. Provancha and Willard (1984) found *C. prolifera* to be the second most common attached constituent in the SAV beds in the Banana River in 1983, with an occurrence of 23.2%. *C. prolifera* appeared to be a major component of the SAV during that year and was often observed in largely monospecific meadows extending across the central, deeper section of the Banana River. Although no quantitative data are available, *C. prolifera* appeared to be the dominant species in deep areas of the Banana River, probably representing 70% of the total SAV between Port Canaveral and the NASA causeway in 1983 (Provancha and Willard 1984). *C. prolifera* extended well past the range of *S. filiforme* showing up in significant amounts at depths exceeding 3 m in 1983 (Provancha, unpublished 1998). Our surveys for the present study showed that in 1986–1987, *C. prolifera* was widely distributed and abundant but showed a "boom and bust" pattern. *C. prolifera* essentially crashed from the system in 1989, after which it was only rarely seen.

In the BRN area, diver observations beyond the terminus of one transect, BRN15, indicated that during many years *Halodule wrightii*, *R. maritima*, and *S. filiforme* grew in waters greater than 5 m. The plant growth at this depth is not robust and shoots occur sporadically, but this depth of growth is extremely uncommon in the remainder of the IRL. These water depths are anomalies in the IRL where basins were dredged in the 1960s during the development of the space center. The waters in this region (BANNOR) are also locally known for the relatively high water clarity (Provancha and Provancha 1988). We believe this clarity allows for the occasional expansion of seagrasses to these depths.

Temporal variation in the percent cover of *Halodule wrightii* indicates a significant decline. The highest coverages occurred in 1983 followed by a decline and then a long, relatively stable period, decreasing again in 1995. The causes for this decline are not entirely clear, but appear to relate, at least partially, to the remarkable rainfall inputs and shifts in *R. maritima* abundance. Temporal trends in the overall percent coverage for *R. maritima* showed several peaks (1986, 1988, and 1996). It was obvious in 1995 and 1996 that *Halodule wrightii* declined during the increase phase of *R. maritima*, and that these changes were concomitant with significant decreases in salinity associated with increased rainfall. These still may not fully explain the decline in *Halodule wrightii* because there was not a simple species shift (i.e., *Halodule wrightii* to *R. maritima*) as noted by our increases in bare ground at numerous sites. As for *S. filiforme*, the decline noted at several sites appears to be salinity related.

Halophila engelmannii was relatively ephemeral in the KSC area. There was a notable overlap in the specific transects that had both *Halophila engelmannii* and *S. filiforme* with occurrences greater than 5%. Those transects with *Halophila engelmannii* had mean depth ranges (47–168 cm) comparable to *S. filiforme*.

The dominant species, *Halodule wrightii*, has shown a decline over the study period. This decline in *Halodule wrightii* was evident in the Banana River, particularly in BANSOU. Mosquito Lagoon experienced changes in *Halodule wrightii* between years, including declines in 1995–1996, but not an overall decline. In 1994–1996, dramatic changes in species composition were documented not only in our study site, but throughout the IRL (Morris, personal communication 1997). The *R. maritima* "bloom" is considered cyclical and is currently at a peak based on our historical dataset (1983–1996). *C. prolifera* displayed a similar peak and subsequent decline in the late1980s.

Despite the fact that the seagrass coverage was not distributed uniformly and coverage varied over time, the frequency of encountering bare ground markedly increased in 1994–1996, causing concern for the condition of seagrass beds, particularly in BRS. We believe recent declines in water salinity are associated with declines in abundance of some seagrass species and increases in others. Since salinity is one of the most basic physical factors affecting the growth and distribution of seagrasses, we would anticipate that a continued depression of salinity would result in a continued reduction in the frequency and cover of SAV species that are less freshwater tolerant, particularly *S. filiforme*.

Despite the downward trends observed for *Halodule wrightii*, the seagrass beds on KSC are in excellent condition relative to other sections of the Indian River complex. This condition is likely due to low levels of human activity within the waters and minimal land-based run-off. The general lack of public access to the waters immediately around KSC results in a refuge for seagrass habitats and associated communities. Further, because these areas do not receive the heavy anthropogenic impacts so common in other portions of the lagoon, the seagrass community around KSC can be used as a benchmark. This benchmark will be useful to researchers and managers in comparing trends observed elsewhere in the lagoon and determining if these trends are site-specific or regional.

ACKNOWLEDGMENTS

Mike Willard assisted in the design of the original seagrass transect protocols. Numerous individuals assisted in collecting the annual transect data, but the following were a constant source of help over the first decade of this project: M. Provancha, C. Hall, R. Smith, and numerous NASA/SHARP students. We also benefited from the assistance of M. Barkazsi, M. Durette, B. Duncan, R. Lowers, J. Mailander, and D. Oddy. Many thanks to P. Schmalzer, M. Mota, and M. Corsello for assistance with data analyses. Rain data came from L. Maull and J. Drese. Many thanks to Dr. W. Knott, NASA, for his continued support and encouragement over the years. This project was part of the Ecological Monitoring Program funded by the Biomedical Office at NASA's Kennedy Space Center under Contracts NAS10-12180 and NAS10-11624. We are grateful to Dr. R. Virnstein and an anonymous reviewer for editorial improvements to the manuscript.

REFERENCES

Daubenmire, R. 1968. *Plant Communities, A Textbook of Plant Synecology.* Harper & Row, New York, 300 pp.

Dawes, C. D., D. Hanisak, and W. J. Kenworthy. 1995. Seagrass biodiversity in the Indian River Lagoon. *Bulletin of Marine Science* 57:59–66.

Down, C. 1978. Vegetation and other parameters in the Brevard County bar-built estuaries. Project Report 06-73, Brevard County Health Department, Environmental Engineering, Brevard County, FL.

Drese, J. H. 1997. A summary of ambient air at John F. Kennedy Space Center with a comparison to data from the Florida Statewide Monitoring Network (1983–1992). NASA Technical Memorandum 97-206236, National Aeronautics and Space Administration, Kennedy Space Center, FL.

Ehrhart, L. M. 1976. Marine turtle studies: a study of a diverse coastal ecosystem on the Atlantic coast of Florida. Final Report submitted to National Aeronautics and Space Administration, Kennedy Space Center, FL.

Gilmore, R. G., C. J. Donohoe, D. W. Cooke, and D. J. Herrema. 1981. Fishes of the Indian River Lagoon and adjacent waters, Florida. Technical Report 41:36, Harbor Branch Foundation, Inc., Fort Pierce, FL.

Irvine, B., M. Scott, and S. Shane. 1979. A study of the West Indian manatee in the Banana River and associated waters, Brevard County, Florida. NASA Contract CC-63426A, Final Report submitted to National Aeronautics and Space Administration, Kennedy Space Center, FL.

Kenworthy, W. J. and D. E. Haunert. 1991. *Results and Recommendations of a Workshop Convened to Examine the Capability of Water Quality Criteria, Standards and Monitoring Programs to Protect Seagrasses from Deteriorating Water Transparency.* NOAA Technical Memorandum, NMFS-SEFC-287.

Kenworthy, W. J., M. S. Fonseca, D. E. McIvor, and G. W. Thayer. 1988. The submarine light regime and ecological status of seagrasses in Hobe Sound, Florida. Annual Report, National Marine Fisheries Service, Beaufort Laboratory, Beaufort, NC.

Ledder, D. 1986. Food habits of the West Indian manatee in South Florida. Master's thesis, Department Biological Oceanography, University of Miami, Miami, FL.

Lefebvre, L. W., D. P. Domning, W. J. Kenworthy, D. E. McIvor, and M. E. Ludlow. 1988. Manatee grazing impacts on seagrasses in Hobe Sound and Jupiter Sound, January-February 1988. In *The Submarine Light Regime and Ecological Status of Seagrasses in Hobe Sound, Florida,* W. J. Kenworthy, M. S. Fonseca, D. E. McIvor, and G. W. Thayer (eds.). Annual Report, National Marine Fisheries Service, Beaufort Laboratory, Beaufort, NC, Appendix B.

Lipkin, Y. 1975. Food of the red sea dugong (Mammalia: Sirenia) from Sinai, Israel. *Journal of Zoology* 24:81–98.

Mailander, J. L. 1990. Climate of the Kennedy Space Center and vicinity. Technical Memorandum 103498, National Aeronautics and Space Administration, Kennedy Space Center, FL.

Morris, L. J. 1997. Personal communication. St. Johns River Water Management District, Palatka, FL.

Ogden, J. C. 1976. Some aspects of herbivore-plant relationships on Caribbean reefs and seagrass beds. *Aquatic Botany* 2:103–116.

Orth, R. J. and K. A. Moore. 1983. Chesapeake Bay: an unprecedented decline in submerged aquatic vegetation. *Science* 222:51–53.

Packard, J. M. 1984. Impact of manatees *Trichechus manatus* on seagrass communities in eastern Florida. *Acta Zoologica Fennica* 172:21–22.

Phillips, R. C. 1960. Observations on the ecology and distribution of the Florida seagrasses. Professional Paper Series 2, Florida State Board of Conservation Marine Laboratory, St. Petersburg, FL.

Provancha, J. A. 1998. Unpublished data. Dynamac Corporation, Kennedy Space Center, FL.

Provancha, J. A. and R. M. Willard. 1984. Seagrass bed distribution and composition at Kennedy Space Center, Brevard County, Florida. Abstract in: *Florida Academy of Science* 47:23.

Provancha, J. A. and M. J. Provancha. 1988. Long-term trends in abundance and distribution of manatees (*Trichechus manatus*) in the northern Banana River, Brevard County, Florida. *Marine Mammal Science* 4:323–338.

Provancha, J. A. and C. R. Hall. 1991. Observations of associations between seagrass beds and manatees in east central Florida. *Florida Scientist* 54:87–98.

Provancha, M. J., P. A. Schmalzer, and C. R. Hall. 1986. Effects of the December 1983 and January 1985 freezing air temperatures on select aquatic poikilotherms and plant species of Merritt Island, Florida. *Florida Scientist* 49:199–212.

Short, F. T. 1991. Light limitation on seagrass growth. In *Results and Recommendations of a Workshop Convened to Examine the Capability of Water Quality Criteria, Standards and Monitoring Programs to Protect Seagrasses from Deteriorating Water Transparency,* W. J. Kenworthy and D. E. Haunert (eds.). NOAA Technical Memorandum, NMFS-SEFC-287, pp. 37–40.

Snelson, F. F. 1976. A study of a diverse coastal ecosystem on the Atlantic coast of Florida. Ichthyological Studies, Final Report, Grant NGR 10-019-004, National Aeronautics and Space Administration, Kennedy Space Center, FL, Vols. 1 and 2.

Thompson, M. J. 1978. Species composition and distribution of seagrass beds in the Indian River Lagoon, Florida. *Florida Scientist* 41:90–96.

Virnstein, R. W. 1995. Seagrass landscape diversity in the Indian River Lagoon, Florida: the importance of geographic scale and pattern. *Bulletin of Marine Science* 57:67–74.

Virnstein, R. W. and L. J. Morris. 1996. Seagrass preservation and restoration diagnostics plan for the Indian River Lagoon. Technical Memorandum 14, St. Johns River Water Management District, Palatka, FL.

Zieman, J. C. 1982. The ecology of the seagrass communities of south Florida: a community profile. United States Fish and Wildlife Service, Biological Service Program, FWS/OBS-82/25.

Section III

Restoration

15 Reciprocal Transplanting of the Threatened Seagrass *Halophila johnsonii* (Johnson's Seagrass) in the Indian River Lagoon, Florida

W. Schlese Heidelbaugh, Lauren M. Hall, W. Judson Kenworthy, Paula Whitfield, Robert W. Virnstein, Lori J. Morris, and M. Dennis Hanisak

Abstract Successful relocation of *Halophila johnsonii* (Johnson's seagrass) may be essential for its long-term survival. *H. johnsonii* is one of the smallest and rarest seagrass species. It has a limited geographic range, and within its range, it is discontinuous and patchy. Male flowers of *H. johnsonii* have never been found; populations may be maintained exclusively by vegetative growth and reproduction. Based on these characteristics, *H. johnsonii* has been listed as threatened by the National Marine Fisheries Service.

Reciprocal transplants were performed in May 1998 to determine the feasibility of relocation and restoration of local populations in the Indian River Lagoon, Florida. Three different sites were selected which varied primarily in mean depth, light penetration, and degree of tidal exchange. Intact plugs of *H. johnsonii*, composed of leaf pairs, roots, rhizomes, and associated sediment, were harvested, put into peat pots, and transplanted within several hours of removal. Transplant and control plots were monitored biweekly for 30 weeks.

The peat pot method was initially successful; however, all three recipient sites had different problems which ultimately caused the loss of all transplant materials. Lagoon Shallow site increased by week 4 to a maximum of 71% cover, followed by an abrupt decline when plots were exposed due to low water levels in June/July. Lagoon Deep site initially increased to a maximum of 88% cover, followed by a gradual decline as light decreased to only 1.3% of surface incident light, well below seagrass light requirements. Sebastian Inlet site also showed an initial increase to a maximum of 95% cover, followed by a gradual decrease associated with sediment burial and/or erosion. Although coverage declined at the transplant sites, beds of *H. johnsonii* at all three donor sites similarly declined. The mortality in the transplant plots was not necessarily a result of the methodology, but more likely a reflection of the inherent dynamic nature of the species.

INTRODUCTION

Seven species of seagrass occur in the tropical Atlantic, Gulf of Mexico, and Caribbean Basin, forming a vital component of the living marine resources in many coastal systems. All seven species are found in the Indian River Lagoon (IRL) (Dawes et al. 1995; Kenworthy 1993). Of these, *Halophila johnsonii* (Johnson's seagrass) is known to occur only along 200 km of inland marine and estuarine waters between Sebastian Inlet (27° 51′ N lat.) and Virginia Key (25° 45′ N lat.),

Florida (Eiseman and McMillan 1980), making it one of the most limited geographical distributions of any seagrass in the world. Within this geographic range, the distribution of Johnson's seagrass is discontinuous, patchy, and temporally variable (Virnstein et al. 1997). Relative to the other six seagrasses, Johnson's seagrass comprises only 0.4% of the total bottom coverage (Virnstein et al. 1997). Despite its small physical stature, the genus *Halophila* is one of the most diverse and widely distributed of all seagrasses, with 12 species worldwide (Phillips and Menez 1988).

Due to its small rhizome structure, high rate of turnover and short life span, *H. johnsonii* has limited energy storage capacity and depends on rapid meristem division and new vegetative growth to maintain populations (Kenworthy 1993; Bolen 1997). Yet despite its rarity and diminutive size, *H. johnsonii* is found in a wide range of environments in both monotypic stands as well as in beds mixed with other seagrasses and macroalgae (Kenworthy 1993; Gallegos and Kenworthy 1996; Virnstein et al. 1997). It can be found growing in subtidal sandy sediments, in high energy tidal channels associated with inlets, on soft mud in deep water near the mouths of freshwater drainage canals, and in the intertidal zone (Virnstein et al. 1997). This local distribution is consistent with its relatively wider physiological tolerance to varying temperature, salinity, and light than, for example, its close congener, *H. decipiens* (Dawes et al. 1989).

Although female plants with mature flowers are routinely found in abundance near inlets, male flowers of *H. johnsonii* have never been found anywhere (Jewitt-Smith et al. 1997). Since no seeds or seedlings have ever been reported, it is assumed that ovaries produced by females remain unfertilized. Bolen (1997) examined female flowers by fluoroscopy and found no evidence of pollen, supporting the idea that fertilization may not occur. The current scientific consensus is that populations of this species are maintained exclusively by vegetative growth and asexual reproduction. Even though rates of growth and asexual reproduction indicate that meristem division may result in the formation of a new leaf pair or lateral branch every 3–4 days (Kenworthy 1993; Bolen 1997), the absence of sexual reproduction may restrict the ability of *H. johnsonii* to recover from disturbances causing the mortality of apical meristems. Unlike the congeneric, *H. decipiens*, which produces 10^3 seeds m^{-2} and re-establishes populations from seedlings, *H. johnsonii* is only capable of recruiting and dispersing vegetatively (Kenworthy 1993).

Interest in developing viable restoration methods and a better understanding of the population dynamics of *H. johnsonii* was heightened recently when the U.S. National Marine Fisheries Service listed it as a threatened species (Federal Register 1998). The agency based the listing of this species on its limited geographic range, disjunct local distribution, and small size. Of particular concern was the inability of *H. johnsonii* to produce viable recruits from seed. This limits its ability to recover from disturbance and makes it vulnerable to extinction. Experiments have shown that some seagrasses can be transplanted using "bare root" techniques which approximate natural vegetative dispersal (Fonseca et al. 1998). Alternatively, less disruptive "core" and "plug" methods, which retain the sediments and rhizosphere intact, have been used successfully with smaller plants. During spring 1998, pilot studies at Harbor Branch Oceanographic Institution revealed for the first time that *H. johnsonii* could be successfully transplanted from the field into seawater mesocosms using both bare root and peat pot methods. These preliminary mesocosm experiments suggested that restoration of damaged *H. johnsonii* populations by transplanting is possible and that field testing was needed to verify the feasibility.

The specific objectives of this study were to (1) determine the feasibility of transplanting *H. johnsonii* with a peat pot method; (2) assess whether the site location and source of the donor plants affect the success of the transplant; and (3) determine if the depth of transplanting affects the success, and if so, to what degree.

MATERIALS AND METHODS

SELECTION OF SITES

The study sites were located on the east coast of Florida in the southern half of the Indian River Lagoon (IRL) (Figure 15.1). Small isolated patches of *H. johnsonii* grow on the flood tidal delta at

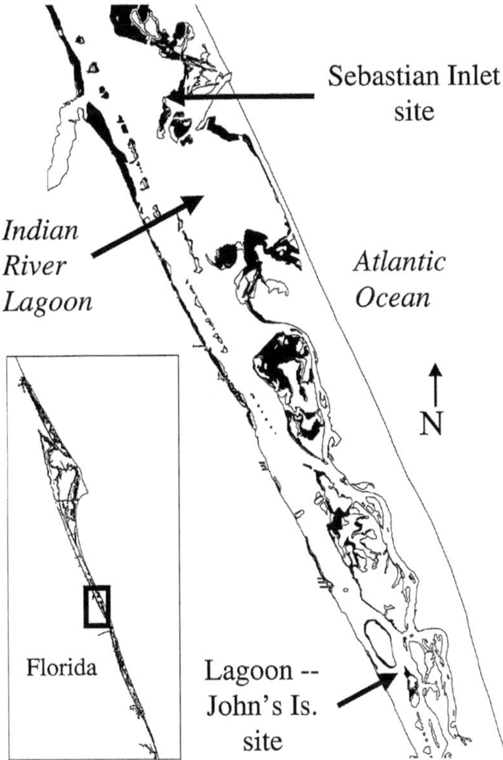

FIGURE 15.1 Location of study sites on the Indian River Lagoon, Florida. The Sebastian Inlet (SI) site was located on the northern side of the flood tidal delta. The Lagoon Deep (LD) and Lagoon Shallow (LS) sites were located near John's Island, on the western shore of a dredge spoil island adjacent to the Intracoastal Waterway.

Sebastian Inlet (27° 51′ 23.7″ N/80° 27′ 90.3″ W) near the northern limit of its geographic range (Kenworthy 1993). This site was chosen as one of the transplant sites because it is subjected to short-term salinity fluctuations due to semidiurnal tidal exchange with the open ocean, thus having high light. A shallow site with a mean water depth of 0.42 m (range 0.25 to 0.54 m) was selected on the northern side of the entrance channel into the lagoon. The second site was in the vicinity of John's Island (27° 42′ 46.9″ N/80° 23′ 78.9″ W), located 17 km south of Sebastian Inlet on the western shore of a dredge spoil island adjacent to the Intracoastal Waterway. This area receives little saltwater exchange due to its distance from Sebastian and Fort Pierce Inlets and represents conditions typical of much of the interior of the IRL. The second site was divided into two strata: (1) a shallow site with a mean water depth of 0.16 m (range 0.05 to 0.35 m), and (2) a deep site, 50 m offshore, with a mean water depth of 1.08 m (range 0.80 to 1.28 m). *H. johnsonii* was observed growing in small isolated patches at the shallow site and in a wide range of patch sizes in the deep site.

Experimental Design and Transplanting Method

Reciprocal transplanting was performed between the three sites. The three recipient sites were labeled as Sebastian Inlet (SI), Lagoon Deep (LD), and Lagoon Shallow (LS). *H. johnsonii* donor plants were taken from natural beds located in close proximity to the transplant recipient sites. The donor plants were collected on May 8, 1998, using a lawn plugger measuring 8 cm × 8 cm × 8 cm (Fonseca 1994). The plugger was used to harvest intact plugs containing *H. johnsonii* leaf pairs, roots, rhizomes, and sediment, with minimal disturbance. These plugs were extruded into peat pots measuring 8 cm on a side. They were transported in plastic containers filled with water and covered

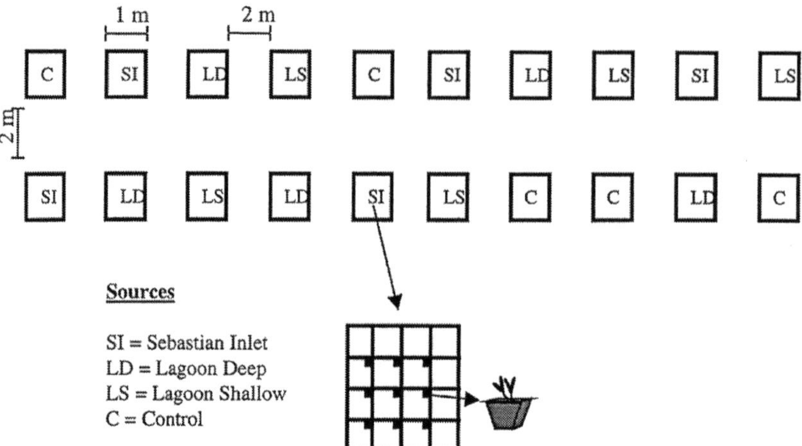

FIGURE 15.2 Planting design of 20 1-m^2 plots at each of the three recipient sites, consisting of five replicates from each source (SI, LD, and LS) and five replicate control plots, with no peat pots planted. Placement of each source was randomly selected.

with wet burlap. The peat pots were transplanted into the recipient sites within several hours of collection. Ten replicate plugs were collected at each donor site, washed free of sediment, and returned to the lab where apical and branch meristems, leaf pairs, and flowers were counted.

Fifteen 1-m^2 plots were planted at each of the recipient sites, five replicates from each source (SI, LD, and LS) and five control plots (no plant material) (Figure 15.2). Two rows of 10 1-m^2 plots were planted with 2 m separating adjacent plots. The placement of each source into the experimental plots was randomly selected. Nine peat pots were planted in a 1-m^2 PVC quadrat frame, divided into 16 squares (Figure 15.2). Holes, approximately 8 cm on a side, were dug and peat pots were placed flush into the sediment. The top edges of the peat pots were broken away to allow plants to spread by vegetative growth.

Transplant Monitoring

One week after transplanting, the plots were visually examined for *H. johnsonii* survival and evidence of bioturbation by rays, crabs, or other burial activity. Non-destructive monitoring of percent cover and number of leaf pairs occurred every 2 weeks between May 19 and September 23, 1998 and monthly thereafter until November 19. Final inspection of the transplant sites was made on December 2 (week 30). Percent cover was determined using the 1-m^2 quadrat frame divided into 16 cells. The quadrat was examined for presence of any *H. johnsonii* blades. The percent cover was calculated based on the number of cells containing blades. The number of leaf pairs was determined by placing a small metal frame, divided into 25 subcells, into three randomly chosen cells from the large quadrat. The total number of leaves present in one randomly chosen subcell was counted and leaf pairs m^{-2} were calculated.

At week 24 (October 21), three replicate plugs of each of the three surviving donor materials were destructively harvested from the Sebastian Inlet transplant site and 10 replicate plugs from the Sebastian Inlet donor site. At week 30 (December 2), all three donor sites were inspected for survival. Sediment accumulation was also recorded at the Sebastian Inlet site at week 30 by measuring the distance between the buried peat pots and sediment surface.

Environmental Monitoring

Water depth, salinity, and water temperature were monitored at all three recipient sites biweekly using a Hydrolab Scout 2. Profiles of photosynthetically active radiation (PAR) were collected biweekly at the Lagoon Deep site using a LI-COR datalogger and 4π (spherical) quantum sensors. Water depths at the Lagoon Shallow and Sebastian Inlet sites were less than 0.5 m; therefore, PAR was collected at the Lagoon Deep site only. PAR was simultaneously recorded at 20 and 50 cm below the water surface and 30 cm from the bottom. This procedure was repeated three times at 1-min intervals. The light attenuation coefficient (Kd) was estimated from the slope of the regression of light vs. depth. The percentage of surface PAR reaching the bottom (Kenworthy and Fonseca 1996) was calculated with the equation:

$$I_z = I_0 * e^{-(kz)}$$

where I_z is the percent surface PAR at depth z (m), I_0 is PAR at surface, z is depth (m), and k is light attenuation coefficient or Kd.

RESULTS

Environmental Conditions

Salinity varied throughout the study, ranging from 16.8 to 27.0 ppt at the Lagoon Shallow and Lagoon Deep recipient sites. Temperatures at these two sites were 27.4°C in May and a high value of 34.5°C in August. At the Sebastian Inlet recipient site, salinity fluctuated between 23.7 and 35.6 ppt, while the temperature ranged from 25.2°C to 28.7°C. Light attenuation (Kd) at the Lagoon Deep site was lowest on week 6 (Kd = 0.33) and highest on week 24 (Kd = 3.41) (Figure 15.3). For the first 16 weeks, PAR exceeded 10% surface irradiance, but there was a steady decline from a peak surface irradiance of 65% on week 6 until week 18, where it declined to ≤ 5% for the next three sampling periods.

Transplants

Initial Transplant Monitoring

Population characteristics of the donor plants are shown in Table 15.1. The mean number of leaf pairs m^{-2} ranged from approximately 11,000 to 13,000 at the time of planting, and the total number

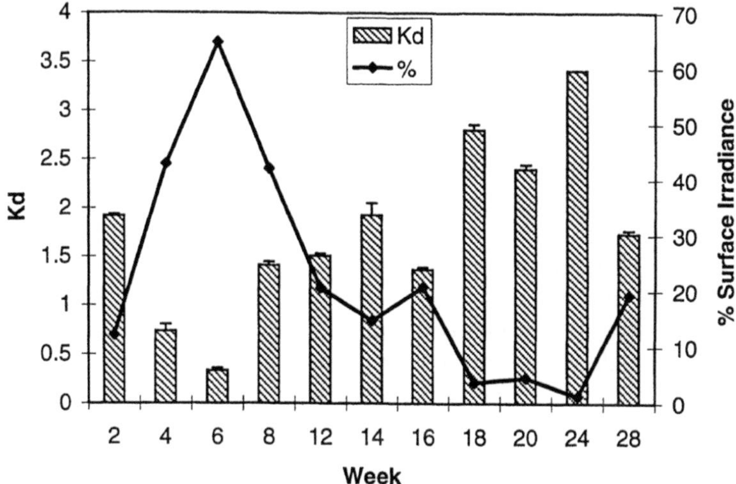

FIGURE 15.3 Biweekly measurements of light attenuation (Kd) and % surface irradiance at the Lagoon Deep recipient site.

TABLE 15.1
Population characteristics of *Halophila johnsonii* collected on May 8, 1998 from Lagoon Deep, Lagoon Shallow, and Sebastian Inlet donor sites. Data are presented as mean ± standard error (n = 10).

	Donor Material		
Plant Characteristics	Lagoon Deep	Lagoon Shallow	Sebastian Inlet
Leaf pairs per pot	85.9 (± 5.7)	80.1 (± 12.9)	72.2 (± 5.8)
Leaf pairs m^{-2}	13,211 (± 879)	12,319 (± 1977)	11,104 (± 892)
Apicals per pot	54.0 (± 3.5)	61.2 (± 9.1)	48.8 (± 5.8)
Apicals m^{-2} (apicals + branch apicals)	8,305 (± 545)	9,413 (± 1402)	7,505 (± 884)
Flowers per pot	0.1 (± 0.1)	1.8 (± 0.6)	0.2 (± 0.2)
Flowers m^{-2}	15.4 (± 15.2)	276.8 (± 96.7)	30.8 (± 30.8)

of apical meristems ranged from approximately 7500 to 9400. At the first monitoring, week 2, all planting plots had 9 of 16 cells (56%) occupied, indicating all planting units survived (Figures 15.4, 15.5, and 15.6).

Lagoon Shallow Recipient Transplant Site

By week 4, percent cover of the Lagoon Deep and Lagoon Shallow source material had increased to approximately 70%, while the Sebastian Inlet donor material decreased to 37% (Figure 15.4A). Initially, all plots were placed at least 1 m away from any naturally occurring seagrass beds, yet *H. johnsonii,* from a patch nearby, was observed within one of the control plots on week 4. A decrease in cover was observed in all plots on week 6, and by week 8 and all subsequent visits, no *H. johnsonii* transplants were found. All natural *H. johnsonii* beds in this area also died. A final visit to this site on week 30 (December 2) confirmed that none of the *H. johnsonii* transplants survived. However, a few small patches (< 0.05 m^2) of naturally occurring *H. johnsonii,* consisting of approximately 5 to 10 leaf pairs each, were found in the surrounding area.

The mean number of leaf pairs m^{-2} (± standard error) on week 4 was 1200 (± 673.9), 960 (± 436.1), and 53 (± 36.3) for donor materials from Lagoon Deep, Lagoon Shallow, and Sebastian Inlet, respectively (Figure 15.4B). On week 6, the means had dropped to 400 (± 237.4), 347 (± 206.1) and 0 leaf pairs m^{-2} for donor materials from Lagoon Deep, Lagoon Shallow, and Sebastian Inlet, respectively.

Lagoon Deep Recipient Transplant Site

Initially all plots showed an increase in percent cover with peaks on weeks 6 and 8 (Figure 15.5A). The Lagoon Deep and Lagoon Shallow donor materials appeared to exhibit better growth than the Sebastian Inlet donor material throughout the first half of the study. By week 20, percent cover declined to about 5–12% for all donor materials. On week 24, all *H. johnsonii* material (transplant as well as naturally occurring patches) had disappeared. Again, natural *H. johnsonii,* adjacent to the treatment, had invaded one of the control plots. The control increased slightly with a peak at week 10, followed by a decrease, corresponding with trends seen in the transplant material.

Mean number of leaf pairs m^{-2} for Lagoon Deep donor material peaked at week 8 at 933 (± 109) and fluctuated until week 18, at which time no plant material was recorded (Figure 15.5B). Sebastian Inlet donor material also peaked on week 8 at 826.7 leaf pairs m^{-2} (± 189.1), but persisted at lower numbers until week 20. The mean number of leaf pairs m^{-2} for Lagoon Shallow donor material peaked at week 6 at 453.3 (± 140), and fluctuated throughout the study until week 24, when no

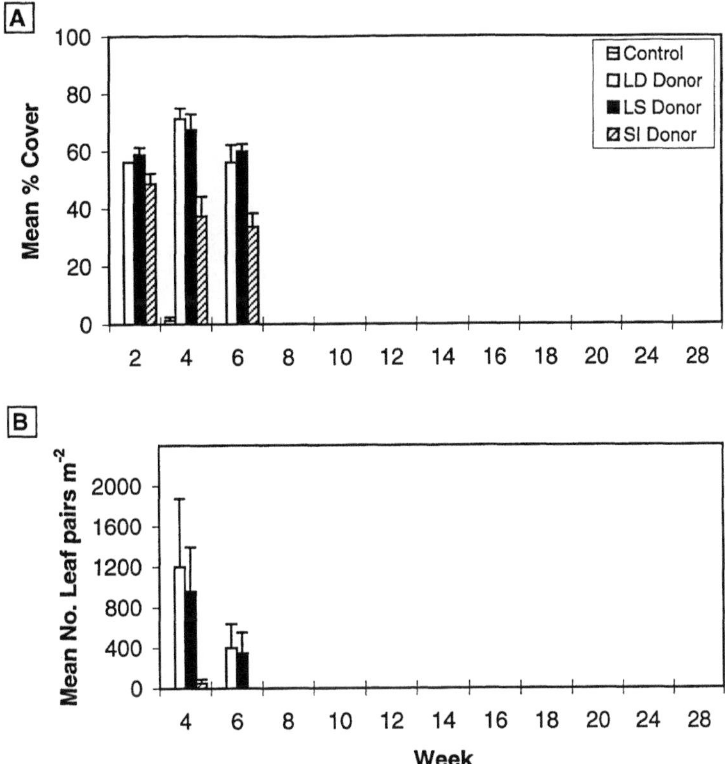

FIGURE 15.4 Biweekly trends within Lagoon Shallow recipient plots: (A) Mean % cover (± SE), (B) Mean number of leaf pairs m^{-2} (± SE). Donor materials consist of Lagoon Deep (LD), Lagoon Shallow (LS), and Sebastian Inlet (SI) *Halophila johnsonii* plugs. Co-occurring trends in natural (control) *H. johnsonii* coverage are also depicted.

plant material was found. Additionally, no *H. johnsonii* was found upon inspection of the Lagoon Deep donor site on week 24.

Sebastian Inlet Recipient Site

Lagoon Deep donor material peaked at approximately 95% cover on week 8, then declined slowly through week 14, and more dramatically by week 16 (Figure 15.6A). By week 20, Lagoon Deep and Sebastian Inlet donor materials had 20% cover, with Lagoon Shallow at 5% cover. At some plots, transplanted material, primarily the Lagoon Deep source, spread into adjacent bare sand outside the 1-m^2 grids. The spread was pronounced and was still evident at week 24, although thinning had occurred. At week 24, Lagoon Deep, Lagoon Shallow, and Sebastian Inlet donor materials decreased to 6, 0, and 5% cover, respectively. No transplanted material was observed on week 28. Mean number of leaf pairs m^{-2} peaked weeks 10 and 12 and fluctuated through week 20 (Figure 15.6B). On week 30 the peat pots were buried under a mean of 12.5 cm (± 0.3) of sediment and no transplant material was surviving.

Despite 100% mortality of the transplants, natural patches of *H. johnsonii* and the donor patches in the area were surviving and flowering. The population characteristics of the Sebastian Inlet donor site are shown in Table 15.2. Leaf pairs m^{-2} dropped from the May collection to 5000 (± 561) and apicals m^{-2} decreased to 4640 (± 711).

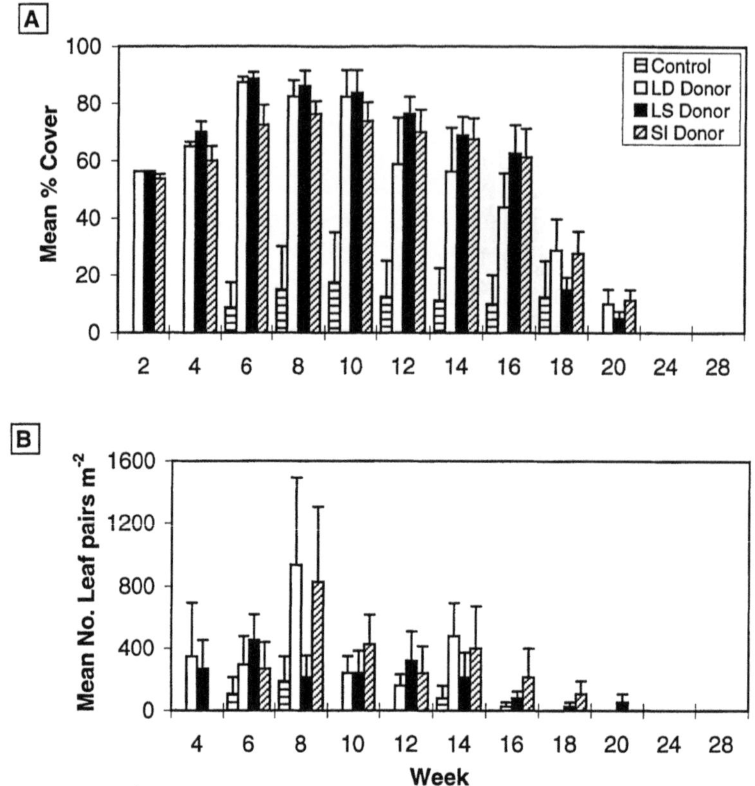

FIGURE 15.5 Biweekly trends within Lagoon Deep recipient plots: (A) Mean % cover (± SE); (B) Mean number of leaf pairs m^{-2} (± SE). Donor materials consist of Lagoon Deep (LD), Lagoon Shallow (LS), and Sebastian Inlet (SI) *Halophila johnsonii* plugs. Co-occurring trends in natural (control) *H. johnsonii* coverage are also depicted.

Sebastian Inlet — Examples of Survival and Growth

Three representative plots from the Sebastian Inlet recipient site, which showed maximum growth from each of the three donor materials, were chosen to illustrate potential growth in percent coverage and number of leaf pairs m^{-2} (Figure 15.7). Lagoon Shallow donor material (replicate #5) increased gradually to 87.5% cover on week 10 and persisted at 75 to 87.5% through week 20. At week 24 no donor plant material was found (Figure 15.7A). Mean number of leaf pairs m^{-2} ranged from 400–800 m^{-2}, peaking at weeks 6 through 8 (Figure 15.7A).

The Lagoon Deep donor material (replicate #1) increased dramatically to 100% cover by week 4 and remained near 100% through week 20 (Figure 15.7B). A peak in mean number of leaf pairs m^{-2} (6000 m^{-2}) was observed by week 12, fluctuating through week 20 (≈1500–3200 m^{-2}; Figure 15.7B). By week 24, the transplant material was still evident although no leaf pairs were counted due to random selection of subcells. A large decrease in cover to 31% was observed on week 24 and all transplant material disappeared by week 28.

A gradual increase was observed for the Sebastian Inlet donor material (replicate #2) with a peak at 100% cover on week 10, which persisted through week 20; a decrease to 25% cover was recorded on week 24 (Figure 15.7C). The mean number of leaf pairs m^{-2} were maintained at high levels (≈1000–3700 m^{-2}) through week 20. No leaf pairs were counted on week 24, and no plant material from the Sebastian Inlet donor source remained by week 28 (Figure 15.7C).

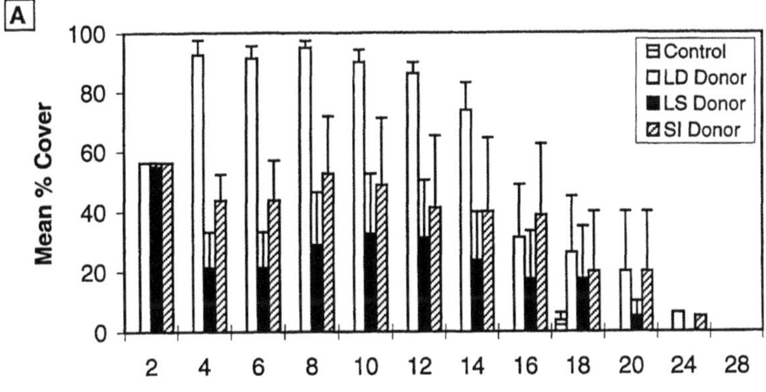

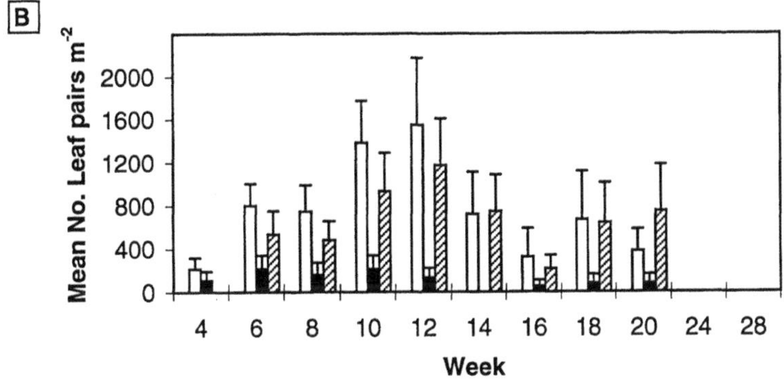

FIGURE 15.6 Biweekly trends within Sebastian Inlet recipient plots: (A) Mean % cover (± SE), (B) Mean number of leaf pairs m^{-2} (± SE). Donor materials consist of Lagoon Deep (LD), Lagoon Shallow (LS), and Sebastian Inlet (SI) *Halophila johnsonii* plugs. Co-occurring trends in natural (control) *H. johnsonii* coverage are also depicted.

TABLE 15.2
Population characteristics of the Sebastian Inlet donor site collected on October 21, 1998. Data are presented as mean ± standard error (n = 10).

Plant Characteristics	Sebastian Inlet Donor Site
Leaf pairs per pot	32.5 (± 3.7)
Leaf pairs m^{-2}	5,000 (± 561)
Apicals per pot	30.2 (± 4.6)
Apicals m^{-2} (apicals + branch apicals)	4,640 (± 711)

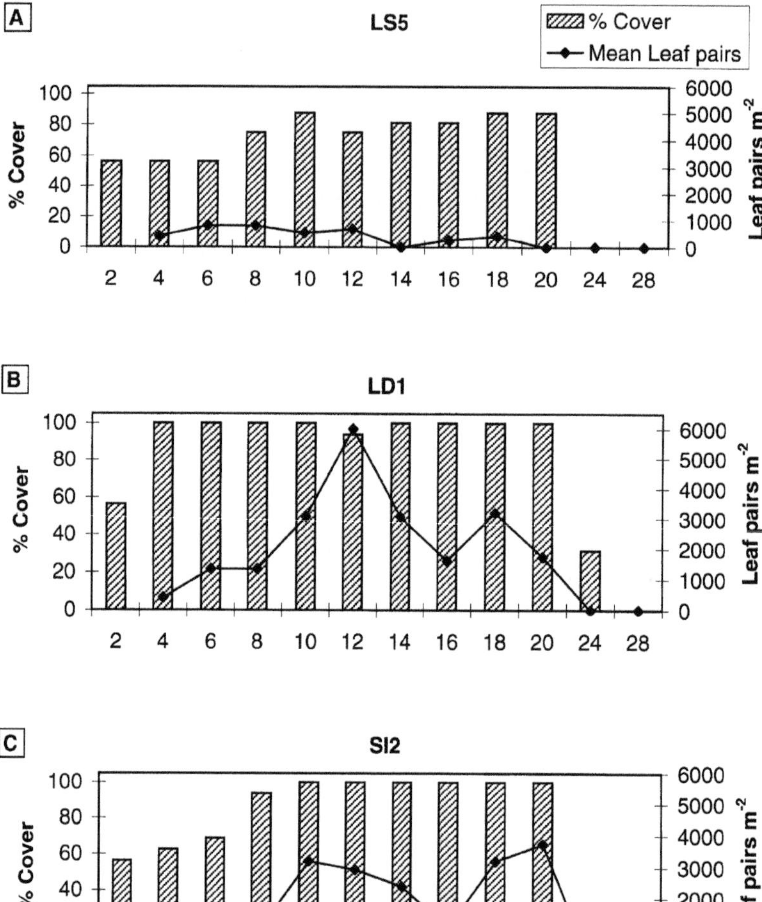

FIGURE 15.7 Mean percent cover and mean number of leaf pairs m^{-2} for three Sebastian Inlet recipient plots exhibiting optimum survival and growth. Donor materials consist of: (A) Lagoon Shallow replicate #5 (LS5); (B) Lagoon Deep replicate #1 (LD1); and (C) Sebastian Inlet replicate #2 (SI2).

From the original nine plantings, the Lagoon Deep (replicate #1) and Sebastian Inlet (replicate #2) donor materials expanded at least 1 m in all directions outside the 1-m^2 plot on weeks 8 through 20. These two plots were adjacent to one another and formed a 3 by 5-m patch of transplanted *H. johnsonii*.

Analyses of plant material collected in October from the Sebastian Inlet transplant site are shown in Table 15.3. Flowers were observed in Sebastian Inlet transplant material collected in October from the Lagoon Deep, Lagoon Shallow, and Sebastian Inlet donor plants.

DISCUSSION

Several transplanting techniques, each with its own advantages and disadvantages, have been used with seagrasses with varying degrees of success (Davis and Short 1997; Fonseca et al. 1998). Some of the most commonly used techniques include the staple method (Derrenbacker and Lewis 1982;

TABLE 15.3
Population characteristics of survivors at the Sebastian Inlet recipient site collected October 21, 1998. Data are presented as mean ± standard error (n = 3).

Plant Characteristics	Donor Material: Lagoon Deep	Lagoon Shallow	Sebastian Inlet
Leaf pairs per pot	54.3 (± 0.9)	13.9 (± 1.8)	69.1 (± 11.3)
Leaf pairs m^{-2}	8,358 (± 137)	2,135 (± 275)	10,625 (± 1739)
Apicals per pot	69.4 (± 2.8)	14.6 (± 3.2)	68.1 (± 4.4)
Apicals m^{-2} (apicals + branch apicals)	10,677 (± 426)	2,239 (± 496)	10,468 (± 681)
Flowers per pot	0.05 (± .01)	0.002 (± 0.0007)	0.05 (± 0.008)
Flowers m^{-2}	9.0 (± 2.0)	0.3 (± 0.1)	8.0 (± 1.3)

Fonseca et al. 1985; Davis and Short 1997), peat pots (Fonseca et al. 1990; Phillips 1990; Harrison 1990), and seeds (Harrison 1991; Orth et al. 1994). Results of these studies clearly show that selection of the appropriate technique depends on a combination of factors including, but not restricted to, the species, transplanting site, and logistics. Due to their small size, *Halophila* spp. are most suitable for the peat pot method (Fonseca 1994). The peat pot plugs are least disruptive to the sediment and plant and decrease the possibility of *H. johnsonii* being eroded by wave action and bioturbation. Because *H. johnsonii* is shallow rooted, this method increases the stability of the newly planted plot. It allows the roots and rhizomes to remain intact while minimally disturbing the sediment nutrient sources. Initial survival and growth responses in this study suggest that the peat pot method worked well for *H. johnsonii.*

Although early survival at all sites was high, later survival varied widely among locations and for different reasons. The Lagoon Shallow site had the lowest survival with 100% mortality by early July (week 8). In June and July, water levels were exceptionally low, dropping from 0.57 to –0.35 ft NGVD, 1929 (Indian River Mosquito Control District 1998). Transplanted and natural *H. johnsonii* materials at the Lagoon Shallow site were subjected to extreme desiccation and high temperatures. On weeks 6 and 12 several plots at this site were completely exposed. Natural patches at this location also disappeared during this time period. Based on photosynthesis and respiration measurements, *H. johnsonii* can tolerate wider temperature fluctuations (10–30°C) than *H. decipiens* as reported by Dawes et al. (1989). It seems reasonable to assume that *H. johnsonii* would not survive prolonged exposure during the hot summer months. On December 2, 1998 seven small patches of *H. johnsonii* were identified in the vicinity of the Lagoon Shallow transplant plots. This suggests that either some very small survivors were missed during the exposure period or new vegetative recruits were established after water levels returned. Water levels in December were similar to levels at the time of transplanting in May, but none of the transplants were recovering.

During May and June, central Florida experienced a severe drought. Rainfall amounts deviated approximately –24 cm from normal during this period (NOAA 30-yr normal, 1961–1990) (SJR-WMD Division of Hydrologic Data Services 1998). The lack of rainfall during May and June may help explain the initial success of the transplant material at the Lagoon Deep recipient site. Light attenuation (Kd) was low during these months, and between 42% and 65% of the surface light was able to penetrate to the bottom (Figure 15.3). The peak *H. johnsonii* percent cover corresponded to peak periods of light penetration. Decreases in growth in late July corresponded to the return of normal summer rainfall patterns and increased light attenuation. Water clarity decreased through week 24 to a low of 1.3% of incident light, well below the minimum requirements for many seagrasses (Kenworthy and Fonseca 1996). This sharp decrease in light may partially explain the loss of all Lagoon Deep transplant material. Additionally, large mats of drift algae (*Gracilaria* spp.) were found covering all of the transplant plots at this site, contributing to a further decrease in the amount of light reaching the plants. Naturally occurring patches of *H. johnsonii* growing in the control plot also disappeared after week 18, suggesting that diminished light levels were affecting all of the populations. No surviving *H. johnsonii* beds were found during a survey of the Lagoon Deep donor site on weeks 24 and 30.

The Sebastian Inlet recipient site is subjected to semi-diurnal tidal fluctuations which maintain good water quality on the flood delta. Rapid fluctuations in water clarity, temperature, and salinity can occur; however, these conditions do not last for extended periods. The eventual loss of all plant material at this site may be explained by two scenarios: (1) peat pots and plant material were eroded due to a severe storm event, and/or (2) peat pots and plant material were buried due to sediment movement. The greatest loss of donor material from this site was seen following Hurricane Bonnie on August 24. Although the hurricane did not hit Sebastian Inlet directly, it produced high winds, waves, and tidal currents from the southeast, which may have eroded a portion of the remaining transplant material. On August 27 (week 16) exposed rhizomes were observed in several plots, possibly due to sediment erosion. The transplant material remaining from all three donor types continued to decline throughout the remainder of the study. Inspection on week 30 found peat pots buried to a mean of 12.5 cm. Burial of peat pots may have occurred due to large volumes of sediment being transported and deposited by strong currents in the Inlet.

Sebastian Inlet donor sites were examined on October 21 and December 2 to determine survival of the original *H. johnsonii* patches. The patches appeared smaller than in May and had sections with exposed rhizome mats, similar to results seen in the transplant plots on week 16. Although the donor patch sizes decreased and leaf pair density decreased from an initial 11,104 m^{-2} to 5,000 m^{-2}, plants persisted, appeared healthy, and were flowering. Flowers recorded in both the May and October sampling events suggest that *H. johnsonii* may flower continuously. Flowering was also observed in all three donor materials collected from the Sebastian Inlet transplant site on October 21. Flowering has been observed in shallow, high light areas, but has rarely been reported in deep populations of *H. johnsonii* growing in the interior of the lagoon.

Other seagrasses which have been used in transplant experiments, for example, *Syringodium filiforme* (171 apicals m^{-2}; Kenworthy and Schwarzschild 1998) and *Thalassia testudinum* (23 apicals m^{-2}; Kenworthy and Schwarzschild 1996) have 1–2 orders of magnitude fewer apical meristems than *H. johnsonii* (7505 to 9412 apicals m^{-2}). This relatively higher number of apical meristems suggests that this species relies heavily on vegetative reproduction for growth and population maintenance. Due to its rapid vegetative reproduction, large numbers of apical meristems m^{-2}, and high potential for spreading, *H. johnsonii* may be a suitable candidate for restoration by transplanting.

Results from this transplant experiment indicate that the peat pot method was initially successful for *H. johnsonii*; however, all three recipient sites had different problems which ultimately caused extinction of all transplant materials. Lagoon Shallow showed an initial increase by week 4, followed by an abrupt decline when water levels receded and exposed the plots. Lagoon Deep initially increased, followed by a gradual decline as light decreased to only 1.3% of incident surface levels, well below normal seagrass light requirements. Sebastian Inlet also showed an initial increase, followed by a decrease associated with erosion and burial of the plants. Although declines in coverage were recorded at the transplant sites, similar declines were also noted at donor beds of *H. johnsonii* at the Lagoon Deep and Lagoon Shallow sites. Thus the mortality observed was not necessarily a result of the transplanting methodology, but more likely a reflection of the inherent dynamic nature of this species in response to environmental changes.

Because no male flowers of *H. johnsonii* have ever been found, vegetative reproduction and dispersal by vegetative fragments are the only known mechanisms for colonization and expansion of this species. In addition to its small size, rarity, and limited geographical distribution, this lack of sexual reproduction illustrates the need for developing a successful transplanting methodology to assist in the conservation and protection of the species. Results from this study indicate that it is feasible to transplant this species, aiding in the relocation and restoration of local *H. johnsonii* populations.

ACKNOWLEDGMENTS

This project was jointly funded by the South Florida Ecosystems Office of the U.S. Fish and Wildlife Service, the St. Johns River Water Management District, and the Recovered Protected Species Program National Marine Fisheries Service. Our thanks to Harbor Branch Oceanographic Institution and the anonymous reviewers for their comments and suggestions and to the many volunteers who assisted us in the field.

REFERENCES

Bolen, L. E. 1997. Growth dynamics of the seagrass *Halophila johnsonii* from a subtropical estuarine lagoon in southeastern Florida, USA. Master's thesis, Florida Atlantic University, Boca Raton, FL.

Davis, R. C. and F. T. Short. 1997. Restoring eelgrass, *Zostera marina* L., habitat using a new transplanting technique: the horizontal rhizome method. *Aquatic Botany* 59:1–15.

Dawes, C. J., C. S. Lobban, and D. A. Tomasko. 1989. A comparison of the physiological ecology of the seagrasses *Halophila decipiens* Ostenfeld and *Halophila johnsonii* Eiseman from Florida. *Aquatic Botany* 33:149–154.

Dawes, C. J., D. Hanisak, and W. J. Kenworthy. 1995. Seagrass biodiversity in the Indian River Lagoon. *Bulletin of Marine Science* 57(1):59–66.

Derrenbacker, J. A. and R. R. Lewis. 1982. Seagrass habitat restoration in Lake Surprise, Florida Keys. In *Proceedings, 9th Annual Conference on Wetlands Restoration and Creation,* R. H. Stoval (ed.). Hillsborough Community College, Tampa, FL, pp. 132–154.

Eiseman, N. J. and C. McMillan. 1980. A new species of seagrass, *Halophila johnsonii*, from the Atlantic coast of Florida. *Aquatic Botany* 9:15–19.

Federal Register. 1998. Vol. 63, No. 177, pp. 49035–49041.

Fonseca, M. S. 1994. A guide to planting seagrasses in the Gulf of Mexico. Sea Grant College Program TAMU-SG-94-601, Texas A & M University, College Station, TX.

Fonseca, M. S., W. J. Kenworthy, G. W. Thayer, D. Y. Heller, and K. M. Cheap. 1985. Transplanting of the seagrasses *Zostera marina* and *Halodule wrightii* for sediment stabilization and habitat development of the east coast of the United States. U.S. Army Corps of Engineers Technical Report EL-85-9.

Fonseca, M. S., W. J. Kenworthy, D. R. Colby, K. A. Rittmaster, and G. W. Thayer. 1990. Comparisons of fauna among natural and transplanted eelgrass *Zostera marina* meadows: criteria for mitigation. *Marine Ecology Progress Series* 65:251–264.

Fonseca, M. S., W. J. Kenworthy, and G. W. Thayer. 1998. Guidelines for the conservation and restoration of seagrasses in the United States and adjacent waters. NOAA Coastal Ocean Program Decision Analysis Series No. 12, NOAA Coastal Ocean Office, Silver Spring, MD.

Gallegos, C. L. and W. J. Kenworthy. 1996. Seagrass depth limits in the Indian River Lagoon (Florida, USA): application of an optical water quality model. *Estuarine Coastal Shelf Science* 42:267–288.

Harrison, P. G. 1990. Variations in success of eelgrass transplants over a five-year period. *Environmental Conservation* 17(2):157–163.

Harrison, P. G. 1991. Mechanisms of seed dormancy in an annual population of *Zostera marina* (eelgrass) from Netherlands. *Canadian Journal of Botany* 69:1972–1976.

Indian River Mosquito Control District. 1998. Tide charts from Grand Harbor, Vero Beach, FL.

Jewitt-Smith, J., C. McMillan, W. J. Kenworthy, and K. Bird. 1997. Flowering and genetic banding patterns of *Halophila johnsonii* and conspecifics. *Aquatic Botany* 59:323–331.

Kenworthy, W. J. 1993. The distribution, abundance and ecology of *Halophila johnsonii* Eiseman in the lower Indian River, Florida. Final report submitted to the Office of Protected Resources, NMFS, Silver Spring, MD.

Kenworthy, W. J. and M. S. Fonseca. 1996. The light requirements of seagrasses *Halodule wrightii* and *Syringodium filiforme* derived from the relationship between diffuse light attenuation and maximum depth distribution. *Estuaries* 19:740–750.

Kenworthy W. J. and A. C. Schwarzschild. 1996. Scientific documentation supporting the development of population dynamics models for predicting the recovery of *Syringodium filiforme* and *Thalassia testudinum*. Final Report submitted to NOAA Sanctuaries and Reserves Division, Damage Assessment Center and NOAA Restoration Center, Silver Spring, MD.

Kenworthy W. J. and A. C. Schwarzschild. 1998. Vertical growth and short-shoot demography of *Syringodium filiforme* in outer Florida Bay, USA. *Marine Ecology Progress Series* 173:25–37.

Orth, R. J., M. Luckenbach, and K. A. Moore. 1994. Seed dispersal in a marine macrophyte: implications for colonizaton and restoration. *Ecology* 75:1927–1939.

Phillips, R. C. 1990. Transplant methods. In *Seagrass Research Methods,* R. C. Phillips and C. P. McRoy (eds.). UNESCO, Paris, pp. 51–54.

Phillips, R. C. and E. G. Menez. 1988. *Seagrasses.* Smithsonian Contributions to the Marine Sciences, No. 34, Smithsonian Institution Press, Washington D.C.

St. Johns River Water Management District, Division of Hydrologic Data Services. May/June 1998. Hydrologic conditions report: summary of rainfall, surface water and ground water levels. Palatka, FL.

Virnstein, R. W., L. J. Morris, J. D. Miller, and R. Miller-Myers. 1997. Distribution and abundance of *Halophila johnsonii* in the Indian River Lagoon. Technical Memorandum No. 24, St. Johns River Water Management District, Palatka, FL, 14 pp.

16 Setting Seagrass Targets for the Indian River Lagoon, Florida

Robert W. Virnstein and Lori J. Morris

Abstract Seagrass protection and restoration targets are established because seagrass is used as a barometer of the Indian River Lagoon's health. Criteria for targets are that they must be scientifically defensible; ecologically achievable; objective, i.e., based on actual data, such as historical distribution; simple; and bounded by natural variability. Three levels of quantitative seagrass coverage targets are set: (1) potential = to a depth of 1.7 m, based on "healthy" areas of the lagoon; (2) historical = maximum extent of where seagrass has been mapped, largely based on 1943; and (3) critical minimum = maximum extent of where seagrass has been mapped in at least half of non-1943 mappings. These quantitative targets will be used to set water quality and pollutant loading targets. Qualitative targets include seagrass and animal density, diversity, and productivity.

INTRODUCTION

Protecting and restoring seagrass beds is important because they provide a multitude of ecological services and functions. This chapter describes the basis and methods used for setting quantitative preservation and restoration seagrass targets (goals) for the Indian River Lagoon (IRL), Florida.

The underlying assumption is that if we restore water quality to historical conditions, seagrass will recover to historical conditions. Of primary importance are those factors affecting light penetration to the seagrass blades (Morris and Tomasko 1993; Virnstein and Morris 1996). These factors include turbidity, chlorophyll *a*, color, and suspended solids.

BACKGROUND

Why is Seagrass Important?

Seagrass beds are valuable because they provide habitat and nursery areas for estuarine animals (Gilmore 1995; Lewis 1984; Virnstein et al. 1983), enhance water quality by removing nutrients and stabilizing sediments, and are highly productive ecosystems (Dawes 1981; Zieman 1982). Seagrass meadows have been described as the marine equivalent of tropical rain forests because of their high structural complexity, biodiversity, and productivity (Simenstad 1994). In the IRL basin, seagrasses provide the ecological basis for a fishery industry worth approximately a billion dollars a year (Virnstein and Morris 1996). Seagrass beds may thus provide about $30,000 per hectare per year of fisheries benefit alone.

Why Seagrass was Chosen as a Barometer of IRL Health

A host of associated benefits (above) depend on the presence of seagrass. Seagrass is sensitive to water quality conditions (Dennison et al. 1993; Stevenson et al. 1993), integrating previous conditions over the past hours, days, months, or even years. Seagrasses are essential species, with many

0-8493-2045-3/99/$0.00+$.50
© 2000 by CRC Press LLC

other species depending on them. Seagrass beds have been identified in the Indian River Lagoon Surface Water Improvement and Management (SWIM) Plan as the most critical habitat in the lagoon (Steward et al. 1994).

Seagrass is considered an appropriate barometer because:

1. Seagrass reflects water quality conditions.
2. Seagrass integrates water quality conditions over time (from hours to years).
3. Seagrass habitat generally harbors high animal density and diversity.
4. Remote sensing methods can be used to map seagrass system-wide. Segments can then be analyzed to measure seagrass changes over time.
5. In addition, small-scale changes in seagrass density, species composition, and maximum depth can easily be monitored using fixed transects.

THE BASIS FOR TARGETS

Management Need and Intended Use of Seagrass Targets

There are currently between 24,000–28,000 hectares of seagrass in the lagoon — about 15–20% less than in decades past. Although seagrass has declined in many areas of the lagoon (Fletcher and Fletcher 1995; Woodward-Clyde 1994; unpublished data), other large areas still have healthy and abundant seagrass beds. Therefore, the goal is to protect existing seagrass in some areas and to restore lost seagrass in other areas.

The numeric seagrass targets described here will provide the basis for water quality targets (especially light attenuation) (Kenworthy and Haunert 1991; Gallegos 1993, 1994; Kenworthy 1993; Gallegos and Kenworthy 1996). Water quality targets will then be used to set pollutant load reduction goals (PLRGs) (Steward et al. 1996).

Criteria for Seagrass Coverage Targets

Criteria for seagrass targets for the IRL were established. These criteria are that targets must be:

1. Scientifically defensible;
2. Ecologically achievable, without regard to whether the targets are economically achievable;
3. Objective, not arbitrary (non-subjective), i.e., based on data (e.g., distribution, maximum depth distribution, and depth contours);
4. Based on actual historical seagrass coverage and depth;
5. Simple, understandable, and explainable; and
6. Bounded by natural variability (ideally in the absence of anthropogenic impacts).

One rationale is that if seagrass has occurred there before, it can occur there again. However, seagrass maps merely provide "snapshots" of past distributions. Without a predictive model, we have no way of knowing how extensive seagrass may have been under unimpacted conditions in the past. For example, seagrass coverage in the IRL in 1943, although documented generally as the most extensive known, was *not* unimpacted. Some impacts, such as dredging, with open-water spoil disposal, were ongoing in 1943 and earlier (e.g., construction of three causeways and dredging of the Intracoastal Waterway).

Other factors could be considered in setting seagrass targets. These possible factors include sediment type and stability, wave dynamics, and shellfish lease sites. While we acknowledge that such factors likely influence whether targets can be reasonably achieved, they are not incorporated here because they are too uncertain.

Assumptions and Implications for Using Seagrass Coverage Targets

Although the targets are primarily simple coverage (area of seagrass), adopting the targets carries the following implications and assumptions:

1. Area (hectares) of seagrass is a valid measure of the "amount" of seagrass.
2. More seagrass is better.
3. Denser seagrass is better. However, some degree of patchiness may provide more "edge" than continuous beds.
4. Higher diversity of seagrass is better.
5. The amount (maximum depth) of seagrass is dependent primarily on water clarity; therefore, more light (above the minimum maintenance level) is better.
6. If we restore water quality conditions, then seagrass will be restored, but the lag time for recovery is not known. Observed rapid increases (within 1 year) in a few areas are promising that initial recovery may proceed rapidly.
7. If we provide seagrass, animals will use it. Lag time and dependence on density and other metrics are unknown.
8. Higher density and diversity of animals is good.
9. Abundance of seagrass represents amount of animal habitat and nursery area, but also important are seagrass density, diversity, location (e.g., adjacent to an inlet or a wetland creek), habitat heterogeneity, and landscape patterns (Virnstein 1995).
10. Amount of seagrass is a surrogate for animal integrity, diversity, and productivity.

THE TARGETS

Quantitative Targets

Three levels of quantitative targets are set (Figure 16.1). These levels are:

1. Potential — to a depth of 1.7 m, to which seagrass might be expected to occur based on distribution in other areas of the lagoon (Figure 16.1A). Healthy, unimpacted areas of the lagoon have seagrass beds extending down to a depth of 1.7 m based on the average of the deep edge of seagrass from transects in "healthy" areas of Hobe Sound, northern Indian River, southern Mosquito Lagoon, and Northern Banana River. Therefore, under ideal conditions we might expect seagrass to extend from near shore to a depth of 1.7 m throughout the lagoon.
2. Historical — maximum extent of wherever seagrass has ever been mapped in the past (Figure 16.1B). Ideally, this target should be met in most (> 50%) years, but some variation is expected.
3. Critical minimum — maximum extent of where seagrass has been mapped in at least half of non-1943 historical mappings (Figure 16.1C). While not strictly a target, this minimum level sets the lowest bounds below which seagrass should never occur, thus allowing for natural variability.

The above quantitative targets apply lagoon-wide, but specific coverage targets vary among segments of the lagoon. In some segments, little or no seagrass has been lost, and thus the historical and minimum targets are met and should be maintained, i.e., the target is to protect and preserve

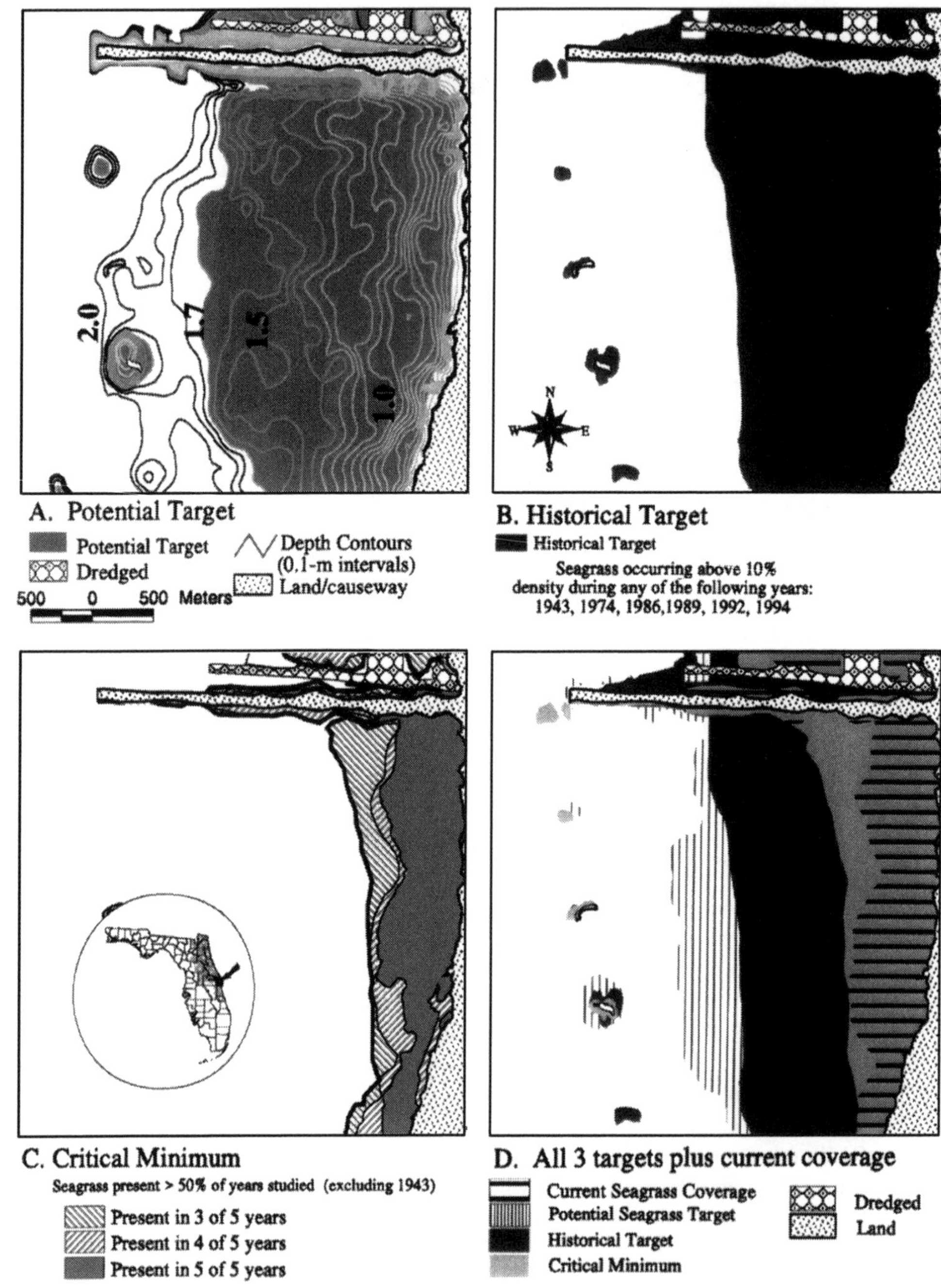

FIGURE 16.1 Maps of a sample segment of Indian River Lagoon just south of State Road 405 (NASA causeway) showing the three target levels (A–C) and a composite of all three (D). (A) Potential Target = shaded area < 1.7 m deep, with 0.1-m depth contours shown. (B) Historical Target, showing where seagrass has been mapped at any time in the past, based on overlapping of maps from 1943, 1974, 1986, 1989, 1992, and 1994. (C) Critical Minimum, where seagrass has occurred at least half the time, excluding 1943. (D) The three targets combined, plotted in the order above, with recent (1994) coverage also indicated.

existing seagrass. In central lagoon segments near Melbourne, 90% of historical beds have been lost since 1943, and restoration to at least historical levels is the primary target.

These targets are considered preliminary. Although the last two targets (historical and minimum) are considered ecologically achievable, we do not yet know whether they are reasonably achievable, given considerations of management capabilities (e.g., for stormwater treatment), costs, and the resolve of society to achieve these targets. Such considerations will come into play later when final targets are recommended and specific plans are made for implementation. At that time, two aspects will be considered in setting final, reasonably achievable seagrass targets. The predictive Pollutant Load Reduction model (Steward et al. 1996) will be used to provide "what if" predictions, e.g., if we reduce loadings of pollutant y by $x\%$, what is the expected improvement in water quality and light attenuation and the subsequent increase in seagrass? Additionally, economics will be considered. The increased value from predicted seagrass recovery will be balanced against the predicted cost of implementation and the willingness of the public to pay.

Qualitative Targets

Maintenance or increase in *area of seagrass coverage* will be the primary quantitative aim. Qualitative aspects are also important and incorporated: (1) seagrass density; (2) seagrass diversity; and (3) animal density, diversity, and productivity. These qualitative aspects of the targets are described as:

1. Successfully meeting any of the three acreage targets assumes a density > 10% out to the outer (deep edge) boundary of seagrass targets. However, no other density criteria are specifically incorporated in these targets.
2. All seagrass species should be protected and preserved. Diversity should include not just number of seagrass species, but also aspects of landscape diversity, such as the number of species within given segments and depth zones (Virnstein 1995). Special consideration should be given to protecting climax species such as *Thalassia testudinum* that probably take decades or longer to recover.
3. In addition to seagrass itself, the targets have the implied goal of protecting, preserving, and restoring animal diversity and abundance, secondary productivity (especially of valued sport and commercial species), functionality (e.g., sediment stability, nutrient uptake, habitat, and nursery), and ecological integrity appropriate to healthy, fully functioning seagrass beds.

METHODS FOR DETERMINING THE THREE LEVELS OF QUANTITATIVE TARGETS

Steps in Drawing Boundaries of Quantitative Targets

1. All existing seagrass coverages were mapped in geographical information system (GIS) Arc/Info. Complete coverage of the lagoon within St. Johns River Water Management District boundaries was done for 1943, 1992, 1994, and 1996. All but Mosquito Lagoon was mapped in 1986 and 1989. Most of Brevard County in the central lagoon was mapped in 1974 (Conrad White, Brevard County ONRM, unpublished). Historical and minimum maps were created in GIS ArcView by overlapping coverages. Areas of < 10% cover were excluded because of high uncertainty of many of these outer boundaries. Also, these targets do not address the small *Halophila* species. Deep beds of *Halophila johnsonii* and *H. decipiens* can occur in deep water in the southern third of the lagoon, but are not mapped because they are not visible in aerial photographs.

2. Seagrass distribution was compared to bathymetry corrected to mean sea level, i.e., actual water depth.
3. Maps were compiled that showed all areas that had seagrass in only 1 year, any 2 years, any 3 years, etc., with 1943 coverage plotted separately. From these maps, the following targets were created:
   A. Potential = to a depth of 1.7 m, mean sea level
   B. Historical = maximum extent of wherever seagrass has occurred in any year in the past, including 1943 and all recent coverages
   C. Minimum = maximum extent of wherever seagrass has occurred in at least half the documented coverage years, excluding 1943.
4. Areas were subtracted from targets that are now dredged areas (channels and borrow pits), filled areas (e.g., causeways), or intertidal shoals. These areas are considered not available for seagrass growth and are *not* incorporated in the coverage targets.
5. Areas of coverage of each target category were then calculated lagoon-wide and for each segment. All map analyses were done in GIS using ArcView.

FINAL PRODUCTS

For each segment in the lagoon, targets will be presented as maps, graphs, and tables of acreage as described below.

- Targets for the entire lagoon and for each segment will be presented as (1) lines and shaded/colored areas on maps; (2) tables of acreage; and (3) bar graphs, comparing acreage of past distribution to target coverage (see Figures 16.1D and 16.2).
- The target (depth, coverage, and line on map) has three components (Figures 16.1 and 16.2):
  (1) potential, where, under ideal conditions, seagrass might extend down to a depth of 1.7 m
  (2) historical, where it has been before; to be met the majority (> 50%) of the time
  (3) critical minimum, below which seagrass should not go.
- Recent coverages are shown as a percent of target as (1) table; (2) coverage on a map (Figure 16.1); and (3) bar graph (Figure 16.2).

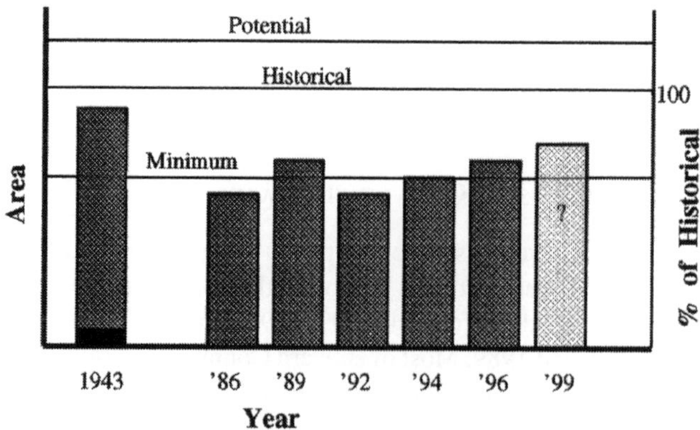

FIGURE 16.2 Plot for a hypothetical segment, illustrating the three target levels. Area of seagrass (of > 10% density) for the segment is plotted for each year that map coverage is available. The black area of the 1943 bar indicates the number of acres permanently lost to dredging and filling.

The primary measure of success is area of seagrass, mapped using aerial photographs at 1:24,000 scale (Virnstein and Morris 1996), usually every 2–3 years. A secondary measure would be the maximum depth of seagrass based on biennial monitoring of 80 fixed seagrass transects in the lagoon (see maps and methods in Virnstein and Morris 1996; Morris et al. 2000).

Seagrass targets can be valuable tools, fulfilling a human need for goals and moving society to action, as has been done for Tampa Bay and Chesapeake Bay.

ACKNOWLEDGMENTS

Ron Brockmeyer, Joel Steward, Bob Day, Becky Robbins, Peter Doering, and others provided thoughtful comments. Ed Carter skillfully provided all GIS analyses and Figure 16.1. This study was funded by the St. Johns Water Management District.

REFERENCES

Dawes, C. J. 1981. *Marine Botany.* John Wiley & Sons, New York, 612 pp.

Dennison, W. C., R. J. Orth, K. A. Moore, J. C. Stevenson, V. Carter, S. Kollar, P. W. Bergstrom, and R. Batiuk. 1993. Assessing water quality with submerged aquatic vegetation. *BioScience* 43:86–94.

Fletcher, S. W. and W. W. Fletcher. 1995. Factors affecting changes in seagrass distribution and diversity patterns in the Indian River Lagoon complex between 1940 and 1992. *Bulletin of Marine Science* 57:49–58.

Gallegos, C. L. 1993. Development of optical models for protection of seagrass habitats. In *Proceedings and Conclusions of Workshops on Submerged Aquatic Vegetation and Photosynthetically Active Radiation,* L. J. Morris and D. A. Tomasko (eds.). Special Publication SJ93-SP13, St. Johns River Water Management District, Palatka, FL, pp. 77–90.

Gallegos, C. L. 1994. Refining habitat requirements of submersed aquatic vegetation: role of optical models. *Estuaries* 17:187–199.

Gallegos, C. L. and W. J. Kenworthy. 1996. Seagrass depth-limits in the Indian River Lagoon (Florida, USA): application of an optical water quality model. *Estuarine Coastal Shelf Science.* 42:267–288.

Gilmore, R. G. 1995. Environmental and biogeographic factors influencing ichthyfaunal diversity: Indian River Lagoon. *Bulletin of Marine Science* 57:153–170.

Kenworthy, W. J. 1993. Defining the ecological light compensation point of seagrasses in the Indian River Lagoon. In *Proceedings and Conclusions of Workshops on Submerged Aquatic Vegetation and Photosynthetically Active Radiation,* L. J. Morris and D. A. Tomasko (eds.). Special Publication SJ93-SP13, St. Johns River Water Management District, Palatka, FL, pp. 195–210.

Kenworthy, W. J. and D. E. Haunert. 1991. *The Light Requirements of Seagrasses: Proceedings of a Workshop to Examine the Capability of Water Quality Criteria, Standards, and Monitoring Programs to Protect Seagrasses.* NOAA Technical Memorandum, NMFS-SEFC-287, pp. 85–94.

Lewis, F. G., III. 1984. Distribution of macrobenthic crustaceans associated with *Thalassia, Halodule,* and bare sand substrata. *Marine Ecology Progress Series* 19:101–113.

Morris, L. J. and D. A. Tomasko (eds.). 1993. *Proceedings and Conclusions of Workshops on Submerged Aquatic Vegetation and Photosynthetically Active Radiation.* Special Publication SJ93-SP13, St. Johns River Water Management District, Palatka, FL, 311 pp.

Morris, L. J., R. W. Virnstein, J. D. Miller, and L. M. Hall. 2000. Monitoring seagrass changes in Indian River Lagoon, Florida using fixed transects. In *Seagrasses: Monitoring, Ecology, Physiology, and Management,* S. A. Bortone (ed.). CRC Press, Boca Raton, FL, pp. 167–176.

Simenstad, C. A. 1994. Faunal association and ecological interactions in seagrass communities of the Pacific Northwest coast. In *Seagrass Science and Policy in the Pacific Northwest: Proceedings of a Seminar Series* (SMA 94-1), S. Wyllie-Echeverria, A. M. Olson, and M. J. Hershman (eds.). EPA 910/R-94-004, Seattle, WA, pp. 1–17.

Stevenson, J. C., L. W. Staver, and K. W. Staver. 1993. Water quality associated with survival of submerged aquatic vegetation along an estuarine gradient. *Estuaries* 16:346–361.

Steward, J., R. W. Virnstein, D. E. Haunert, and F. Lund. 1994. Surface Water Improvement and Management (SWIM) Plan for the Indian River Lagoon, FL. St. Johns River Water Management District and South Florida Water Management District, Palatka and West Palm Beach, FL, respectively, 120 pp.

Steward, J. S., F. Morris, R. W. Virnstein, L. J. Morris, and G. Sigua. 1996. Indian River Lagoon pollution load reduction model and recommendations for actions. Technical Memorandum, St. Johns River Water Management District, Palatka, FL, 23 pp. plus appendices.

Virnstein, R. W. 1995. Seagrass landscape diversity in the Indian River Lagoon, Florida: the importance of geographic scale and pattern. *Bulletin of Marine Science* 57:67–74.

Virnstein, R. W. and L. J. Morris. 1996. Seagrass preservation and restoration: a diagnostic plan for the Indian River Lagoon. Technical Memorandum No. 14, St. Johns River Water Management District, Palatka, FL, 18 pp. plus appendices.

Virnstein, R. W., P. S. Mikkelsen, K. D. Cairns, and M. A. Capone. 1983. Seagrass beds versus sand bottoms: the trophic importance of their associated benthic invertebrates. *Florida Scientist* 46:363–381.

Woodward-Clyde Consultants. 1994. Historical Imagery Inventory and Seagrass Assessment, Indian River Lagoon. Final Technical Report, Indian River Lagoon National Estuary Program, Melbourne, FL, 106 pp.

Zieman, J. C. 1982. The ecology of seagrasses of south Florida: A community profile. U.S. Fish and Wildlife Service, Office of Biological Services, Washington, D.C., FWS/OBS-82-85. 185 pp.

17 Seagrass Bed Recovery after Hydrological Restoration in a Coastal Lagoon with Groundwater Discharges in the North of Yucatan (Southeastern Mexico)

Jorge A. Herrera-Silveira, Javier Ramírez-Ramírez, Nelly Gómez, and Arturo Zaldivar-Jimenez

Abstract In the Yucatan Peninsula the lagoons are the major geomorphological features of the coast. The shallow water (< 2.5 m) of these ecosystems favors coverage of submerged aquatic vegetation (SAV) dominated by seagrasses. The hydrological characteristics of Chelem Lagoon were modified as a consequence of the reduction of freshwater discharge through the springs. Due to harbor and road construction and natural events such as hurricanes, seagrasses almost disappeared and the benthos was occupied by macroalgae. Mean salinity was 42‰ and the SAV coverage, reduced to 20–30‰, was dominated by *Laurencia microcladia, Batophora oerstedii, Acetabularia* sp. In 1994 a hydrological restoration program was initiated, specifically targeting the removal of sediments in the springs thereby increasing freshwater discharges to the lagoon. After 5 years of restoration effort, the mean salinity dropped from 42‰ to 35‰, and the SAV coverage increased to 80%, dominated by *Halodule wrightii* with patches of *Ruppia maritima* and *Thalassia testudinum*. More recently, eutrophication and turbidity caused by urbanization and dredging have reduced SAV growth and production.

INTRODUCTION

Seagrass communities have important ecological functions. They serve as habitats and nurseries for a variety of invertebrates, fish, and mammals (Hartman 1979; Bjorndal 1980; Sheridan and Livingston 1983). They help stabilize the substrate, enhance sedimentation, and reduce currents thus improving water clarity and reducing erosion. Seagrasses are capable of recycling sediment nutrients in the water column (Livingston 1984; Tomasko et al. 1996). With these important ecological functions, seagrasses benefit fisheries and improve water quality. However, seagrass communities are seriously threatened due to many factors such as dredging, eutrophication, oil pollution, salinity, and thermal stresses. In addition, the impact that natural events such as hurricanes have on seagrasses has received little attention (Onuf 1994; Hicks et al. 1998). The changes in submerged vegetation associated with different impacts are documented by ample historical and experimental evidence. Studies conducted thus far indicate that changes favor the dominance of faster-growing autotrophs, leading to the replacement of seagrasses and slow-growing macroalgae by blooming, fast-growing macroalgae and phytoplankton (Stevenson et al. 1993; Lapointe et al. 1994; Duarte 1995). The increase of human activities along the coasts forces the increase in development infrastructure, such as harbors, increasing the stress placed on the submerged plant communities. In areas where the seagrass meadows are seriously or potentially impacted, restoration and conservation programs are being implemented (Fonseca et al. 1998). Unfortunately, in developing countries it is difficult to design management programs because of lack of ecological information.

The objective of this study was to evaluate the response of the seagrass community to a hydrological restoration program in Chelem Lagoon (Yucatan Peninsula, southeastern Mexico), including long-term trends of coverage and total seasonal biomass subjected to salinity and temperature stressors. However, during the monitoring program, the region was affected by hurricane activity; therefore, it was possible to evaluate the response of the submerged aquatic vegetation (SAV) after these events. Moreover, to achieve a better understanding of the coastal lagoon seagrass communities subjected to groundwater discharges, the biomass and productivity of *Thalassia testudinum* was studied relative to the hydrological and sediment variables.

STUDY LOCATION

The coastal lagoons of Yucatan are marginal marine depressions parallel to the coast and separated from the sea by a sand barrier (Herrera-Silveira et al. 1999). The karst nature of the Yucatan Peninsula with no rivers favors the formation of sink holes (cenotes) that constitute the typical freshwater ecosystems of the region (Marín and Perry 1994). The freshwater inputs to coastal lagoons are made via precipitation and groundwater discharges in a diffuse array of springs distributed along the lagoons. The freshwater inputs tend to reduce salinity and temperature and increase nutrients in the coastal lagoons, due to the low salinity concentrations (2–5‰), low water temperature (17–22°C), and high NO_3^-, and soluble reactive silica (SRSi) from the springs (Herrera-Silveira 1994).

Chelem Lagoon is located in the northern part of the Yucatan Peninsula (Figure 17.1), 32 km from Merida City. It is long (14.7 km), narrow (1.8 km), and shallow (mean depth = 1.5 m) with an area of 13.6 km^2 connected to the sea through Yucalpeten Port.

The weather in the region is hot-semiarid; the annual mean temperature is 26.2°C, varying from 20°C in January to 35°C in May. The mean annual rainfall is 45 cm and the evaporation rate is 130 cm. Two main seasons are recognized, the dry with low rainfall (March–May, 0–5 cm) and the rainy (June–October) with high rainfall (>30 cm). Furthermore in this part of the Gulf of Mexico, the period from November to February is known locally as *nortes* season and is characterized by strong northern winds (>80 km/h), little rainfall (2–6 cm), and low temperatures (<22°C), imposed by cold air masses from the north

This lagoon supports the largest portion of the human population on Yucatan coast. Along the barrier island, the settlements are Chuburna, Chelem, Yucalpeten, and Progreso. With 90% of the Yucatan fishing fleet, urban development along the Chuburna–Progreso corridor has decreased the dune and mangrove vegetation. To facilitate urban and industrial development, Yucalpeten Harbor was constructed 30 years ago. The highways and railroad embankments have reduced water circulation favoring the intrusion of seawater and marine sediments. The results in Chelem Lagoon are an increase in mean salinity and a shift of the seagrasses to a macroalgae community (Zizumbo 1987; Herrera-Silveira et al. 1995). In addition to these cultural impacts, the area has been subjected to the effects of natural events such as hurricanes. During the last 10 years, three large hurricanes have crossed Chelem Lagoon: Gilbert in 1988 and Opal and Roxanne in 1995.

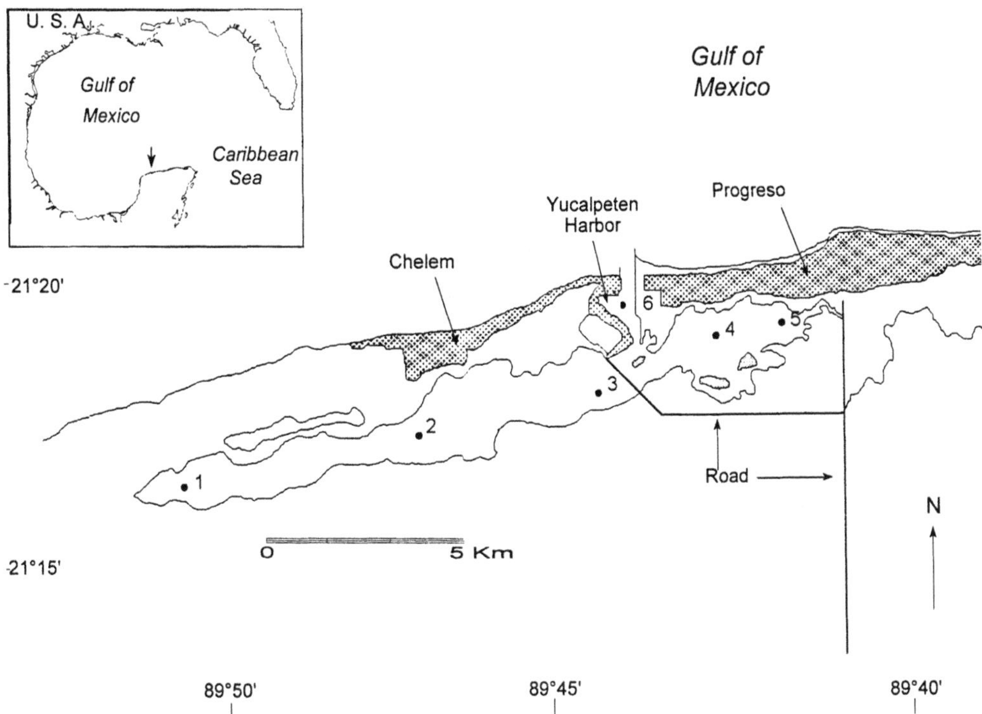

FIGURE 17.1 Location of the Chelem Lagoon, long- and short-term monitoring sampling sites.

MATERIALS AND METHODS

In each dry, rainy, and *nortes* season from 1986 to 1998, six stations (Figure 17.1) were sampled for salinity and temperature. The biomass of the submerged aquatic vegetation was determined from three replicate samples (above- and below-ground to a depth of 30 cm) with a 35-cm diameter core sampler. Each sample was washed on a 2 mm sieve and dried at 60°C for 24 hr to constant weight (Wetzel 1965; Westlake 1974). The distribution and cover of the submerged aquatic vegetation in the lagoon was estimated once a year by snorkeling along 50 m transects spaced 250 m. In 1994, a program to restore the freshwater inputs to the lagoon was initiated; 52 springs were cleaned of sediments to permit increased freshwater flow.

Monthly from September 1997 to September 1998, measurements of biomass, leaf productivity, and shoot density were estimated at two sites (stations 3 and 4). In each site, three replicate samples were collected with a 35-cm diameter core device driven 15–20 cm into the sediment. Samples were thoroughly cleaned of epiphytes and sediments, and dried at 60°C for 24 hr to constant weight. Temperature and salinity were measured with a YSI-85 oxymeter and the light attenuation coefficient with a LI-COR irradiometer (LI-1000 spherical quantum sensor 1939). At both stations, leaf productivity was obtained using the blade marking technique (Zieman 1974). Three sampling areas were marked with 20 × 20 cm PVC pipe quadrats in a monotypic *Thalassia testudinum* bed.

RESULTS AND DISCUSSION

The hydrology of Chelem Lagoon was characterized by a seawater flow through the Yucalpeten Harbor, reduced fresh groundwater due to sedimentation, and poor water interchange between the west and east zones of the Lagoon because of road construction. This increased the water residence

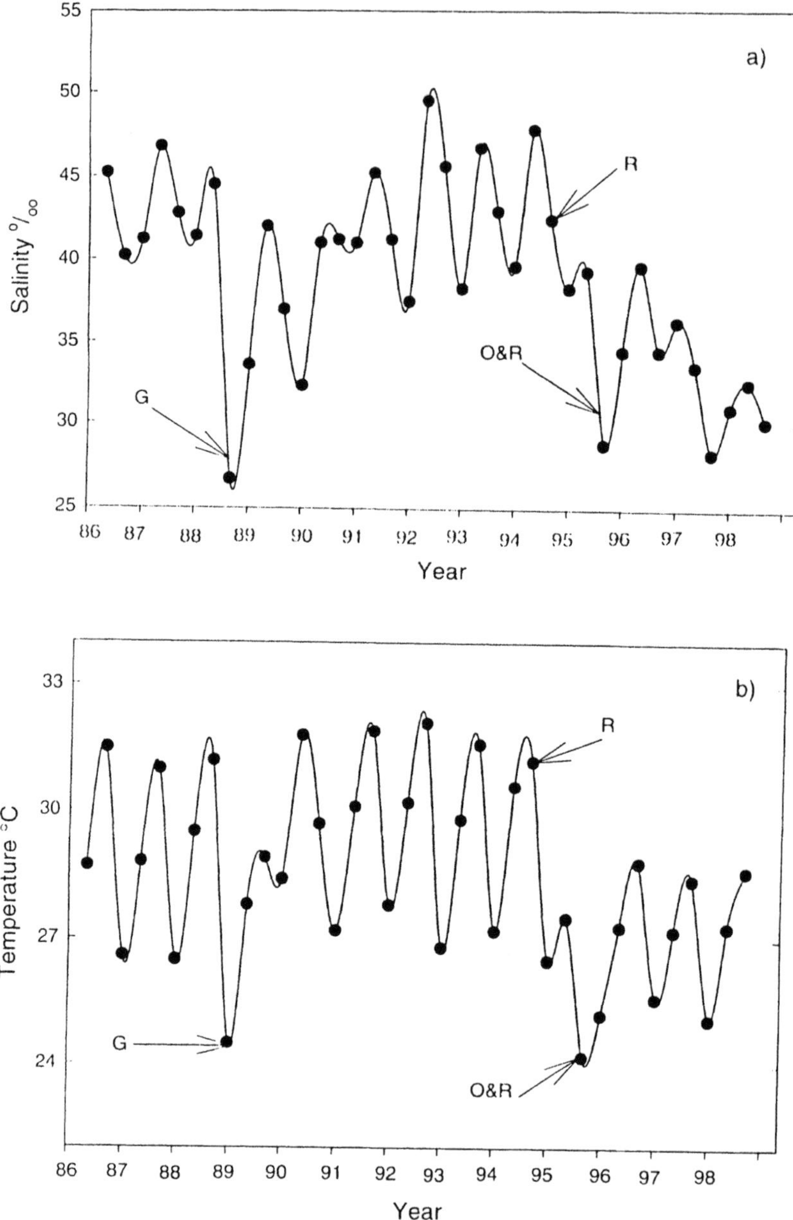

FIGURE 17.2 Long-term seasonal variation of a) salinity and b) temperature from 1986 to 1998 in Chelem Lagoon; Gilbert (G), Opal and Roxanne (O&R) hurricanes, (R) begin of the restoration program.

time (> 60 days) in the west zone as deduced from the high salinity (38–45‰) levels of this area (Herrera-Silveira et al. 1999). Between 1986 and 1988, the mean salinity was 42‰; however after Hurricane Gilbert, the mean salinity dropped to 28‰ (Figure 17.2a). During the following years, it varied from 38 to 48‰, higher than the mean salinity reported as optimal (25–35‰) for tropical seagrasses (Fonseca et al. 1998). Lower values were observed during each season following Hurricanes Opal and Roxanne in 1995 (Figure 17.2a). During the restoration program at the end of 1994, the mean salinity varied from 32 to 36‰, better for seagrass growth. The mean temperature before the restoration program varied from 29 to 31.5°C. After it was 25 to 28.5°C (Figure 17.2b).

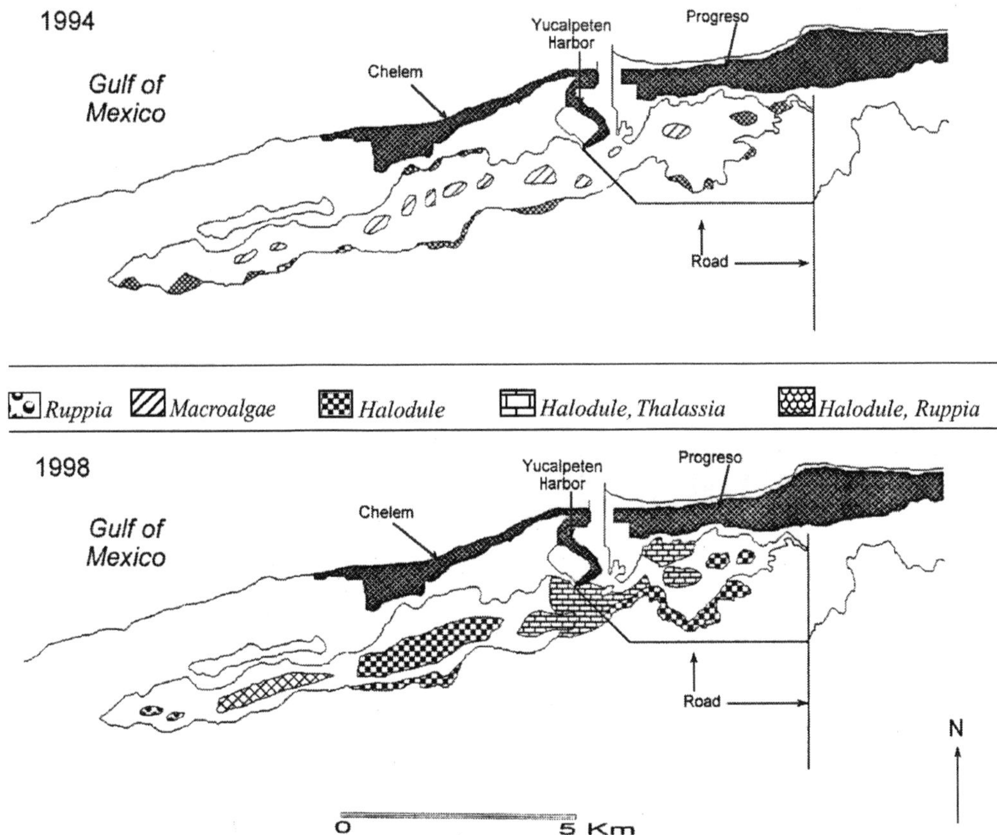

FIGURE 17.3 Long-term variation of the composition and distribution of the submerged aquatic vegetation in 1994 and 1998 in Chelem Lagoon.

The SAV coverage in Chelem Lagoon during 1994 was 35% (Figure 17.3a) dominated by macroalgae *(Laurencia microcladia, Batophora oerstedii, Acetabularia* sp.). The seagrass community was reduced about 10% at the borders of the lagoon and dominated by *Ruppia maritima* with a few patches of *Halodule wrightii* and *Thalassia testudinum.* During the 1998 survey (Figure 17.3b) (5 years after the restoration program), the SAV coverage increased to 70%, with a shift in structure. The dominant species in the SAV group in 1998 were *H. wrightii, T. testudinum,* and *R. maritima,* respectively. This species dominance shift of the submerged vegetation community has also been observed in Laguna Madre (Quammen and Onuf 1993).

The seasonal seagrasses and macroalgae biomass (Figure 17.4) between 1986 and 1994 were dominated by macroalgae with an annual mean of 260 g dry wt m^{-2} and were strongly seasonal with higher biomass during the rainy season (430 g dry wt m^{-2}) than during dry and *nortes* seasons (125–250 g dry wt m^{-2}). However, the annual mean seagrass biomass was low (< 150 g dry wt m^{-2}). After the restoration of the fresh groundwater discharge, the seagrass biomass increased to an annual mean of 400 g dry wt m^{-2}, though in some sites, during a rainy season, the biomass was > 1500 g dry wt m^{-2}.

The monitoring program in stations 3 and 4 was conducted because field observations suggested that the differences between these two sites were due to *T. testudinum* growth patterns. The temperature and salinity were similar in both places, showing a seasonal pattern (Figure 17.5). The mean temperature drops from 27.3°C in September to 23.3°C in February, reaching the highest values in June (29.8°C) (Figure 17.5a). The salinity was lowest at the end of the rainy season (32‰) and highest at the end of the dry season (41‰) (Figure 17.5b).

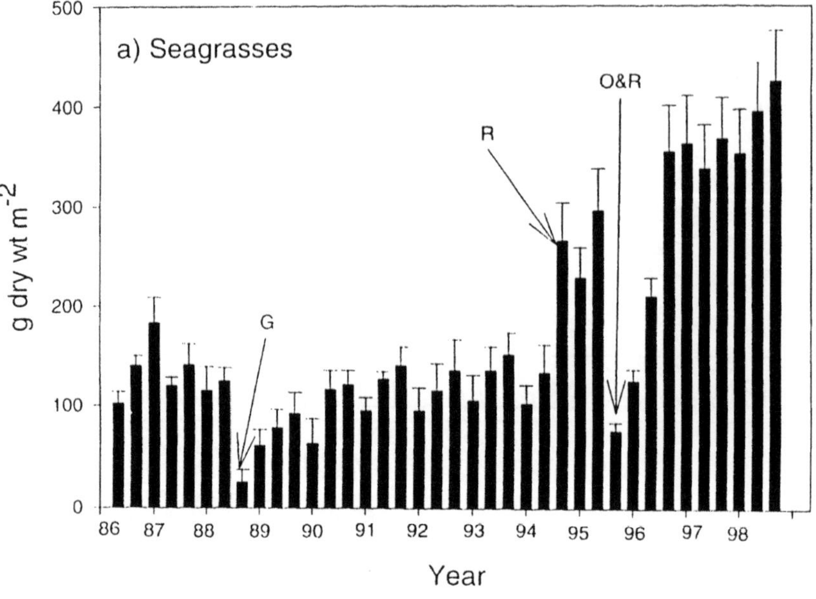

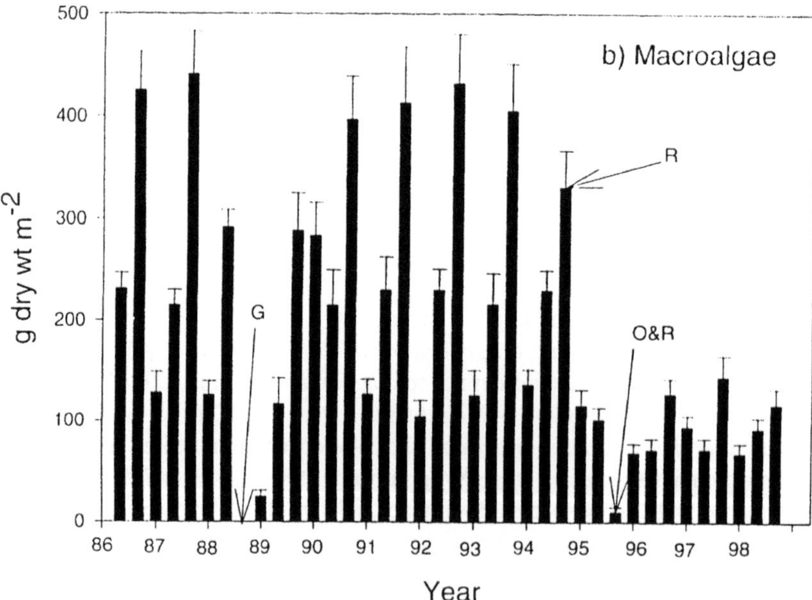

FIGURE 17.4 Long-term seasonal variation of mean biomass of a) seagrasses and b) macroalgae from 1986 to 1998 in Chelem Lagoon. Vertical bars show ±SE. Gilbert (G), Opal and Roxanne (O&R) hurricanes, (R) begin of the restoration program.

The light attenuation coefficient (k) had a seasonal pattern with small differences between sites (Figure 17.5c). The highest k (3 m^{-1}) was observed during *nortes* season. In this period of the year, the resuspension of sediments takes place due to greater wind velocities (>80 km h^{-1}) and the shallowness of the lagoon (<1.5m)(Hicks et al. 1998; Herrera-Silveira et al. 1998). The lowest k (<1 m^{-1}) was recorded during the rainy season, when 5 days per month have velocities higher than 20 km h^{-1} (Comisión Nacional del Agua 1998). During this season, no forcing function is present to resuspend the sediments. Station 4 had higher values of k than station 3. This is attributed to

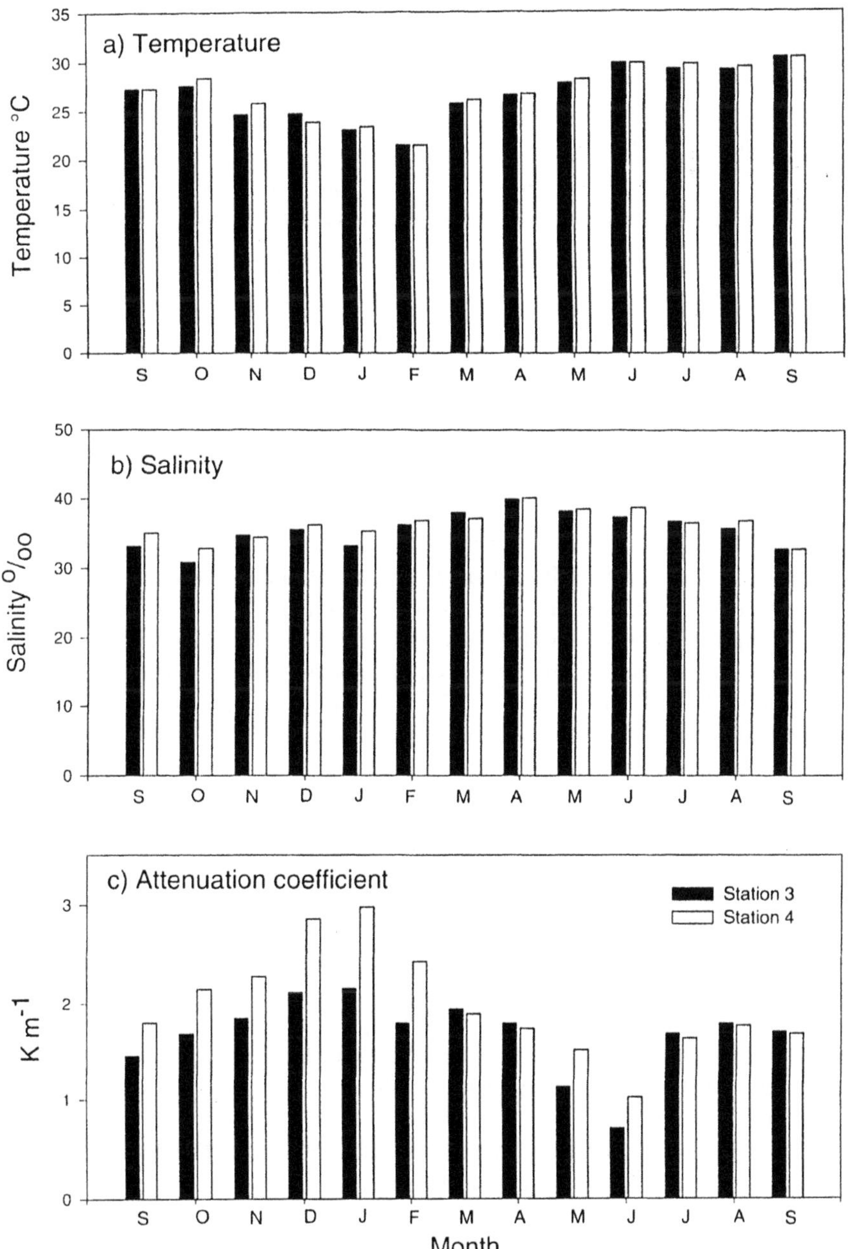

FIGURE 17.5 Seasonal variation of the mean a) temperature, b) salinity, and c) attenuation coefficient from September 1997 to September 1998 in Chelem Lagoon.

the different types of sediments (Table 17.1); site 3 is dominated by sand while site 4 is silt. If an eventual resuspension occurs, the sand settles faster than silt.

Figure 17.6 shows the seasonal trends of shoot density, total biomass (above-below ground), and leaf productivity at the two stations. The three variables of *T. testudinum* show higher values at station 4.

The mean shoot density in the latter varied from 115 shoots m^{-2} in December to 285 shoots m^{-2} in September, with no clear seasonal pattern. However, at station 3 the pattern shows fewer

TABLE 17.1
Characteristics of the sediments in the sampling sites of *Thalassia testudinum* in Chelem Lagoon (data from Valdés and Real 1994).

Site	Porosity	Organic Matter (%)	Sand (%)	Silt (%)	Clay (%)	Sediment Type
Station 3	0.4	3.4	57.5	8.5	34	Sand/clay
Station 4	0.7	9.2	31.5	56	12.5	Silt/sand

shoots from September (480 shoots m^{-2}) to February (330 shoots m^{-2}) in agreement with the temperature pattern (Figures 17.5a and 17.6a). It is also in agreement with the increase of shoots during the dry and rainy season which reached their highest mean density in June (675 shoots m^{-2}), followed by a decrease at the end of the rainy season. This pattern can be related to the available light; the lowest k (1 m^{-1}) was registered during the same period. Light has been described as the principal factor affecting the growth of the seagrasses (Pérez and Romero 1992; Dunton 1994).

The total biomass at station 3 did not show a clear pattern; however, at station 4, two peaks of high biomass were observed: the first during *nortes* season (1480 g dry wt m^{-2} in December), and the second during the rainy season (1783 g dry wt m^{-2} in June). These two peaks are related to different portions of the plants. In the first case, the biomass was dominated by the below-ground components (above-below-ground ratio = 0.8), while during the June peak, this ratio was 1.3 and the above-ground component was important. This pattern suggests that during low temperatures and high light attenuation coefficients, *T. testudinum* tend to cover a greater area, while during the rainy season as the temperature and light availability increase, *T. testudinum* reaches canopy height and shoot density (Dennison 1978). This pattern has been observed in *H. wrightii* in coastal lagoons of Texas and Florida (Zieman et al. 1989).

Leaf productivity (Figure 17.6c) follows a similar seasonal pattern as shoot density (Figure 17.6a), with low mean productivity during the *nortes* season (1 and 9 g dry wt m^{-2} d^{-1} in stations 3 and 4, respectively), and a maximum during June (4 and 18 g dry wt m^{-2} d^{-1}). As with shoot density, the leaf pattern should be related to the covariation of temperature and light; the lowest rates were observed during the season of lowest temperatures and highest k (Figure 17.5).

In Bojorgez Lagoon, located on the Caribbean coast of Mexico, shoot density (mean 647 shoots m^{-2}) was similar to that observed at station 4 (675 shoots m^{-2}); however, leaf production in Chelem Lagoon was higher (8 dry wt m^{-2} d^{-1}) than in the coral reef lagoon (1.17 dry wt m^{-2} d^{-1}). This is probably due to differences in sediment nutrient concentrations. The Caribbean sediments are predominantly carbonate and sand, while in Chelem Lagoon they are silt and clay (Valdés and Real 1994; Short 1987; Erftemeijer 1994; Tussenbroek 1995). These sediment characteristics may explain the differences observed in growth variables (Figure 17.6) between stations 3 and 4. Station 3 is characterized by sand/clay sediments and low organic matter (3.4%), while at station 4, the sediments are silt/sand with higher organic matter concentration (9.2%) (Table 17.1).

While there are no data on the light attenuation coefficient for Chelem Lagoon from 1986 to 1996, the long-term and seasonal observations of this work suggest that salinity is an important factor to control. It is important to control salinity in conservation and restoration programs of tropical seagrasses over the long-term context, mainly in semiarid coastal zones with shallow water ecosystems where evaporation rates are conducive to their salinization. However, temperature and light attenuation are variables that act in a seasonal context. Moreover, light could play an important long-term role in the growth of seagrass beds as higher attenuation is related to eutrophication (Valiela et al. 1992), due to the phytoplankton, epiphytes, and macroalgae growth (Sand-Jensen and Borum 1991).

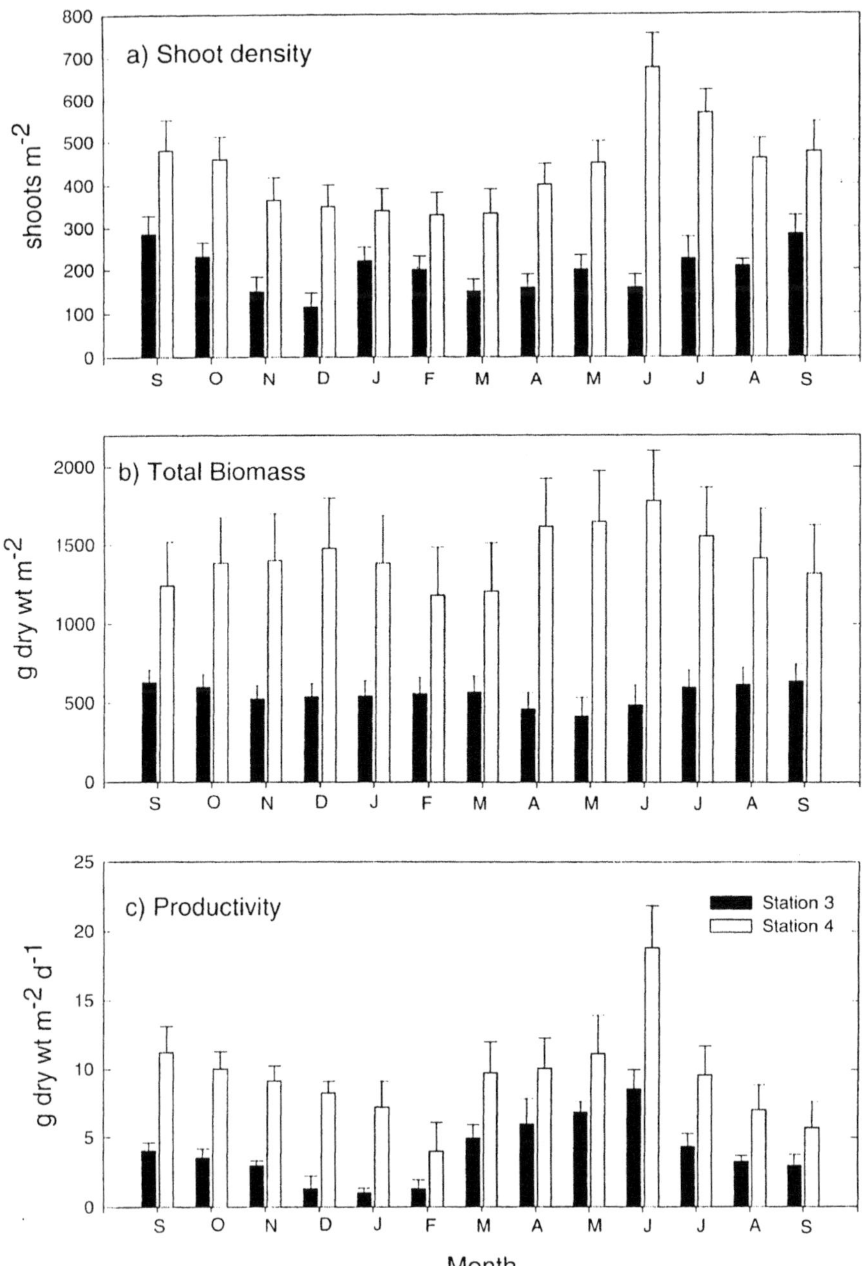

FIGURE 17.6 Seasonal variation of the mean a) shoot density, b) total biomass, and c) leaf productivity from September 1997 to September 1998 in Chelem Lagoon. Vertical bars show ± SE

The distribution and growth of SAV in Chelem Lagoon after the passages of hurricanes suggests that this community has the capacity to naturally recover from high energy, short-duration, low frequency events. However, anthropogenic impacts such as nutrient inputs (which can be of low energy but long term and high frequency) can initiate the total and definitive disappearance of seagrasses.

Conservation and restoration programs must consider the role of the forcing functions that could negatively affect the growth and development of seagrasses. Their effects must be considered in a space-time context as each variable may act at a different scale according to covariation with other variables. This relationship was observed in Chelem Lagoon where salinity appeared as a long-term variable, while temperature and light were short term. However, sediment characteristics are important in a spatial context. This behavior has been observed in other autotrophic communities such as the mangroves (Twilley et al. 1996).

The long- and short-term observations of the SAV in Chelem Lagoon, show: (1) capacity for recovering after natural events; (2) salinity reduction as a result of hydrological restoration favoring the recovery of seagrass beds; (3) light and temperature play an important role in the seasonal growth; and (4) organic matter and texture of the sediments can be responsible for the differences observed in the biomass and productivity of *T. testudinum* between sites in Chelem Lagoon.

ACKNOWLEDGMENTS

This project was funded by the National Council for Biodiversity (CONABIO B019) and National Council for Science and Technology (CONACyT 4147P-T). The authors are grateful to L. Capurro and two reviewers for their valuable comments and suggestions.

REFERENCES

Bjorndal, K. A. 1980. Nutrition and grazing behavior of the green turtle *Chelonia mydas*. *Marine Biology* 56:147–154.

Comisión Nacional del Agua. 1998. Datos climatológicos del Norte de Yucatán. Reporte Anual, 18 pp.

Dennison, W. C. 1978. Effects of light on seagrass photosynthesis, growth and depth distribution. *Aquatic Botany 27*:15–26.

Duarte, C. M. 1995. Submerged aquatic vegetation in relation to different nutrient regimes. *Ophelia* 41:87–112.

Dunton, K. H. 1994. Seasonal growth and biomass of the subtropical seagrass *Halodule wrightii* in relation to continuous measurements of underwater irradiances. *Marine Biology* 120(3):479–489.

Erftemeijer, P. L. 1994. Differences in nutrient concentrations and resources between seagrass communities on carbonate and terrigenous sediments in south Sulawesi, Indonesia. *Bulletin of Marine Science* 54(2):403–419.

Fonseca, M. S., W. J. Kenworthy, and G. W. Thayer. 1998. Guidelines for the conservation and restoration of seagrasses in the United States and adjacent waters. NOAA Coastal Ocean Program, Decision Analysis Series No. 12, 222 pp.

Hartman, D. S. 1979. Ecology and behavior of the manatee (*Trichechus manatus*) in Florida. American Society of Mammalogists, Special Publication No. 5.

Herrera-Silveira, J. A. 1994. Nutrients from underground discharges in a coastal lagoon Celestún, Yucatán, México. *Verhandlungen Internationale Vereingung Limnologie* 25:1398–1401.

Herrera-Silveira, J. A., J. R. Ramírez, F. A. Comín, I. M. Sánchez, M. T. Martín, and J. C. Soto. Octubre 1995. Biodiversidad de Productores Primarios de Lagunas Costeras del Norte de Yucatán CONABIO-B019, Informe Final, 60 pp.

Herrera-Silveira, J. A., J. R. Ramírez, and A. Zaldivar. 1998. Overview and characterization of the hydrology and primary producers communities of selected coastal lagoons of Yucatan, Mexico. *Aquatic Ecosystem Health & Management* 1:353–372.

Hicks, D. W., C. P. Onuf, and J. W. Tunnell. 1998. Response of shoal grass, *Halodule wrightii,* to extreme winter conditions in the Lower Laguna Madre, Texas. *Aquatic Botany* 62:107–114.

Lapointe, B. E., D. A. Tomasko, and W. R. Matzie. 1994. Eutrophication and trophic state classification of seagrass communities in Florida. *Bulletin of Marine Science* 54(3):696–717.

Livingston, R. J. 1984. The relationship of physical factors and biological response in coastal seagrass meadows. *Estuaries 7*(4A):377–390.

Marín, L. E. and E. C. Perry. 1994. The hydrogeology and contamination potential of northwestern Yucatan, Mexico. *Geofísica Internacional* 33(4):619–623.

Onuf, C. P. 1994. Seagrasses, dredging and light in Laguna Madre, TX, USA. *Estuarine Coastal Shelf Science* 39:75–91.

Pérez, M. and J. Romero. 1992. Photosynthetic response to light and temperature of the seagrass *Cymodosea nodosa* and the prediction of its seasonality. *Aquatic Botany* 43:51–62.

Quammen, M. L. and C. P. Onuf. 1993. Laguna Madre: seagrass changes continue decades after salinity reduction. *Estuaries* 16:302–310.

Sand-Jensen, K. and J. Borum. 1991. Interactions among phytoplankton, periphyton, and macrophytes in temperate freshwaters and estuaries. *Aquatic Botany* 41:137–175.

Sheridan, P. F. and R. J. Livingston. 1983. Abundance and seasonality of infauna and epifauna inhabiting a *Halodule wrightii* meadow in Apalachicola Bay, Florida. *Estuaries* 6:407–419.

Short, F. T. 1987. Effects of sediments nutrients on seagrasses: literature review and mesocosm experiment. *Aquatic Botany* 27:41–57.

Stevenson, J. C., L. W. Staver, and K. W. Staver. 1993. Water quality associated with survival of submersed aquatic vegetation along an estuarine gradient. *Estuaries* 16(2):346–361.

Tomasko, D. A., C. J. Dawes, and M. O. Hall. 1996. The effects of anthropogenic nutrient enrichment on turtle grass (*Thalassia testudinum*) in Sarasota Bay, Florida. *Estuarine Research Federation* 19(2B):448–456.

Tussenbroek, B. I. V. 1995. *Thalassia testudinum* leaf dynamics in a Mexican Caribbean coral reef lagoon. *Marine Biology* 122:33–40.

Twilley, R. R., R. Chen, and M. Koch-Rose. 1996. The significance of nutrient redistribution and regeneration to the recovery of mangrove ecosystems of South Florida in response to Hurricane Andrew. Final Report, Everglades National Park and University of Southwestern Louisiana, Cooperative Agreement CA5280-4-9019, 139 pp.

Valdés, L. D. and E. Real. 1994. Flujos de amonio, nitrito, nitrato y fosfato a través de la interfase sedimento-agua, en una laguna tropical. *Ciencias Marinas* 20(1):65–80.

Valiela, I., K. Foreman, M. LaMontagne, D. Hersh, J. Costa, P. Paulette, B. DeMeo-Anderson, C. D'Avanzo, M. Babione, C. Sham, J. Brawley, and K. Lajtha. 1992. Couplings of watersheds and coastal waters: sources and consequences of nutrient enrichment in Waquoit Bay, Massachusetts. *Estuaries* 15(4):443–457.

Westlake, D. F. 1974. Macrophytes. In *A Manual of Methods for Measuring Primary Production in Aquatic Environments,* R. A. Vollenweider (ed.). IBP Handbook No. 12, Blackwell, London, pp. 32–41.

Wetzel, R. G. 1965. Techniques and problems of primary productivity measurements in higher aquatic plants and periphyton. *Memorias Italianas Idrobiología* 18:249–267.

Zieman J. C. 1974. Methods for the study of the growth and production of turtle grass, *Thalassia testudinum* Köning. *Aquaculture* 4:139–143.

Zieman, J. C., J. W. Fourqurean, and R. L. Iverson. 1989. Distribution, abundance and productivity of seagrasses and macroalgae in Florida Bay. *Bulletin of Marine Science* 44(1):292–311.

Zizumbo, D. 1987. El deterioro del sistema ecológico, Cienega de Progreso, Yucatán, México. Secretaria de Ecología, Gobierno del Estado de Yucatán, Mérida, Yucatán, 66 pp.

18 Observations on the Regrowth of Subaquatic Vegetation Following Transplantation: A Potential Method to Assess Environmental Health of Coastal Habitats

William P. Davis, Michelle R. Davis, and David A. Flemer

Abstract In 1991, experimental transplantings of *Vallisneria americana* (tapegrass, vallisneria, or wild celery) were initiated at selected sites which lacked grass beds along the north shore of Perdido Bay, located on the Alabama–Florida border. Abatement of organic and color-staining components had been implemented to improve the water quality of effluent discharged by a pulp mill into the headwaters of Elevenmile Creek, a stream entering this low salinity estuary. This study was designed to assess whether previous *in situ* habitat conditions (e.g., light exclusion, water, or sediment toxicity) had prevented natural recruitment of aquatic grasses or if other factors, (e.g., propagule transport) existed which might limit or delay *V. americana* colonization or growth.

Different transplanting configurations were employed in our experimental designs to observe success in establishment of beds and assess our ability to measure plant growth among the varying micro-habitats and substrates. The initial transplanting, in 1991, consisted of two plants each, spaced at 40 cm centers in four 6 × 1 m parallel row-plots. Subsequently these plants spread rapidly by runners merging the rows into a continuously expanding grass bed. Second and third trials conducted in 1995 were planted in a cross-shaped configuration, which has emerged as our preferred design. The growth of these transplants indicated *V. americana* grass beds were recruitment limited, rather than constrained by prevailing conditions of water quality/toxicity, light reduction, or unsuitable substrate during the study period. Our experience may represent a fundamental method for routine utilization of the responses of submerged aquatic vegetation (SAV) to assess a broad range of questions concerning habitat and water quality of potential sites for habitat restoration.

INTRODUCTION

Observations of significant reductions in submerged aquatic vegetation (SAV) populations have been reported worldwide, attributed especially to increased light attenuation (Dennison et al. 1993). Decline of SAV has been pronounced in coastal waters and estuaries of the Gulf of Mexico (Duke and Kruczynski 1992). In some locations the declines are 20% of historical levels (e.g., Tampa Bay, Florida, Lewis et al. 1985), whereas other areas have losses accruing to 40–50% of historical distribution and abundance. These submerged vascular plant communities are recognized as one of the most important ecological communities in shallow coastal waters (Zieman and Zieman 1989). Coastal populations of SAV exhibit high primary production, often exceeding highly manipulated tropical croplands (Zieman and Wetzel 1980). Furthermore, SAV habitats support numerous important ecological functions for a variety of aquatic organisms, including invertebrate, fish, turtle, bird, and mammal species (McRoy and Helfferich 1980). In addition to contributions to maintenance of food webs, SAV functions in nutrient cycling, stabilizing fine sediments and mitigating shoreline erosion dynamics (Phillips and McRoy 1980; Korschen and Green 1988). Therefore, related to the importance in these diverse ecological functions, SAV beds have become widely recognized as sentinels of the ecological status for estuarine ecosystems (Thayer et al. 1975; Livingston 1984; Doust et al. 1994; Neckles 1994). It is important to environmental and living resource managers to minimize SAV losses, as previously reported in nearby Escambia Bay (Olinger 1976), and foster re-establishment of SAV communities. Thus natural resource agencies have increased their interest in protection, restoration, and creation of wetlands in general with seagrass and estuarine SAV habitats, and communities are receiving increased attention (Morris and Tomasko 1993).

Two species, *Vallisneria americana* (tapegrass, vallisneria, or wild celery) and *Ruppia maritima* (widgeon grass), are the most common SAV species in tidal fresh and brackish waters of northern Gulf of Mexico estuaries (Zieman and Zieman 1989). *V. americana* will tolerate salinities ranging from freshwater to 12–18‰ (parts per thousand). *R. maritima* is an even more euryhaline species occurring more often in this region in shallow brackish waters. *V. americana* was the dominant species of SAV that we observed in upper Perdido Bay.

Since our initial transplanting (1991), we have conducted field measurements, including plant density and general bed morphology, characterizing shape, size, and sculpture to document the progress and success of the SAV beds. In this study we summarize our observations during the period 1991 through early 1998.

DESCRIPTION OF THE STUDY SITE

Perdido Bay (Figure 18.1) has been influenced by changing land use activities, which presumably affected historical distribution and abundance of SAV. Beginning in the fall of 1989, water clarity in inshore waters of upper Perdido Bay increased concurrently with the implementation of water treatment and color reduction of the kraft mill effluent (KME) discharged by the paper mill forming the headwaters of Elevenmile Creek. Secchi disk observations conducted on another project increased from 1–10 cm before color reduction, to 20–40 cm after in the mouth of Elevenmile Creek. This increased clarity eventually allowed one to discern the bottom in 30–50+ cm along the north shore of the bay. During this period we also observed other aquatic plants, such as *Najas* and *Ceratophyllium*, colonizing the upper reaches of Elevenmile Creek, presumably also responding to increased water transparency.

Prior to summer 1988, *V. americana* beds were limited to specific shallow locales in upper Perdido Bay (e.g. Lillian Bridge, Grassy Point, Millview, etc.) which, potentially, could have served as sources of propagules. However, there was no evidence of any plants colonizing the north shore, between Elevenmile Creek, Ramsey Beach to Marcus Bayou. Because of this distribution, we speculated that perhaps colonization of *V. americana* was more recruitment-limited than habitat (i.e., substrate)-limited. Prompted by questions about what parameters actually limited spread of grasses from nearby sites, in 1991 we initiated experimental plantings of *V. americana.* Our study site, which then lacked any grass beds, was located adjacent to a private dock at Ramsey Beach (Figure 18.1), on the north shore of upper Perdido Bay.

The Ramsey Beach study site is bordered on the east by a 111 m pier, oriented north–south, with pilings set approximately at 3 m intervals. The north shoreline is a lawn edged by rip-rap. The adjacent intertidal zone is characterized by shallow waters with a few old cypress stumps

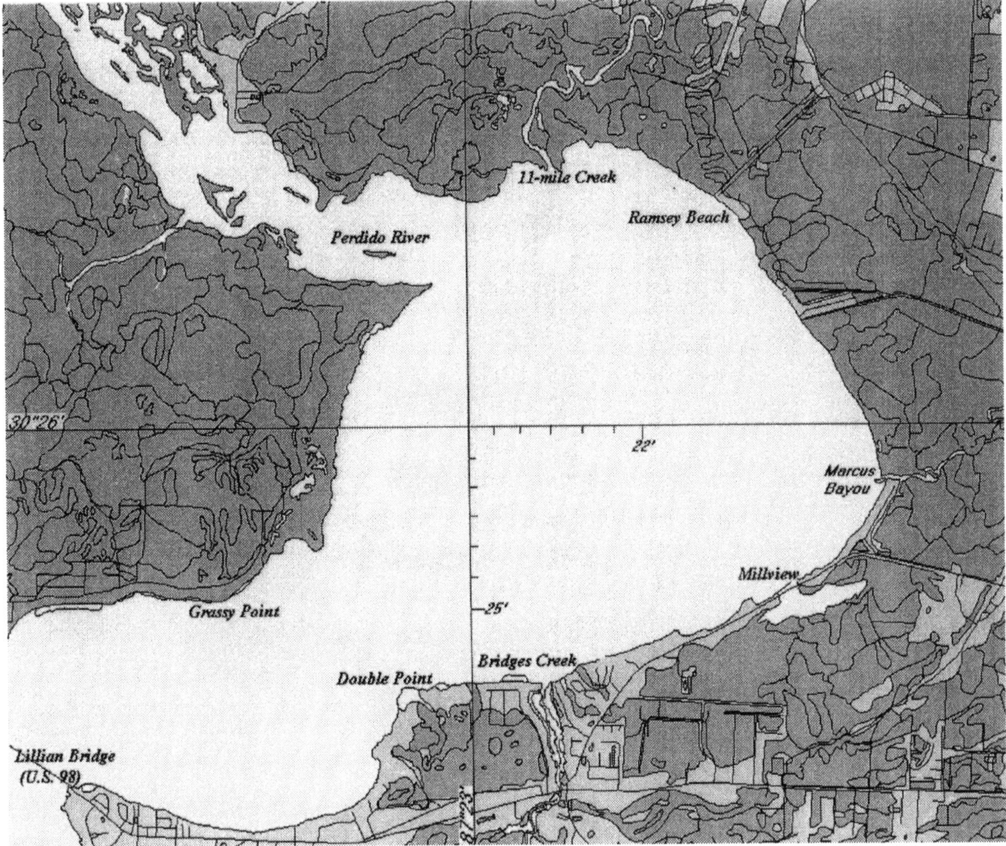

FIGURE 18.1 Upper or northern section of Perdido Bay indicating the locations of the Ramsey Beach study site relative to other sites in the estuary.

(Figure 18.3). The substrate is composed of a fine sandy mud, very gently sloping toward open water. Water depth increases only 5–8 cm from inshore to the offshore part of the study area. The prevailing warm season winds blow from the southwest. Salinity ranged from 0 to12‰. The bottom substrate is uniformly fine quartz and soft sediments in which one's footprint is typically about 5 cm deep. Bottom sediments included numerous juvenile and adult clam shells (*Rangia*). Living *Rangia* vary in density and age among years but data on benthic organisms were not quantified.

METHODS AND INITIAL OBSERVATIONS

Different planting configurations for *V. americana* "sets" were used during the study to compare growth responses in different substrates and micro-habitats. In the first experimental transplant, plants were trimmed by removal of old foliage and limited to a plant with a maximum of one runner. This initial planting used plants moved from a bed adjacent to Lillian Bridge (Figure 18.1). Sets consisted of two trimmed *V. americana* planted 40 cm apart in four parallel 6 × 1 m row/plots. Each row was separated by a 1 m strip (Figure 18.2). Turbidity during the summer months of 1991 prevented any practical visual observation of the transplants. During our first attempt to count plants in the row/plots, we were astonished at the extent of growth and spread of runners that had occurred during the 5-month period, April 5 to August 28, 1991 (Figure 18.2). Rapid growth of plants continued, so that by June 30, 1993, enumeration of individual plants and runners was impossible. Thereafter, we adopted an alternate method to assess and characterize the bed.

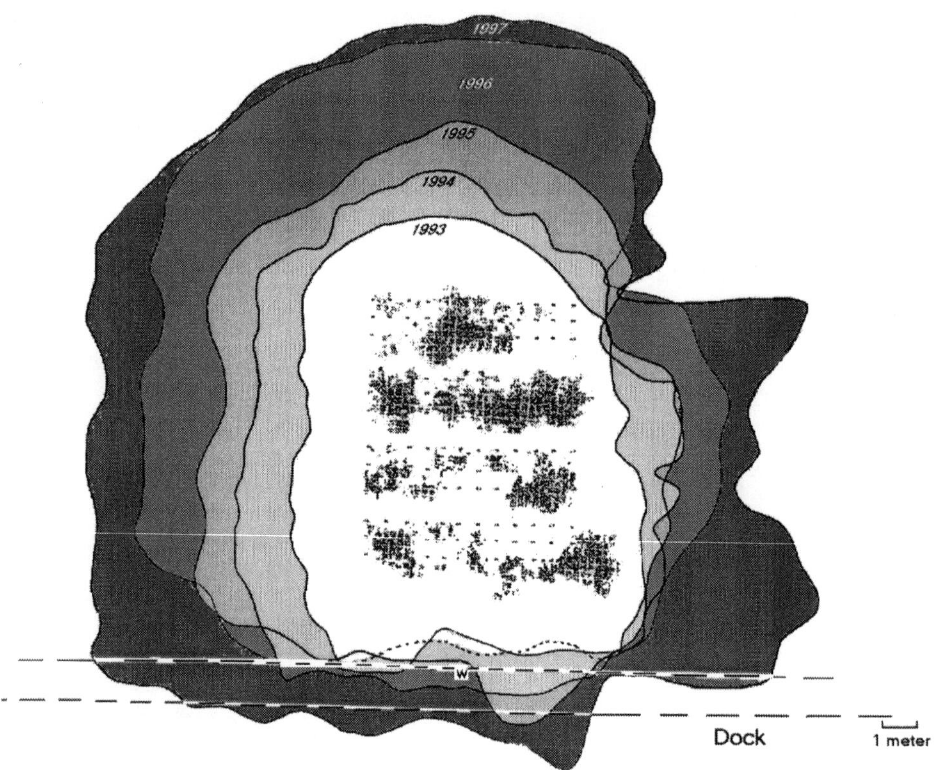

FIGURE 18.2 Configuration of initial 6 × 1 m transplant plots represented by paired dots. Growth during first (1991) year represented by stippled zones. Growth of the *V. americana* bed from 1992 through November 1997 as measured around the perimeter of the SAV bed. The site is adjacent to a privately owned dock at Ramsey Beach. The position of the north–south dock along the east border of the bed and 1 m scale bar are indicated at the base of the figure.

Forming a reference square over the bed, using lines stretched from the dock pilings to stakes, we measured, at 1 m intervals, the distance from the line to the edge of bed. These measurements are translated into the outline perimeters of the grass bed illustrated for annual assessments for 1993–1997 (Figure 18.2).

RESULTS

Extremely low tides and excellent water clarity in November 1993 afforded an opportunity to observe and photograph the entire bed that had grown from our initial planting (Figure 18.3). In addition to the shape of the bed, our photographic and visual observations revealed typical growth and morphological differences between the shoreward plants and offshore plants of each SAV bed. In the inshore half, plants tended to be more densely spaced with broader and/or longer leaf blades, whereas plants comprising the offshore half of the bed appeared to be less densely spaced, with fewer leaves reaching the water surface. Although this feature varied somewhat during seasonal observations, the Ramsey Beach inshore beds appeared bicolored in color photographs, with a reddish cast inshore but more green offshore.

Additional transplantings were conducted during May and June of 1995. Our observations from the 1991 transplants led us to select a new plot design for the next experimental trials. The new configuration consisted of four 1 m^2 blocks, each planted 1 m out from a central stake in four cardinal directions (Figure 18.4a) forming a simple cross-shaped plot. Three "cross-plots" were

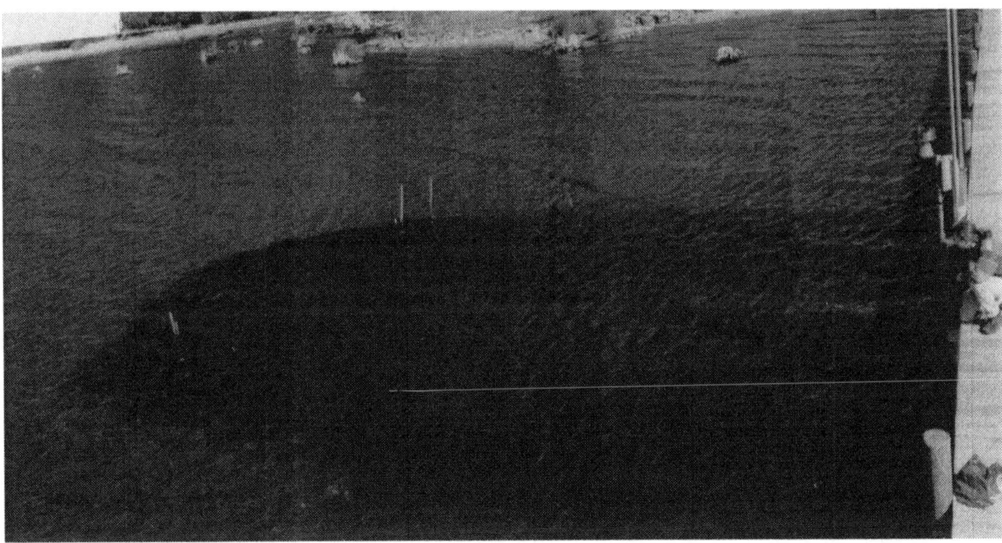

FIGURE 18.3 Composite photograph from a pier at Ramsey Beach (November 1993) looking north–northwest at the bed development of *V. americana* from initial planting (1991).

planted spaced 50 m apart along two parallel transects (spaced ca 18–20 m apart) extending approximately 120 m out from the shoreline, west of the 1991 planting. On December 9, 1995, we observed that the inshore cross-plot of each transect exhibited excellent growth, with the open 2 m space around the center stake filled in with new growth and runners extending into the area between the parallel transects (Figure 18.4b). In comparison, the middle transplant plots of each transect demonstrated only moderate growth, the meter blocks still evident. Only six plants remained in one of the two most offshore plots. Our assessment in winter 1997–1998 revealed that plants had

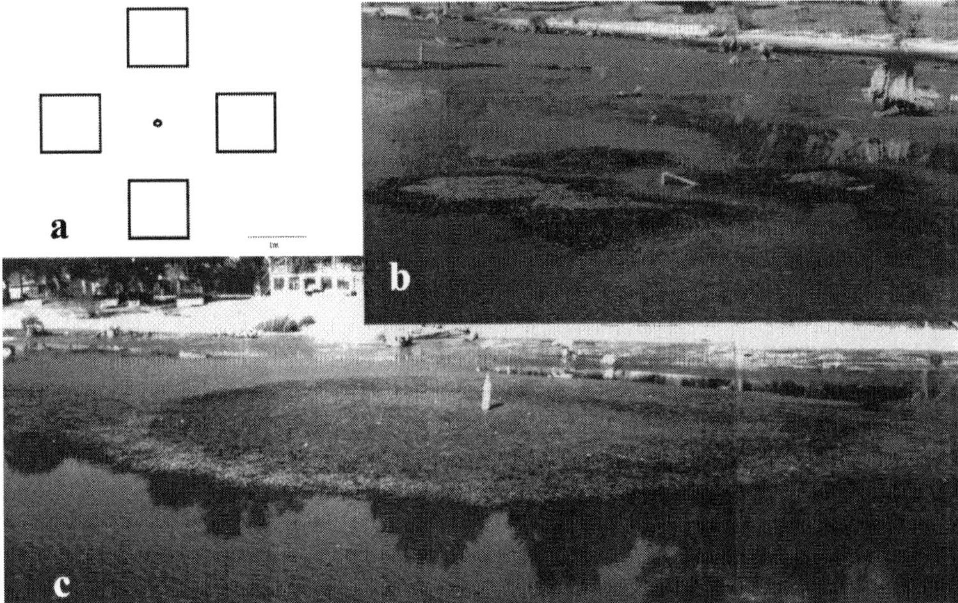

FIGURE 18.4 (a) Diagram of the cross-plot planting configuration. (b) Indicates growth of *V. americana* at the end of the first season (winter 1995) in an inshore cross-plot. (c) By winter 1997, the inshore cross-plot demonstrates the extensive growth and expansion of the beds which occurred in both inshore and middle plots.

significantly spread and filled in the space between the middle plots of the transects (Figure 18.4c); the single surviving offshore plot had expanded to a small bed over 1 m^2.

The bed perimeter outlines in Figure 18.2 illustrate the dramatic changes in the north shoreward border of the 1991 SAV bed, reflecting dynamic extension and regressions of this edge. The frequency of our assessments was unfortunately too far apart to permit finer resolution of the change dynamics of shape of the north edge of the bed. However, it is evident that dynamic factors were affecting the shoreward border differently than the other boundary edges. Hurricanes Erin (August) and Opal (October) both passed through the region in 1995. Perimeters of all transplant beds were measured and photographed emersed during an extremely low tide in December 1995. The low water level revealed a 15–20 cm berm of sediment along the shoreward margin of the SAV bed planted in 1991, and each inshore planting of the 1995 transects. In December 1997 the inshore berm previously observed was absent from all beds and north-edge plants had grown further shoreward (Figure 18.2). During 1997, Hurricane Danny affected Perdido Bay with only high water, but without significant winds.

The observations of expansion and growth of the plant bed under the dock provide additional clues to some of these inshore dynamics. As recorded in Figure 18.2, *V. americana* plants had spread under the dock during 1994 and 1995. The dock was uplifted in 1995 during hurricane surges (Erin), to be reset later by hydro-jetting (early 1996). In our September 1996 assessment (Figure 18.2), the edge of the bed had discernibly retreated, no longer extending as far beneath the dock. However, in 1997 the bed had regrown with plant runners expanding beneath and significantly eastwardly beyond the dock. These observations represent examples of the interactions of hydrographic processes and dynamics of plant growth that occur in inshore SAV beds.

In July 1997, observations at the Ramsey Beach plantings revealed the distinctive bi-colored differentiation of shoreward and offshore plants noted earlier in each planting of the transect cross-plots. These observations suggest that agitation, probably from the wave actions reflected from the shoreline, affects the inshore perimeters of these plots (perhaps also contributing to the berm). This wave agitation creates a different dynamic on the inshore sides of these plots that affects differences of growth and morphology observed among the plants in varying locations of each plot.

During low water and high visibility periods of winter 1997–1998, we conducted surveys along the east and north shorelines of Perdido Bay. Historically, there had been *V. americana* beds in the north basin of Perdido Bay along the south shore (Millview) and west shore (Lillian) and especially at Grassy Point. These beds continue to exist, but we were surprised to find numerous smaller (< 1–2 m diameter) and moderate (2.5–4 m diameter) beds along the north shoreline, both east, but especially west of the Ramsey Beach site. *V. americana* beds were not observed along the east shore from Bayou Marcus creek toward Ramsey Beach for a distance of approximately 1.6 km, in an area where the shoreline lacks residential development.

DISCUSSION

It is hypothetical whether or not propagules from our experimental plantings accelerated the colonization and expansion of SAV beds along the Perdido north shore from Ramsey Beach to a few hundred meters east of the mouth of Elevenmile Creek. Previous to our transplanting, only one *V. americana* bed existed and it was substantially larger than the smaller patches observed in 1997–1998 (longest axis exceeding 3–4 m). This bed resulted from a single transplanting by other researchers (Livingston and McGlynn, Environmental Planning and Analysis, Inc., Tallahassee, Florida, personal communication) established prior to 1992. All other beds we observed along the north shore of Perdido Bay were significantly smaller, which, based upon our planting observations, would represent 1–3 year growth stage. Collectively, these observations suggest that initial colonization may be limited by propagule recruitment. However, once colonization begins, new beds become additional sources of propagule production with further growth rapidly occurring. Observations made during winter 1997–1998 support this scenario, revealing that numerous new beds appeared along the northern shore of Perdido Bay. Most of these beds were small, resembling the 1 to 2-year-old stage when compared to beds we transplanted previously. Exactly how the "new" beds became established along the north shore remains speculative. The size of the new beds resembles that of our experimental transplantings after 1–2 years of growth. One cannot ignore the occurrence of Hurricanes Erin and Opal (1995) and storm effects as potential mechanisms of dispersal.

Previous to these transplanting experiments and this new colonization, there were no observed or historical records of SAV beds along the north shore of Perdido Bay. Conceptually, in a freshwater-dominated estuarine system, one would expect the upper bay SAV to have received the greatest potential exposure to incoming river-borne contamination from effluents, and thereby act as the best sentinels for exposure effects from that source. In contrast, SAV species with higher salinity adaptation would potentially respond to exposure to contamination sources located further "down" the bay estuarine system. During the period of our observations, since 1986, SAV beds have existed in the inland waterway adjacent to the mouth of Perdido Bay both before and after abatement of deep staining and associated components of KME. We also observed *V. americana* beds occurring along the south and west shorelines of upper Perdido Bay as well. Therefore, we are confident that the success of the transplanted and regrowth of *Vallisneria* beds demonstrates the utility of these SAV species as sentinels and as ecological indicators of the effects of pollution abatement in estuarine systems.

SUMMARY

Observations of growth and spreading of *V. americana* in trial planting plots have led us to believe that recruitment of propagules can represent the limiting factor to initial dispersion and establishment of new *V. americana* beds, at least under the hydrographic conditions in this situation. To assess the response of plants in a situation with multiple interacting factors, plants set out in the cross-plot configuration have revealed interesting and dynamic ecological responses. Inshore plot perimeters have grown more rapidly, maintained more plant and leaf density, and often demonstrated red leaf blade coloration. The inshore edge of the bed appears to capture wave wash and sediments which in some instances resulted in formation of a berm. Anthropogenic sediment disturbance can apparently inhibit extension of the bed perimeter, based upon retreat of the bed from the dock during hydro-jetting. Therefore, agitation, potentially from shore line wave reflection, may be an important factor in the sculpture of SAV beds. Refinement of hydrographic monitoring approaches to test these hypotheses should be factored into strategies using grasses in ecological assessment approaches.

We believe that increased areal coverage of desirable SAV in upper Perdido Bay will provide renewed ecological benefits. Efforts to transplant *V. americana*, or any SAV, to augment or re-establish essential fishery habitat, for example, is consistent with the objectives of the Magnuson-Stevens Fishery Management and Conservation Act. It would be important in future execution of studies of this sort to concurrently assess invertebrate and fish species to better define the effects on grass-bed associated communities.

ACKNOWLEDGMENTS

Numerous people assisted and contributed to the success of the plantings and our observations. We were assisted greatly during transplanting and conducting plant growth assessments by Matthew MacGregor, Barbara Ruth, Margaret Posten. Drs. Emile Lores and David Weber, Eric P. Ericson, Florida Department of Environmental Protection, and Jennifer Jackson assisted us in composition of the figure of Perdido Bay from GIS data and in preparation of the contour figure. Expert advice and composing finalized figures was given by Vic Camargo and Steve Embry. A number of student summer employees assisted in the physical labor during the observation period. This chapter is Contribution No. 1060 of Gulf Ecology Division, United States Environmental Protection Agency, National Health and Environmental Effects Laboratory.

REFERENCES

Dennison, W. C., R. J. Orth, K. A Moore, J. C. Stevenson, V. Carter, S. Kollar, P. W. Bergstrom, and R. A. Batiuk. 1993. Assessing water quality with submerged aquatic vegetation. *Bioscience* 43:86–94.

Doust, L. L., J. L. Doust, and M. Biernacki. 1994. American wild celery, *Vallisneria americana*, as a biomonitor of organic contaminants in aquatic ecosystems. *Journal of Great Lakes Research* 20(2):333–354 .

Duke, T. and W. L. Kruczynski. 1992. Status and trends of emergent and submerged aquatic vegetated habitats, Gulf of Mexico. United States Environmental Protection Agency, Washington, D.C., EPA/800-R-92-003, 161 pp.

Korschen, C. E. and W. L. Green. 1988. American wild celery (*Vallisneria americana*): Ecological considerations for restoration. Fish and Wildlife Service Technical Report 19, U.S. Department of the Interior, Washington, D.C., 24 pp.

Lewis, R. R., M. J. Durako, M. D. Moffler, and R. C. Phillips. 1985. Seagrass meadows of Tampa Bay: a review. In *Proceedings, Tampa Bay Area Scientific Information Symposium* (May 1982), S. F. Treat, J. L. Simon, R. R. Lewis, and R. L. Whitman, Jr. (eds.). Burgess Publishing Company, Minneapolis, MN, pp. 210–246.

Livingston, R. J. 1984. The relationships of physical factors and biological response in coastal sea grass meadows. *Estuaries* 7(4A):377–390.

Livingston and McGlynn. Personal communication. Environmental Planning and Analysis Inc., Tallahassee, FL.

McRoy, C. P. and C. Helfferich. 1980. Applied aspects of seagrasses. In *A Handbook of Seagrass Biology: An Ecosystem Perspective,* R. C. Phillips and C. P. McRoy (eds.). Garland Publishing Incorporated, New York, pp. 297–343.

Morris, L. J. and D. A. Tomasko (eds.). 1993. *Proceedings and Conclusions of Workshops on Submerged Aquatic Vegetation Initiative and Photosynthetically Active Radiation.* Special Publication SJ93-SP13, St. Johns River Water Management District, Palatka, FL.

Neckles, H. A. (ed.). 1994. *Indicator Development: Sea Grass Monitoring and Research in the Gulf of Mexico.* Report of a workshop held at Mote Marine Laboratory, Sarasota, FL, January 28–29, 1992. U.S. Environmental Protection Agency, Washington, D.C., EPA/620/R-94/029, 64 pp.

Olinger, L. W. 1976. Environmental and Recovery Studies of Escambia Bay and the Pensacola Bay System, Florida. Surveillance and Analysis Division, Atlanta, GA. U.S. Environmental Protection Agency, Washington, D.C., EPA-904/9-76-016, 438 pp.

Phillips, R. C. and C. P. McRoy (eds.). 1980. *A Handbook of Seagrass Biology: An Ecosystem Perspective.* Garland Publishing Incorporated, New York.

Thayer, G. W., D. A. Wolfe, and R. B. Williams. 1975. The impact of man on sea grass systems. *American Scientist* 63:288–296.

Zieman, J. C. and R. G. Wetzel. 1980. Productivity in seagrasses: methods and rates. In *A Handbook of Seagrass Biology: An Ecosystem Perspective,* R. C. Phillips and C. P. McRoy (eds.). Garland Publishing Incorporated, New York, pp. 87–116.

Zieman, J. C. and R. T. Zieman. 1989. The ecology of the sea grass meadows of the west coast of Florida: a community profile. U.S. Department of the Interior, Fish and Wildlife Service, Washington, D.C., Biological Report 85 (7.25), 155 pp.

Section IV

Management

19 Scaling Submersed Plant Community Responses to Experimental Nutrient Enrichment

Laura Murray, R. Brian Sturgis, Richard D. Bartleson, William Severn, and W. Michael Kemp

Abstract Detailed mechanistic understanding of how nutrient enrichment leads to losses of seagrasses and related submerged aquatic vegetation (SAV) is still lacking, despite extensive research on the topic. In this study, we compare results from a series of three mesocosm experiments to address how physical and biotic scales influence responses of SAV communities to nutrient enrichment. These experiments, which involved the SAV species (*Potamogeton perfoliatus*) formerly abundant in Chesapeake Bay, considered the following specific ecosystem scales: (1) frequency and timing of nutrient additions; (2) residence time of water within mesocosms; and (3) trophic complexity (food-chain length). Ecosystem model simulations were used to help guide experimental designs and interpretations.

Time scales of response to nutrient enrichment differed for SAV (8–9 wk) and their attached epiphytes (2–6 wk). SAV growth responses to nutrients varied with season; in spring the above-ground plant tissues were most sensitive, while in fall responses were confined to below-ground biomass (roots and rhizomes). In the fall experiment, continuous nutrient input resulted in greater enhancement of epiphytes and inhibition of plant growth than did identical loading rates delivered as pulsed inputs. This may be explained by the higher biomass and kinetic saturation coefficients of the vascular plants, which favored their uptake of higher pulsed nutrient concentrations. In general, longer residence time of water over SAV beds improved the plant's ability to cope with nutrient enrichment, while faster water exchange rates favored epiphyte growth at the expense of SAV. Although herbivorous grazing on epiphytes partially relieved SAV growth inhibition at moderate nutrient loading, grazing did not significantly alter epiphyte or plant responses to enrichment at higher nutrient levels.

A comparison of effects of the three scaling factors suggests that grazing exerted the largest relative influence on the SAV community under moderate nutrient enrichment. However, at high nutrient loading rates, changes from continuous to pulsed nutrient delivery and from high to low water exchange rates both resulted in stronger relative responses than did increased grazing. Results of these studies provide a basis for explaining variability in reported SAV community responses to nutrient enrichment and for extrapolating results from controlled experiments to conditions in nature.

INTRODUCTION

Worldwide losses of seagrasses and other submerged aquatic vegetation (SAV) have been well documented (e.g., den Hartog and Polderman 1975; Orth and Moore 1983; Silberstein et al. 1986).

Many of these losses are associated with over-enrichment of coastal waters with nutrients, which stimulates growth of microalgal communities, resulting in reduction of photosynthetically active radiation (PAR) at SAV leaves (Phillips et al. 1978; Kemp et al. 1983; Twilley et al. 1985; Cambridge et al. 1986; Sand-Jensen and Borum 1991; Duarte 1995). The process by which nutrient enrichment leads to declines in SAV abundance is, however, highly non-linear, and the associated mechanisms are still poorly described.

One essential approach to expand the understanding of these mechanisms involves the use of controlled experimental ecosystems (mesocosms). Such studies are needed to infer causal relationships at the integrated level of whole ecosystems (Odum 1984). Over the last several decades, there has been a growing literature on mesocosm studies of SAV responses to nutrients. The scales of these experiments vary widely in terms of their:

1. size — from 0.01 m^3 chambers (Sturgis and Murray 1997) to 400 m^3 ponds (Kemp et al. 1983);
2. duration — from 2 wk (Philips et al. 1978) to 6 mo (Twilley et al. 1985);
3. water exchange rate — from rapid (e.g., 16 d^{-1}, Neckles et al. 1993) to slow (e.g., 0 d^{-1}, Neundorfer and Kemp 1993);
4. ecological complexity — from simple ecosystems dominated by SAV and algae (Short et al. 1995) to more complex ones also including herbivorous invertebrates and carnivorous fish (Brönmark and Vermaat 1997).

The inherent reduced scales of all experimental ecosystems, compared to natural habitats, may substantially alter their responses to experimental manipulations, including SAV responses to nutrient enrichment (e.g., Kemp et al. 1980; Frost et al 1988). There has been a growing recognition of how ecological processes, in general, vary with scales of time and space (O'Neill 1989; Wiens 1989) as well as ecological complexity (Frost et al. 1988). The reduced scales of mesocosms result in both "artifactual scaling effects" associated with containers (e.g., wall growth, Chen et al. 1997) and "fundamental scaling effects" characteristic of all ecosystems, both natural and experimental (e.g., depth effects on photosynthesis, Petersen et al. 1997). In a sense, mesocosms are a "halfway house" between the abstractions of mathematical models and the reality of natural ecosystems (Kemp et al. 1980; Lawton 1996). Therefore, coupling mesocosm experiments with simulation models can give a better understanding of the internal workings of ecosystems.

In nature, delivery of nutrients to SAV communities occurs through many routes, each of which has characteristic time-scales of variability. These nutrient sources include the relatively continuous flows of groundwater and industrial inputs, as well as seasonally varying inputs from streams and tidally varying exchange with adjacent open waters (Figure 19.1). While input rates and frequencies are of primary importance, nutrient concentrations in SAV communities also vary with relative size of SAV beds and rates of water exchange with adjacent estuarine ecosystems (e.g., Moore et al 1995). In addition, SAV responses to nutrient inputs at different temporal and spatial scales are dependent on the complexity (e.g., food-web structure) of the associated ecological community. Trophic cascade theory suggests that SAV growth and survival may be dependent on both bottom-up (nutrient inputs) and top-down (from upper trophic levels) controls (Carpenter and Kitchell 1988). Indeed, a number of recent studies have demonstrated how SAV community responses to eutrophication can be mediated by algal grazers (e.g., Orth et al. 1984; Hootsmans and Vermaat 1985; Neckles et al. 1993), which in turn may be controlled by their predators (Brönmark and Vermaat 1997).

The objective of this chapter is to compare how ecosystem scales (time, space, and ecological complexity) modify nutrient responses of an estuarine community dominated by the SAV species, *Potamogeton perfoliatus*. Specifically, we compare results of three SAV nutrient enrichment studies in which experimental scales were varied as: (1) the frequency and timing of nutrient additions;

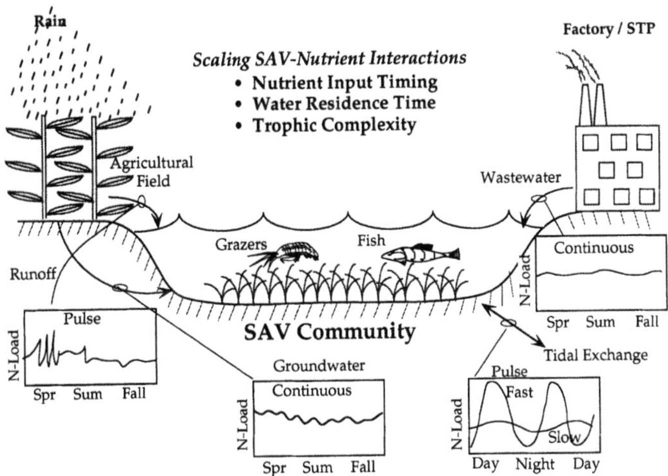

FIGURE 19.1 Schematic depicting the variations of nutrient inputs to submersed plant communities.

(2) the residence time of water within the mesocosm; and (3) the SAV community's trophic complexity. Ecosystem model simulations were used to help guide experimental designs and interpretations, and the question of extrapolating experimental results to conditions in nature is discussed.

METHODS

EXPERIMENTAL DESIGN

Three separate experiments were conducted in a mesocosm facility at the University of Maryland's Horn Point Laboratory in Cambridge, MD. The first, referred to as the nutrient delivery experiment, tested the hypothesis that both variability and timing of nutrient enrichment influence responses of submersed plant communities. The second study, referred to as the trophic complexity experiment, investigated the thesis that grazing on the SAV epiphytes partially mitigates negative SAV effects of nutrient addition. The third investigation, referred to as the water exchange rate experiment, tested the hypothesis that the average water residence time within SAV beds alters community responses to nutrients added in that water. All experimental ecosystems contained the submerged plant *P. perfoliatus*, which was a historically important SAV species in Chesapeake Bay (e.g., Kemp et al. 1983). All experimental protocols and analysis were done in similar fashions. Small variations in experimental designs are described separately for each study.

Nutrient Delivery Experiments

Experiments to investigate effects of varying frequency and timing of nutrient inputs were conducted in an outdoor greenhouse. Two experiments were conducted during different phases of Chesapeake Bay's SAV growing season: one in the early part of the season (May–July), the other during the later portion of the plant growth period (July–October). The experimental design consisted of four replicates of each nutrient enrichment treatment and controls. Treatments included dissolved inorganic nutrients (DIN) delivered in both a pulse (at a rate matching the water retention time) and continuous input mode at a nutrient loading rate of 38 μmol N l^{-1} d^{-1} (NH_4NO_3) and 3.8 μmol P l^{-1} d^{-1} PO_4^{3-} (NaH_2PO_4). These mean nutrient loading rates were similar to those used in many previous experiments with seagrasses and SAV (e.g., Twilley et al. 1985; Neckles et al. 1993; Short et al. 1995). For all experiments, low concentrations of dissolved nutrients in water were obtained using 400 m^3 SAV ponds as "nutrient scrubbers," and water was filtered (2 μm) prior to use. For these experiments water retention time in each mesocosm was 3.5 days in spring and 7 days in fall experiments; the study duration was 12 weeks. Plant material was harvested for analysis at the end of the experiment. Measurements of nutrient levels, light conditions, plant growth, and accumulation of epiphytic material were made at weekly time intervals over the duration of the experiment to assess the time-course of responses to treatments. Details of experimental design are provided in Sturgis and Murray (1997).

Water Exchange Experiments

Experiments to test how variations in water retention time alter SAV community response to nutrients were conducted in laboratory SAV mesocosms under controlled temperature and light conditions. Experiments included a sequential replication (3) of four rates of water exchange (1, 3, 6, and 12 times per day). In this chapter, we report only the results from the slowest water exchange (once per day, referred to as "controls") and from "fast" experiments (exchanged six times per day). Two levels of nutrient input concentrations (low, 2–4 μM DIN and high, 20–24 μM DIN) were used in each experiment. The resulting nitrogen loading rates were (in fast exchange experiments) approximately 18 μmol N l^{-1} d^{-1} for low and 144 μmol N l^{-1} d^{-1} for high treatments. Each replicate experiment was conducted for a period of 6 weeks. Plants and sediments were replaced between experiments to ensure similarity in experimental conditions among treatments. Small plants (5–20 cm stem length) were established in pots with estuarine sediments and transferred to the experimental mesocosms when they reached a height of 20 cm.

Trophic Complexity Experiments

Experiments to address the question of how trophic complexity influenced SAV community response to nutrient enrichment were conducted in the indoor mesocosm facility as described above. These experiments included the following trophic manipulations: one macrophyte species (*P. perfoliatus)*, referred to as control, and one macrophyte species + one grazer species (gammarid amphipod), referred to as grazer. We also used fish treatments with mummichug (*Fundulus* sp.) as amphipod predators (Murray et al. 1997); however, results of these are not included in this document. Two nutrient regimes were established in separate experiments: low (< 10 μmol l^{-1} N); high (> 30 μmol l^{-1} N). Low nutrient water was obtained from experimental ponds as described above and exchanged once every 24 h. The low nutrient experiments were conducted for 6 weeks; the high nutrient experiments were run for 12 weeks. Plant growth and epiphyte mass were measured biweekly; measurements made at the 6-week time period were used from the high nutrient experiments for comparison to the low nutrient experiments.

Experimental Systems

Nutrient delivery experiments were conducted in 10 liter acrylic mesocosms containing *P. perfoliatus* planted in plastic pots (15 cm diameter × 15.2 cm height) with 15 cm of sediment collected from a vegetated area of the Choptank River. Chambers were placed in water baths to maintain temperature to within 2°C of ambient estuarine water conditions. Air stone bubblers maintained the internal mixing of the systems. Light, measured by a LI-COR (Model 100-32) irradiance meter, was 95% of outside light.

Both the water exchange and the trophic complexity experiments were conducted in 100 liter glass aquaria with recirculating pumps for internal mixing. Light levels at the sediment surface were maintained at levels above those considered limiting for this plant species (≥ 200 μ Ein m^{-2} s^{-1}, Harley and Findlay 1994), and artificial lights were programmed on a 12 hour light/dark cycle.

Temperatures ranged between 23–25°C. In all experiments, wall growth was cleaned weekly, and materials removed from the walls were allowed to remain within the mesocosm.

Experimental Measurements

Plant Variables

For all experiments, *P. perfoliatus* rhizome segments were obtained from nearby experimental estuarine ponds (Twilley et al. 1985; Neundorfer and Kemp 1993). Randomly selected rhizome-shoot segments were planted into sediments collected locally from a site which formerly supported SAV beds. At the conclusion of the experiments, above ground portions of the plants were scraped free of epiphytic material, rinsed in de-ionized water, and dried to a constant weight at 60°C. Below-ground tissue was rinsed free of sediments and analyzed as above-ground biomass. Plant growth rate was estimated from weekly measurements of stem density and length. Growth rates were computed utilizing the previous week's measurement and calculated as growth per day for the experimental period.

Epiphytic Community Variables

Epiphytic biomass on the plant leaves was collected differently in the separate experiments. In the nutrient delivery experiments, epiphytes were collected from five leaves (fourth leaf from the plant apex) in each chamber. In the trophic complexity experiments, epiphytes were collected from an average of nine leaves, three each from the top, middle, and bottom portions of the plant (Severn 1998). For the exchange rate experiments, epiphytes were taken from whole plants (two per mesocosm). For all experiments, epiphytic material was scraped into filtered estuarine water (1.2 μm). The resultant algal slurry was filtered onto pre-ashed, weighed filters, which were then dried to a constant weight at 60°C. Once the dry weight was determined, samples were combusted at 500°C for 4 h and the ash-free dry weight (AFDW) calculated by difference.

Simulation Modeling

The dynamic behavior of experimental SAV communities and responses to experimental treatments were simulated using a numerical ecosystem process model. This model was structured around 14 state variables, including dissolved inorganic nitrogen and phosphorus (DIN, DIP) in water column and sediment pore waters; labile and refractory sediment organic matter; algal biomass for phytoplankton, epiphytes, and microphytobenthos; SAV biomass partitioned into leaves/stems and roots/rhizomes; herbivorous amphipod grazers; and predatory fish (Figure 19.2).

The model's finite difference equations included terms to represent nutrient uptake, recycling, and transformations; plant photosynthesis, growth, respiration, and mortality; grazing, predation, and related bioenergetic processes for experimental invertebrates and fish; and physical variables such as water mixing and exchange, light attenuation, and temperature. Equation formulations used standard hyperbolic kinetic relations for photosynthesis-light, nutrient uptake, grazing, and predation (e.g., Kemp et al. 1995, Madden and Kemp 1996). The model was coded in a structured simulation language (Stella, High Performance Systems), with physical forcing functions programmed as continuous functions. Numerical computations were done using fourth-order Runga–Kutta methods and a fixed time-step of 1 h.

The model was used to explore potential experimental design options (e.g., grazer densities, water exchange rates) prior to initiation of some empirical studies. It was also used to analyze empirical observations to test hypothesized data interpretations. Finally, the model is currently being used to extend results from specific experiments to other combinations of treatments not

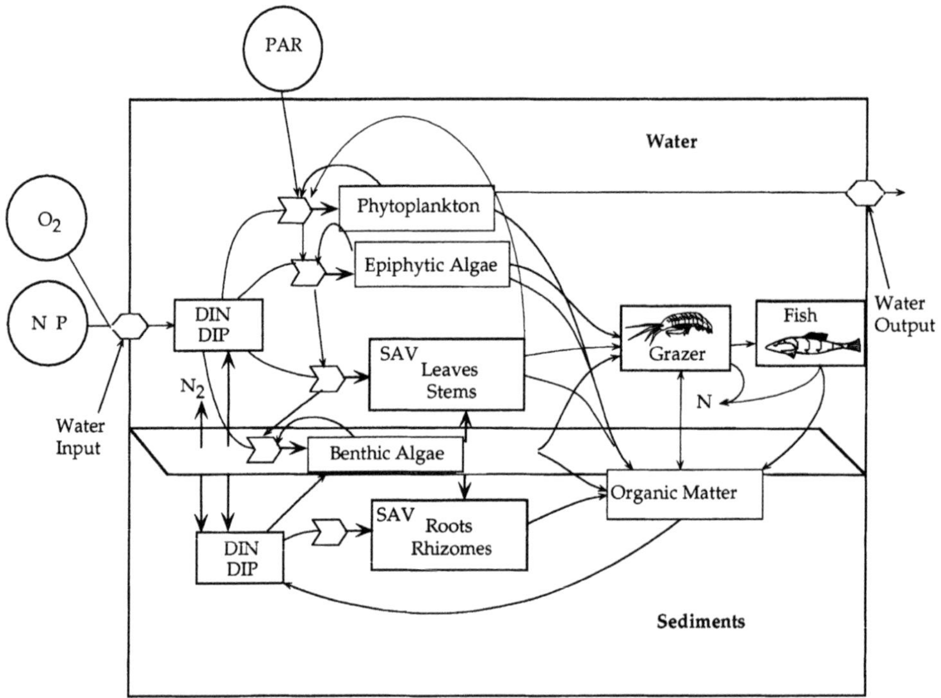

FIGURE 19.2 Diagram of conceptual model of SAV trophic complexity in mesocosms.

explicitly studied empirically, and to extrapolate experimental findings to conditions in natural ecosystems.

RESULTS

Nutrient Delivery Experiments

Although nutrient enrichment resulted in significant reductions in plant biomass during spring regardless of delivery frequency, there was 30–60% less biomass in mesocosms treated with variable (pulsed) nutrient inputs compared to those receiving continuous inputs at the same loading rate (Figure 19.3A). The inability to detect a difference between pulse and continuous delivery was due to relatively large variance and small sample size. For the fall experiment, nutrient enrichment resulted in root/rhizome biomass to be significantly reduced; however, leaf/stem biomass declined significantly only with continuous delivery of nutrients (Figure 19.3B). Furthermore, in the fall experiment, continuous delivery resulted in less (40–70% lower) plant biomass than did the pulse treatment (Figure 19.3B). This difference was significant for both leaf/stem and root/rhizome fractions.

Time-course observations on *P. perfoliatus* growth and epiphyte biomass (as chlorophyll *a*) revealed interesting and complex patterns during the spring experiment (Figure 19.4). Similar patterns were observed for the fall experiment (Sturgis and Murray 1997). Plant growth generally increased slowly for all mesocosms over the first 5–7 wks of the experiment. Although the pulse-treated plants started showing signs of nutrient enrichment effects after 4–5 wks, consistently significant depressions in plant growth for nutrient enriched systems were evident only after 8–9 wks (Figure 19.4A). In contrast to this pattern for plant growth, accumulation of epiphyte biomass in nutrient enriched systems exhibited significant increases over the first 6–8 wks of the experiment, and reached a stable plateau after 4–6 wks in continuous treatments and after 6–8 wks in pulse

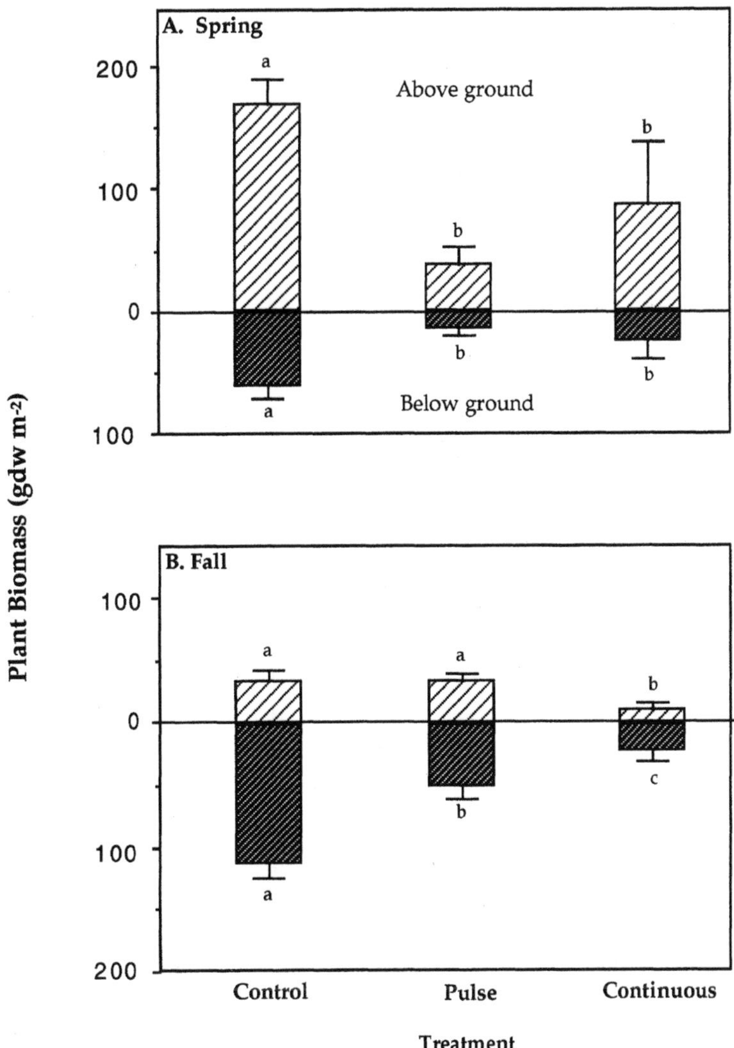

FIGURE 19.3 End of experiment plant biomass for nutrient delivery, spring (A) and fall (B) experiments (mean + SE). Superscript letters show significant differences; $p < 0.05$, ANOVA. (Adapted from Sturgis and Murray 1997.)

treatments (Figure 19.4B). Levels of epiphyte biomass remained steady for the duration of the experiment in both nutrient treatments after week 6. By week 10, responses in plant growth to pulsed and continuous treatments were consistent with the levels of epiphyte biomass accumulation during weeks 2–6, where both reductions in plant growth and increases in epiphyte biomass were greatest with pulse treatments. There was, however, a 4–8 week lag between epiphyte build-up and plant growth responses (Figure 19.4A, B). Thus, epiphyte biomass responses to nutrient addition within the first 2–6 weeks of these experiments served as good predictors of subsequent plant responses, a fact that we exploited in later experiments.

Exchange Rate Experiments

Model simulations were used to predict responses of epiphytic biomass to changes in water exchange rates under different nutrient loading rates (and associated DIN concentrations,

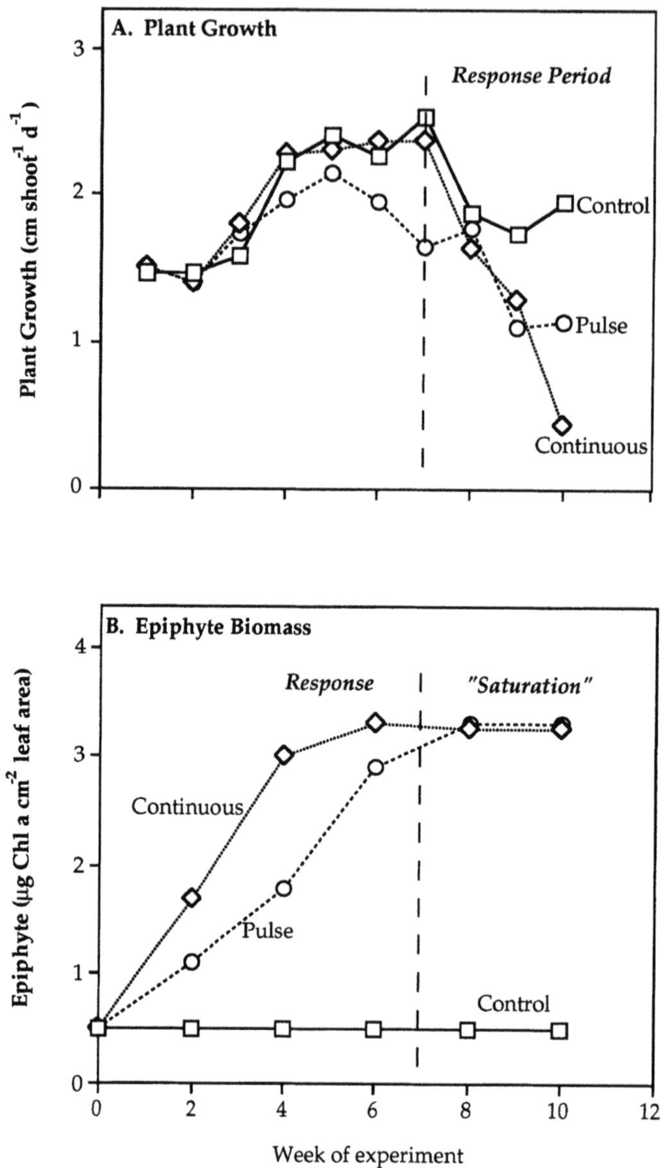

FIGURE 19.4 Short-term response of plant growth (A) and epiphytes (B) to variations in nutrient delivery during spring experiments. (Redrawn from Sturgis and Murray 1997).

Figure 19.5A). In general, epiphyte biomass tends to increase with increasing water exchange rate and with nutrient concentration. Exchange-rate response-functions, however, differ in shape with nutrient loading and concentration. At high nutrient levels, epiphyte biomass increases rapidly with initial increases in water exchange rate, but the response "saturates" at approximately 5 d^{-1} (Figure 19.5A). At relatively low nutrient concentrations (e.g., < 5 μM), epiphyte biomass exhibits a linear response to increasing exchange rate up to 15 d^{-1}. At intermediate nutrient levels, responses are mixed. Based on these model simulations, water exchange rates of 1, 3, 6, and 12 per day were to be used in empirical studies.

Experimental observations on the effects of water exchange rate in modifying SAV responses to nutrient loading can be seen as changes in the biomass (total dry weight) of epiphytic communities

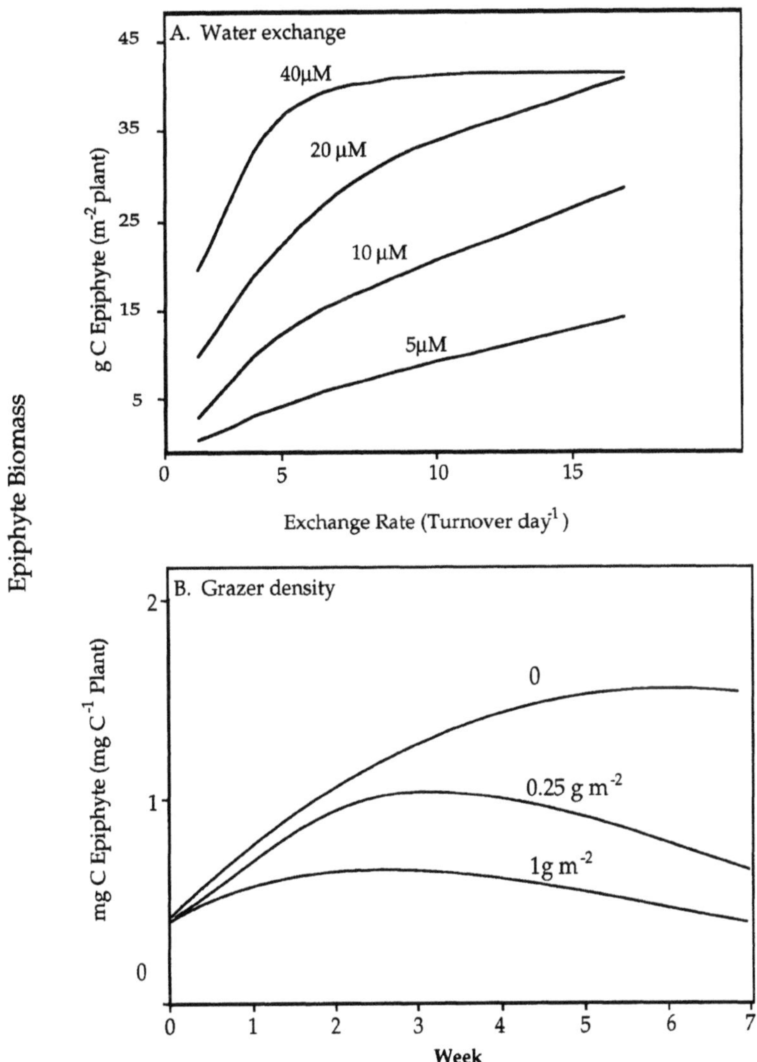

FIGURE 19.5 Model output of the effect of water exchange rate at four nutrient input concentrations of DIN (A) and grazer densities (gm^{-3}) (B) on epiphyte biomass.

(Figure 19.6A). Although epiphytic algal chlorophyll *a* did increase significantly with nutrient enrichment (data not shown, L. Murray, unpublished data), increases in total dry weight with nutrient enrichment were not significant (Figure 19.6A). Epiphytic responses to variations in water exchange rate were, however, significant at both low and high nutrient levels, with 34% increases from control (1 d^{-1}) to fast (6 d^{-1}) exchange at low nutrient levels and approximately 27% increase at high nutrients (Figure 19.6A). These responses are similar to those predicted from model simulations (Figure 19.5A). Although our use of total dry weight to characterize the epiphytic community tends to exaggerate physical rather than nutrient effects, the results do emphasize the importance of this scaling variable.

Trophic Complexity Experiments

Simulated time-courses of epiphyte biomass under nutrient enrichment and differing densities of amphipod grazers, revealed complex patterns (Figure 19.5B). Grazing results in a marked decline

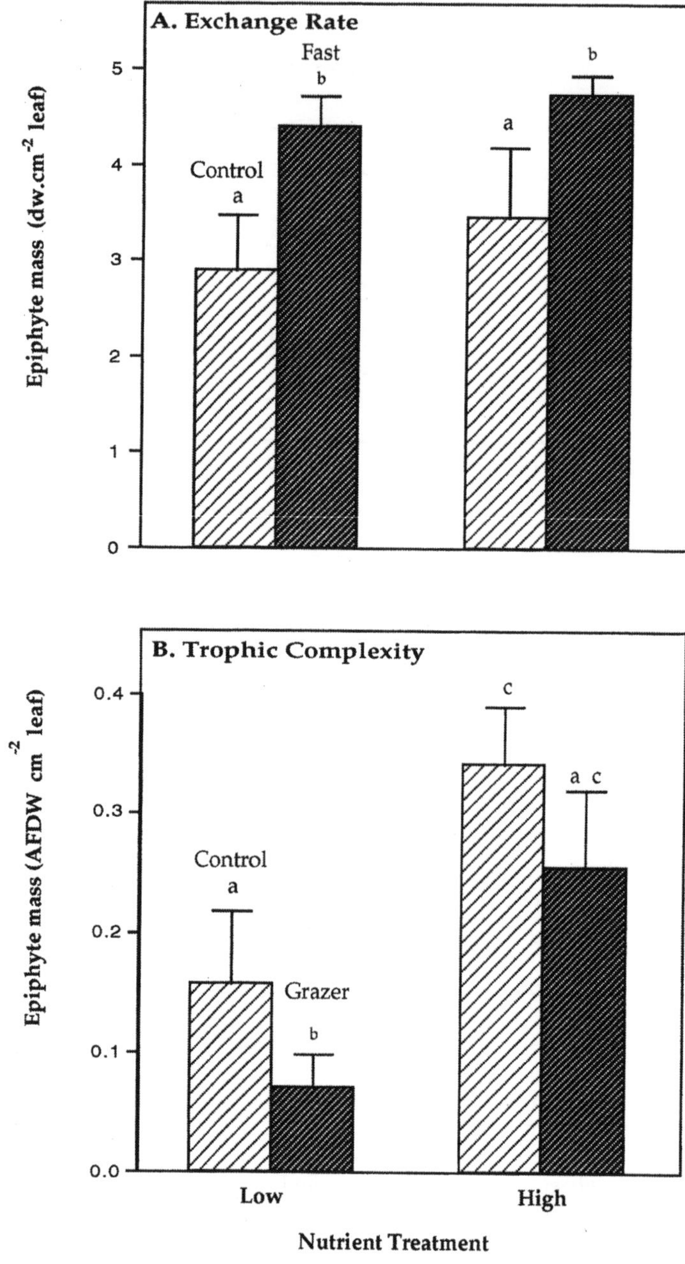

FIGURE 19.6 Epiphyte mass for exchange rate (A) and trophic complexity (B) and experiments. Values (means + SE) are for the end of the experiment. Superscript letters designate significant differences; $p < 0.05$, ANOVA.

in epiphyte biomass after initial increase over 2–3 wks. At initial grazer densities of 0.25 g m^{-2}, epiphyte levels were controlled effectively. This grazer level was selected for the trophic complexity mesocosm experiment.

As with the other two experiments, nutrient enrichment resulted in significant increases in epiphyte biomass (by two- to threefold) with nutrient addition (Figure 19.6B). In this case, we used ash-free dry weight (AFDW) as a measure of the organic fraction of epiphytic material available to grazer consumption. Amphipod grazing caused marked decreases in epiphytic material under both low (56%) and high (25%) nutrient treatments. However, this grazing effect was only significant at low nutrients (Figure 19.6B). Grazing in high nutrient treatments reduced epiphyte material to levels comparable to those under control conditions at low nutrients.

Comparisons Among Experiments

To compare results among these three experiments, all treatment responses for both epiphytes and SAV plants have been calculated in relative terms as a percent of respective control systems (Table 19.1). In all cases, nutrient enrichment resulted in negative plant responses and positive responses of the epiphytic community (reading across, Table 19.1). For comparative purposes, results from the spring nutrient delivery experiments are presented because these correspond best to the timing of the other experiments. While both modes of nutrient delivery resulted in decreased plant growth and elevated levels of epiphytes, effects were more pronounced with pulse delivery (29% reduction in plant growth and an epiphyte increase of 165%). Increases in water exchange rate resulted in decreased plant growth and increased epiphytic biomass under both low and high nutrients. It is interesting to note that epiphyte accumulation increased less in response to nutrient addition (from low to high) than it did in response to exchange rate treatment (from control to fast exchange).

Of the three scaling treatments, herbivorous grazing treatments exerted the largest influence on SAV community responses to experimental nutrient enrichment. For example, under grazing pressure, nutrient enrichment resulted in a 131% decrease in plant growth compared to an 88% decrease without grazing. In contrast, at high exchange rates plant growth decreased by 32% with nutrient

TABLE 19.1
Comparison of relative responses of submersed plants and their epiphytic communities to environmental manipulations.

	Nutrient Input Level			
	Plant Response[d]		Epiphyte Response[e]	
Environmental Manipulation	**Low**	**High**	**Low**	**High**
Nutrient Delivery[a]				
Control	0	–49	0	+85
Pulse	—	–78	—	+82
Trophic Complexity[b]				
Control	0	–88	0	+112
Grazer	+43	–88	–56	+63
Exchange Rate[c]				
Control	0	–26	0	+20
Fast	–16	–48	+52	+64

Note: All comparisons are made for a 6-week experimental period. Relative responses calculated as percent of control [100 (Treatment – Control)(Control)$^{-1}$].

[a] Comparisons from spring experiments. "High" nutrient levels are synonymous with continuous inputs. There was no "pulsed" delivery for low nutrient treatments (Sturgis and Murray 1997).
[b] Herbivorous grazers were the gammarid amphipod, *Gammarus* spp.
[c] Water exchange rates were once per day for "controls" and six times per day for "fast" exchange.
[d] Submersed plant responses measured as above ground mass (g m^{-2}) for "nutrient delivery" experiments and as growth rate (cm m^{-2}) for "trophic complexity" and "exchange rate" experiments.
[e] Epiphyte responses measured as organic ash-free dry weight (AFDW) of epiphytic material for trophic complexity and exchange rate and µg Chl *a* for nutrient delivery experiments.

enrichment compared to a 26% decrease with nutrient addition under slow (control) exchange rate. Increased exchange rate had similar effects (on both plants and epiphytes) at low and high nutrient loading rates. This was not the case for epiphyte grazing treatments. Grazing elicited substantial epiphyte (56% decrease) and plant (43% increase) responses at low nutrient loading. However, at high nutrient loading rates, although grazing significantly reduced epiphyte biomass (49%), epiphyte levels remained high enough so that grazing caused no differences in plant growth (Table 19.1).

DISCUSSION

Results of mesocosm experiments presented here are consistent with conclusions of many previous studies, which indicate that growth of seagrasses and other coastal SAV is inhibited by excessive nutrient enrichment to overlying waters (e.g., Kemp et al. 1983; Neckles et al. 1993; Short et al. 1995). In most cases, nutrient inhibition of SAV growth is related to accumulation of phytoplanktonic and epiphytic algae and the resulting attenuation of light to seagrass leaves (e.g., Borum 1985; Twilley et al. 1985; Tomasko et al. 1996). Although nutrient enrichment of coastal waters tends to cause reduced SAV growth, the actual relationship between nutrient level and submersed plant response appears to vary depending on environmental conditions (e.g., Kemp et al. 1983). Results of experiments such as those presented here suggest that these environmental effects on seagrass-nutrient dynamics are related to the respective scales of time, space, and ecological complexity in both experimental and natural ecosystems.

In part these scaling effects arise from the complex nature of ecological interactions. For example, while nutrient conditions can significantly alter plant growth and survival, the SAV system itself can also alter the nutrient conditions via ecological feedback processes (e.g., Kemp et al. 1984). Indeed, submersed plant communities have the ability to modify nutrient regimes in their surrounding environment via direct uptake and assimilation, as well as through indirect pathways related to sedimentation, recycling, and transformation processes (e.g., Kemp et al. 1984; Caffrey and Kemp 1991; Duarte 1995). Furthermore, in many cases, plant growth may also be limited by inadequate nutrient supplies in sediments of tropical and temperate coastal seagrass beds (e.g., Short 1987; Short et al. 1990; Perez et al. 1991). Even in habitats where declines in SAV abundance were attributed to eutrophication, sediment fertilization experiments revealed that plant growth can still be stimulated by nutrient additions (e.g., Murray et al. 1992; Pedersen and Borum 1993). Hence, the key question is: when does nutrient enrichment become excessive, and how do environmental conditions affect the relationship between nutrient levels and submersed plant growth?

The timing and frequency of nutrient delivery to SAV beds is related to the relative sizes of the bed and its surrounding watershed, as well as the nature of the nutrient sources (Figure 19.1). When the areal ratio of bed size to surrounding watershed is relatively small, spring runoff events can generate large nutrient input pulses. In late summer, convective rainfall events can also deliver nutrient pulses. Where direct inputs from local watersheds are negligible, nutrient delivery associated with tidal exchange or groundwater seepage will tend to be more continual. With respect to timing of nutrient inputs, experimental nutrient additions inhibited SAV growth regardless of whether they were delivered during spring or late summer; however, effects differed with seasons (Sturgis and Murray 1997). In spring, when plants were actively growing, the biomass allocation to leaves and stems was most sensitive to experimental nutrient stress. In the late summer period of stationary growth and incipient senescence, nutrient/light stress caused reductions in below-ground biomass and associated loss of overwintering reserves. Experimental SAV communities exhibited different responses to pulse vs. continuous nutrient delivery, and this difference changed with season (Figure 19.3). Inhibition of plant growth was less pronounced with pulsed (vs. continuous) inputs in fall. This is probably because plant biomass levels were higher, affording them a greater ability to compete with algae for uptake of high nutrient concentrations over short durations. In contrast, during spring experiments, when biomass was relatively low but growth rates high, submersed plants fared better under continuous nutrient inputs and were inhibited more with pulsed loading.

Observed differences in response to short-duration pulsed inputs vs. longer continuous loading of nutrients are directly related to differences in biomass turnover times (biomass/respiration) for submersed plants (2–6 wk) compared to algal cells (1–2 d). When the duration and/or frequency of pulsed inputs approximates algal turnover-times, effects on algal growth and community composition will be more pronounced than they would be in response to continuous inputs or pulsed inputs of much longer or shorter time scales (e.g., Turpin et al. 1981). Furthermore, under all experimental nutrient enrichment treatments, the time required for plant responses tended to be longer than that for algal responses because of differences in respective biomass turnover-times (Figure 19.4). Whereas, SAV growth responses to nutrient addition were delayed until 7–9 wk after initiation of treatment, epiphytic algae exhibited rapid responses, generally within 2 wk. Although the plant response was delayed, it is reasonable to conclude that it was a consequence of the more rapid epiphyte response. It was, thus, concluded that significant community responses to nutrient treatment and scaling would be evident within relatively short time periods. Therefore, the study duration and sampling strategy for subsequent nutrient scaling experiments reflects recognition of these time-scales. In general, the intense sampling during a 6 wk experiment was sufficient to capture the full algal community responses and the initial plant response.

The average residence time of water in a natural submersed plant community is a function of several physical scales of the ecosystem, including water depth, bed size, and tidal range. In addition, increased friction associated with the plant canopy can lengthen the water residence-time (e.g., Rybicki et al. 1997). Incomplete mixing of water on ebb or flood tides can also extend the effective residence time of water over SAV beds (e.g., Sanford et al. 1993). In our experiments, variations in water residence-time (which would result from differences in these physical scales) significantly altered epiphyte and plant responses to nutrient enrichment. In fact, variations in exchange rate had greater effect on epiphyte growth than did variations in nutrient concentration (Figure 19.6A). In general, epiphyte accumulation increased with increases in water exchange rate; however, the responses were non-linear, particularly at higher nutrient concentrations (Figure 19.6A). This effect results largely from the fact that shorter water residence-times (larger exchange rates) effectively increase nutrient input rates, where input is the product of nutrient concentration and exchange rate. In addition, when SAV biomass is relatively high, longer water residence-times allow plants to effectively reduce nutrient levels, thereby precluding overgrowth by epiphytic algae. These explanations are supported by model simulations (Figure 19.5A). The saturation of epiphyte nutrient response, which was predicted by the model at higher nutrient levels (Figure 19.5A), was also evident in mesocosm experiments (Figure 19.6A).

Previous investigators have argued indirectly that natural ecosystems tend to increase in complexity with increases in the characteristic spatial and temporal scales (e.g., Frost et al. 1988). One obvious measure of ecosystem complexity is the number of trophic levels, which tends to increase directly with temporal and spatial scale in aquatic ecosystems (e.g., Sheldon et al. 1972). By adding trophic complexity to our experimental ecosystems, we were, in a sense, increasing the scales of these systems. In these experiments, this resulted in markedly reduced epiphytic biomass (Figure 19.6B). These findings are consistent with other studies which suggest grazing by surface-feeding herbivores reduces epiphytic biomass on submersed plants (Orth et al. 1984; Hootsmans and Vermaat 1985; Howard and Short 1986). Responses to herbivorous grazing treatments differed under low and high nutrient enrichment conditions. Although grazing significantly reduced epiphytic biomass under low nutrient conditions, the response was insignificant at high nutrients. Apparently, the increased algal mortality associated with grazing was insufficient to compensate for the enhanced growth at higher nutrient levels (Figure 19.6B). This is similar to the findings of a previous mesocosm experiment examining interacting effects of nutrient enrichment and grazing on growth of eelgrass and its epiphytic community (Neckles et al. 1993).

The primary objective of this chapter was to illustrate that responses of a submersed plant community to nutrient enrichment differ substantially depending on the physical and biotic scales of that community. Here, we consider three factors interacting to alter community responses to nutrient addition: input frequency and timing, water residence time, and trophic complexity. Although these three factors are measured in different units and are thus difficult to compare, we calculate relative responses to treatments (as a percentage of respective controls) to provide a basis for comparison. While there is general agreement that herbivorous grazing can modify nutrient enrichment effects on SAV (e.g., Jernakoff et al. 1996), there has been little focus on other physical scaling factors, such as those studied here.

A comparison among experimental variables at low nutrients indicates that the relative responses of epiphytes to variations in trophic complexity (grazing) and exchange rate are of similar magnitude but opposite direction (Table 19.1). The reduction in relative epiphyte growth with grazing was, in fact, comparable to the increase in epiphyte growth with higher rates of water exchange. Relative effects on plant growth were; however, substantially larger with grazing than with water exchange (Table 19.1). Note that there was no equivalent low nutrient treatment for the nutrient delivery experiment.

At high nutrient levels, relative responses to scaling treatments were quite different. Changes in time-scales of both nutrient delivery and water exchange caused a greater negative response in plant growth than did experimental variations in trophic complexity. However, relative responses of epiphyte biomass to scaling factors were greatest for trophic complexity and exchange rate treatments, with little differences resulting from changes in nutrient delivery (Table 19.1). Apparently, the epiphyte community response to elevated nutrients was saturated at high nutrient levels, such that additional effects arising from variations in delivery frequency were swamped. On the other hand, relative responses to variations in trophic complexity and exchange rate were similar under low and high nutrient conditions (Table 19.1). These results support our contention that these and other physical and biotic scaling factors strongly modulate seagrass community responses to nutrient enrichment.

The exact nature of these scaling relationships remains to be described; however, the trends observed in these studies should be robust. Ultimately, our ability to extrapolate experimental results to conditions in natural ecosystems depends on understanding regulating mechanisms (Petersen et al. 1997). Once these mechanisms are broadly described, they can be incorporated into numerical models (such as that used in this study) to assist in development of quantitative scaling relationships (e.g., Kemp et al. 1980). Conclusions from our comparative analysis help explain observed variabilities in natural seagrass ecosystems and in their responses to perturbations such as nutrient enrichment. For example, some SAV communities may be able to exist in eutrophic conditions as long as they have relatively low flushing rates and/or have stable populations of herbivorous grazers. Of course, caution should be used in making simplistic extrapolations from mesocosm experiments to conditions in nature (Schindler 1998). The inherent artifacts associated with experimental systems must be considered (e.g., Chen et al 1997), and scaling relationships used for extrapolation must apply to both experimental and natural ecosystems (Cairns 1983, Levin 1992).

Although the problem of eutrophication and SAV losses occurs on many scales — from small coves to large estuaries and from brief episodes to decadal trends — its impact is being felt globally (e.g., Short and Wyllie-Echeverria 1996). Because of the vast scope and daunting mitigation costs for this problem (Dennison et al. 1993), it is of fundamental importance to better understand how physical and ecological scales alter SAV community response to nutrient stress. It is no longer sufficient to say that nutrient enrichment can cause SAV loss; it is now essential that we understand what nutrient levels cause what damage to submersed plant communities under what physical and ecological conditions.

ACKNOWLEDGMENTS

This research was supported by two grants from the U. S. Environmental Protection Agency: Multiscale Experimental Ecosystem Research Center (MEERC, R 819640), and Chesapeake Bay Program (CB 993586-01-1). We are particularly indebted to the useful comments provided by an anonymous reviewer and to summer undergraduate interns (C. Fellows, J. Bryner, L. Taylor, J. Bailey, J. Krut, S. Simmons, O. Thomas, M. Tinkler) who helped with sampling and analyses.

REFERENCES

Borum, J. 1985. Development of epiphytic communities on eelgrass (*Zostera marina*) along a nutrient gradient in a Danish estuary. *Mar. Biol.* 87:211–218.

Brönmark, C. and J. E. Vermaat. 1997. Complex fish-snail-epiphyton interactions and their effects on submerged freshwater macrophytes. In *The Structuring Role of Submersed Macrophytes in Lakes,* E. Jeppesen, Ma. Søndergaard, Mo. Søndergaard, and K. Christoffersen (eds.). Springer-Verlag, New York, pp. 47–68.

Caffrey, J. M. and W. M. Kemp. 1991. Seasonal and spatial patterns of oxygen production, respiration and root-rhizome release in *Potamogeton perfoliatus* L. and *Zostera marina. L. Aquat. Bot.* 40:109–128.

Cairns, J. J. 1983. Are single species toxicity tests alone adequate for estimating environmental hazard? *Hydrobiologia* 100:47–57.

Cambridge, M. L., A. W. Chiffings, C. Brittan, L. Moore, and A. J. McComb. 1986. The loss of seagrass in Cockburn Sound, western Australia. II. Possible causes of seagrass decline. *Aquat. Bot.* 24:269–285.

Carpenter, S. R. and J. F. Kitchell. 1988. Consumer control of lake productivity. *BioScience* 38:764–769.

Chen, C.-C., J. E. Petersen, and W. M. Kemp. 1997. Spatial and temporal scaling of periphyton growth on walls of estuarine mesocosms. *Mar. Ecol. Prog. Ser.* 155:1–15.

den Hartog, C. and P. J. G. Polderman. 1975. Changes in the seagrass populations of the Dutch Waddenzee. *Aquat. Bot.* 1:141–147.

Dennison, W. C., R. J. Orth, K. A. Moore, J. C. Stevenson, V. Carter, S. Kollar, P. W. Bergstrom, and R. A. Batiuk. 1993. Assessing water quality with submersed aquatic vegetation: habitat requirements as barometers of Chesapeake Bay health. *BioScience* 43:86–94.

Duarte, C. M. 1995. Submerged aquatic vegetation in relation to different nutrient regimes. *Ophelia* 41:87–112.

Frost, T. M., D. L. DeAngelis, S. M. Bartell, D. J. Hall, and S. H. Hurlbert. 1988. Scale in the design and interpretation of aquatic community research. *Complex Interactions in Lake Ecosystems,* Springer-Verlag, New York.

Harley, M. T. and S. Findlay. 1994. Photosynthesis-irradiance relationships for three species of submersed macrophytes in the tidal freshwater Hudson River. *Estuaries* 17:200–205.

Hootsmans, M. J. M. and J. E. Vermaat. 1985. The effect of periphyton-grazing by three epifaunal species on the growth of *Zostera marina* L. under experimental conditions. *Aquat. Bot.* 22:83–88.

Howard, R. K. and F. T. Short. 1986. Seagrass growth and survivorship under the influence of epiphyte grazers. *Aquat. Bot.* 24:287–302.

Jernakoff, P., A. Brearley, and J. Nielsen. 1996. Factors affecting grazer-epiphyte interactions in temperate seagrass meadows. *Oceanogr. Mar. Biol. Ann. Rev.* 34:109–162.

Kemp, W. M., M. R. Lewis, J. J. Cunningham, J. C. Stevenson, and W. R. Boynton. 1980. Microcosms, macrophytes, and hierarchies: environmental research in the Chesapeake Bay. U.S. Technical Information Center, U.S. Department of Energy.

Kemp, W. M., W. R. Boynton, R. R. Twilley, J. C. Stevenson, and J. C. Means. 1983. The decline of submerged vascular plants in upper Chesapeake Bay: summary of results concerning possible causes. *Mar. Technol. Soc. J.* 17:78–89.

Kemp, W. M., W. R. Boynton, R. R. Twilley, J. C. Stevenson, and L. G. Ward. 1984. Influences of submersed vascular plants on ecological processes in upper Chesapeake Bay. In *The Estuary as a Filter,* V. S. Kennedy (ed.). Academic Press, New York, pp. 367–394.

Kemp, W. M., W. R. Boynton, and A. J. Hermann. 1995. Simulation models of an estuarine macrophyte ecosystem. In *Complex Ecology,* B. Patten and S. E. Jørgensen (eds.). Prentice-Hall, Englewood Cliffs, NJ, pp. 262–278.

Lawton, J. H. 1996. The Ecotron facility at Silwood Park: the value of "big bottle" experiments. *Ecology* 77:665–669.

Levin, S. A. 1992. The problem of patten and scale in ecology. *Ecology* 73:1943–1967.

Madden, C. J. and W. M. Kemp. 1996. Ecosystem model of an estuarine submersed plant community: calibration and simulation of eutrophication responses. *Estuaries* 19(2B):457–474.

Moore, K. A., J. L. Goodman, J. C. Stevenson, L. Murray, and K. Sundberg. 1995. Chesapeake Bay nutrients, light and SAV: relationships between water quality and SAV growth in field and mesocosm studies. EPA, Chesapeake Bay Program, Annapolis, MD, NTIS, CB003909-02.

Murray, L., W. C. Dennison, and W. M. Kemp. 1992. Nitrogen versus phosphorus limitation for growth of an estuarine population of eelgrass (*Zostera marina* L.). *Aquat. Bot.* 44:83–100.

Murray, L., R. B. Sturgis, R. D. Bartleson, and W. Severn. 1997. Scaling response of SAV community to nutrient loading: effects of trophic complexity. *Scale-Dependent Relations Governing Extrapolation from Mesocosm to Nature in Coastal Ecosystems,* V. S. Kennedy (ed.). Report to MEERC, U.S. Environmental Protection Agency, Washington, D.C.

Neckles, H. A., R. L. Wetzel, and R. J. Orth. 1993. Relative effects of nutrient enrichment and grazing on epiphyte-macrophyte (*Zostera marina* L.) dynamics. *Oecologia* 93:285–295.

Neundorfer, J. V. and W. M. Kemp. 1993. Nitrogen versus phosphorus enrichment of brackish waters: response of *Potamogeton perfoliatus* and its associated algal communities. *Mar. Ecol. Prog. Ser.* 94:71–82.

Odum, E. P. 1984. The mesocosm. *BioScience* 34:558–562.

O'Neill, R. V. 1989. Perspectives in hierarchy and scale. In *Perspectives in Ecological Theory,* J. Roughgarden, R. M. May, and S. A. Levin (eds.). Princeton University Press, Princeton, NJ, pp. 140–156.

Orth, R. J. and K. A. Moore. 1983. Chesapeake Bay: an unprecedented decline in submerged aquatic vegetation. *Aquat. Bot.* 222:51–53.

Orth, R. J., K. L. Heck, Jr., and J. van Montfrans. 1984. Faunal communities in seagrass beds: a review of the influence of plant structure and prey characteristics on predator-prey relationship. *Estuaries* 7:339–350.

Pedersen, M. F. and J. Borum. 1993. An annual nitrogen budget for a seagrass *Zostera marina* population. *Mar. Ecol. Prog. Ser.* 101:169–177.

Perez, M., J. Romero, C. M. Duarte, and K. Sand-Jensen. 1991. Phosphorus limitation of *Cymodocea nodosa* growth. *Mar. Biol.* 109:129–133.

Petersen, J. E., C.-C. Chen, and W. M Kemp. 1997. Scaling aquatic primary productivity: experiments under nutrient- and light-limited conditions. *Ecology* 78:2326–2338

Phillips, G. L., D. Eminson, and B. Moss. 1978. A mechanism to account for macrophyte decline in progressively eutrophicated freshwaters. *Aquat. Bot.* 4:103–126.

Rybicki, N. B., H. L. Jenter, V. Carter, and R. A. Baltzer. 1997. Observations of tidal flux between a submersed aquatic plant stand and the adjacent channel in the Potomac River near Washington, D.C. *Limnol. Oceanogr.* 42:307–317.

Sand-Jensen, K. and J. Borum. 1991. Interactions among phytoplankton, periphyton, and macrophytes in temperate freshwaters and estuaries. *Aquat. Bot.* 41:137–175.

Sanford, L. P., W. C. Boicourt, and S. R. Rives. 1993. Model for estimating tidal flushing of small embayments. *J. Waterway Port Coastal Ocean Eng.* 118:635–654.

Schlindler, D. W. 1998. Replication versus realism: The need for ecosystem-scale experiments. *Ecosystems* 1:323–334.

Severn, W. A. 1998. Physiological and morphological response of *Potamogeton perfoliatus* across scales of depth, nutrient loading and trophic complexity. Master's thesis, University of Maryland, College Park, MD.

Sheldon, R. W., A. Prakash, and W. H. J. Sutcliff. 1972. The size and distribution of particles in the ocean. *Limnol. Oceanogr.* 17:323–340.

Short, F. T. 1987. Effects of sediment nutrients on seagrasses: literature review and mesocosm experiment. *Aquat. Bot.* 27:41–57.

Short, R. T. and S. Wyllie-Echeverria. 1996. Natural and human-induced disturbance of seagrasses. *Environ. Conserv.* 23:17–27.

Short, F. T., W. C. Dennison, and D. G. Capone. 1990. Phosphorus-limited growth of the tropical seagrass *Syringodium filiforme* in carbonate sediments. *Mar. Ecol. Prog. Ser.* 62:169–174.

Short, F. T., D. M. Burdick, and J. E. Kaldy, III. 1995. Mesocosm experiments quantify the effects of eutrophication on eelgrass, *Zostera marina. Limnol. Oceanogr.* 40:740–749.

Silberstein, K., A. W. Chiffings, and A. J. McComb. 1986. The loss of seagrass in Cockburn Sound, western Australia. III. The effect of epiphytes on productivity of *Posidonia australis* Hook F. *Aquat. Bot.* 24:355–371.

Sturgis, R. and L. Murray. 1997. Scaling of nutrient inputs to submersed plant communities: temporal and spatial variations. *Mar. Ecol. Prog. Ser.* 152:89–102.

Tomasko, D. A., C. J. Dawes, and M. O. Hall. 1996. The effects of anthropogenic nutrient enrichment on turtle grass (*Thalassia testudinum*) in Sarasota Bay, Florida. *Estuaries* 19:448–456.

Turpin, D. H., J. S. Parslow, and P. J. Harrison. 1981. On limiting nutrient patchiness and phytoplankton growth: a conceptual approach. *J. Plankton Res.* 3:421–431.

Twilley, R. R., W. M. Kemp, K. W. Staver, J. C. Stevenson, and W. R. Boynton. 1985. Nutrient enrichment of estuarine submersed vascular plant communities. 1. Algal growth and effects on production of plants and associated communities. *Mar. Ecol. Prog. Ser.* 23:179–191.

Wiens, J. A. 1989. Spatial scaling in ecology. *Functional Ecol.* 3:385–397.

20 Seagrass Ecosystem Characteristics, Research, and Management Needs in the Florida Big Bend

Robert A. Mattson

Abstract The seagrass ecosystem present in nearshore waters of the Florida Big Bend coast is the second largest in the eastern Gulf of Mexico, covering about 3,000 km^2. This area (from the St. Marks to the Anclote River) generally receives less research and management attention compared to other seagrass areas in Florida (i.e., Florida Bay, Indian River Lagoon). It represents the northern distributional limit of American tropical seagrasses. The major seagrass species are *Thalassia testudinum*, *Syringodium filiforme*, and *Halodule wrightii. Halophila engelmanni*, *Halophila decipiens*, and *Ruppia maritima* are also present but constitute a lesser component of coverage and standing crop. Benthic green algae in the order Siphonales and drift algae (primarily Rhodophyta) are major floristic components of these beds, in some areas standing crop of algae exceeds that of seagrasses.

Four subregions may be recognized in the northern portion of the region: Apalachee Bay, Deadman Bay, Suwannee Sound and adjacent coastal waters, and Waccasassa Bay. Submerged vegetation coverage was estimated by a 1972 NMFS study: Apalachee Bay 128.0 km^2; Deadman Bay 7.4 km^2; Suwannee Sound 32.3 km^2; and Waccasassa Bay 98.0 km^2. These data are for inshore, shallow-water areas (primarily < 2 m MLW depth). Other submerged vegetation mapping studies conducted in the region misclassified vegetation coverage in some areas or only mapped in a spatially defined region.

Limited amounts of field data are available to describe the seagrass/benthic algal communities in the Big Bend, with few studies reporting standing crop, short-shoot densities, or productivity data. Differences between seagrass community characteristics in the Big Bend vs. Florida Bay have been attributed in part to climatic differences. Studies conducted in Apalachee Bay indicate that reduction in light quantity and quality associated with discharge of bleached kraft mill effluent have resulted in significant reductions in seagrass coverage and standing crop.

Because seagrasses are influenced by water clarity, which can be affected by a variety of water quality factors, they can serve as biological indicators to evaluate the effectiveness of efforts to protect water quality. A key component leading to this would be determining minimum water quality characteristics needed to maintain suitable water clarity. Given the current high ecological integrity of this portion of the Big Bend coast (perhaps the least disturbed in Florida), it is desirable to collect additional descriptive data (both mapping and field data) at an adequate number of sample sites to document current seagrass community conditions and form the basis for a monitoring program to assess this condition on an ongoing basis. The results from studies conducted in this area and other estuaries could be used to derive protective water quality criteria to maintain existing seagrass coverage and community composition. An important research question is whether the

0-8493-2045-3/99/$0.00+$.50
© 2000 by CRC Press LLC

seagrasses in this area are more sensitive to environmental perturbation than those in areas in the southern part of the state, since they reside at the northern limit of their distribution and are subject to natural climatic stresses.

INTRODUCTION

The Big Bend region of the Florida coast has been described as "... one of the least polluted coastal regions of the continental United States." (Livingston 1990). The seagrass ecosystem present along this coast is the second largest in the eastern Gulf of Mexico, encompassing an estimated 3000 km^2 (Iverson and Bittaker 1986). While there has been research performed in this area which has contributed greatly to our understanding of seagrass ecology (Virnstein 1987), overall the area remains poorly studied and receives much less management attention than other seagrass areas in Florida. This chapter reviews and summarizes the available data and literature on the seagrasses in the region. Zieman and Zieman (1989) provided an extensive review of Florida west coast seagrasses, although much of the data discussed came from other west coast regions, such as Tampa Bay. This chapter might be considered a supplement of that earlier report, incorporating more recently gathered information from the Big Bend.

The Big Bend coast (Figure 20.1) has long been recognized as a low energy coastline (McNulty et al. 1972; Davis 1997); a broad (approximately 150 km wide) shallow coastal shelf combined with prevailing winds results in low wave and wind energy. The coast is dominated by intertidal marsh with no barrier islands or extensive sandy beach areas. The Cedar Key area represents the northernmost extent of mangrove occurrence on the Florida west coast. Mattson (1997) and Livingston (1990) provided general overviews of the region's ecological characteristics. The region may be classified as an "estuarine realm" (after Levinton 1982), a coastal area bounded by the sea on one side and freshwater inflow (from both surface and groundwater sources) on the landward side. The low wave energy and shallow depths, combined with low sediment loads and generally high contributions of clear groundwater from the Florida Aquifer System in the rivers draining to the region, create a physical environment highly conducive to the survival and growth of seagrasses in the Big Bend.

FLORISTIC CHARACTERISTICS

Six species of seagrasses are reported in the region (Table 20.1): three in the Hydrocharitaceae (*Thalassia testudinum*, *Halophila engelmannii*, and *Halophila decipiens*), two in the Cymodoceaceae (*Halodule wrightii* and *Syringodium filiforme*), and one in the Ruppiaceae (*Ruppia maritima*). Eiseman (1980) stated that *Halophila baillonis* does not occur in Florida, and that identifications of this plant should be referred to *Halophila decipiens*. *Halophila baillonis*, however, continues to be listed in major floristic monographs of the region (e.g., Godfrey and Wooten 1979; Clewell 1985). Clewell (1985) listed the plant as occurring in deep grassbeds off the coast of Franklin County, although this is likely based on herbarium records, not personal observation. Additional systematic attention to this issue may be in order.

The major associate species in seagrass beds in the region are benthic green algae in the order Siphonales (Table 20.1). This includes 7 species of *Caulerpa*, two species each of *Udotea*, *Penicillus*, and *Codium*, and *Halimeda incrassata*. A few species in the Dasycladales are also present (*Acetabularia* spp. and *Batophora oerstedii*). The siphonaceous species tend to be found in areas of higher salinity (≥ 15–20 ppt), while the Dasycladaceous algae are able to tolerate lower salinities and may be found in inshore estuarine areas, as well as offshore marine habitats (Mattson 1995). In addition, three species of attached brown algae are components of the grassbeds where suitable areas of rocky substratum are present: *Sargassum filipendula*, *S. pteropleuron*, and *Padina*

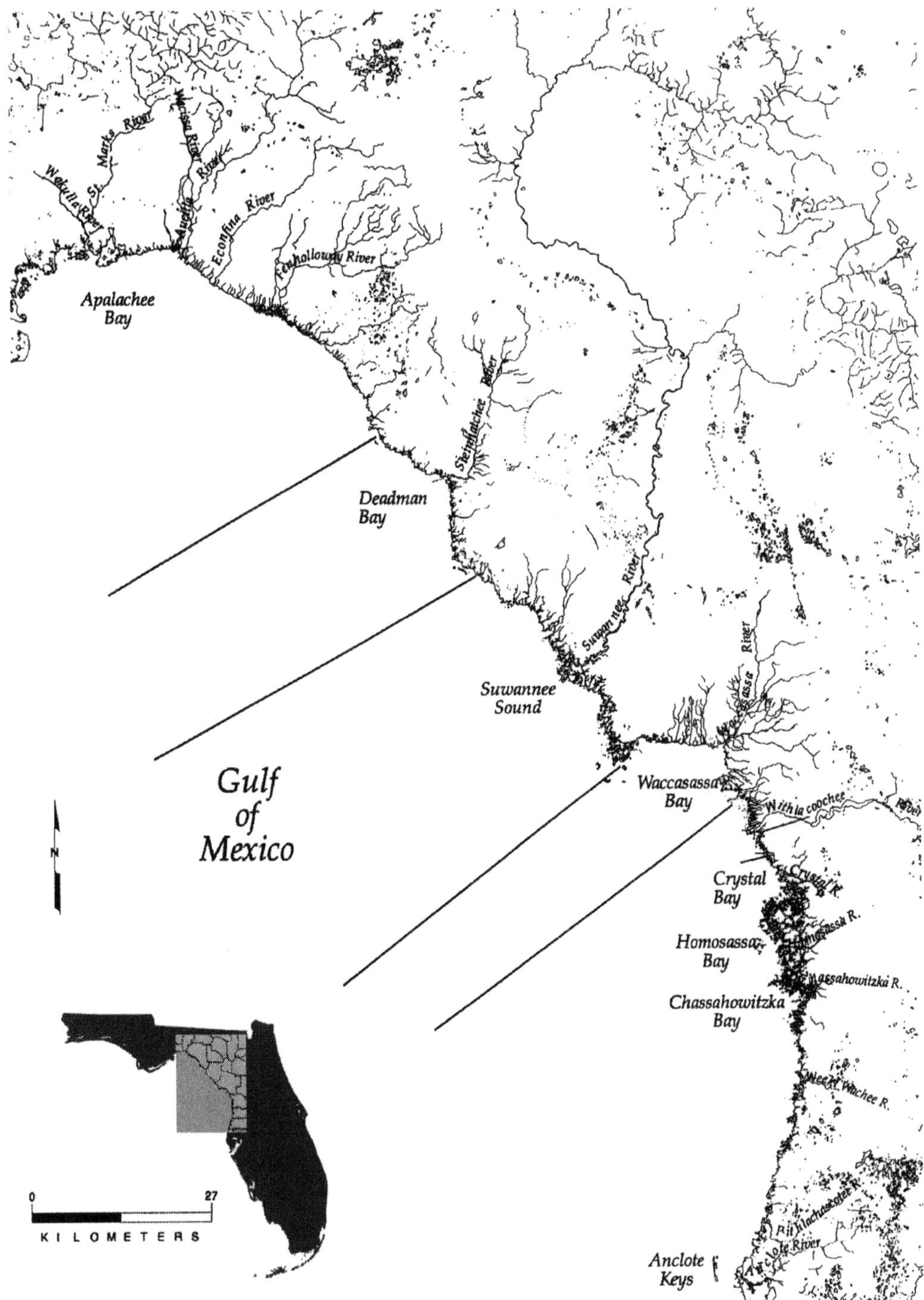

FIGURE 20.1 Map of the Big Bend, showing subregions in the northern half of the region.

TABLE 20.1
Seagrasses and common macroalgal taxa reported to occur in the Florida Big Bend. (?) = questionable record of occurrence.

SEAGRASSES

Hydrocharitaceae	Cymodoceaceae	Ruppiaceae
Thalassia testudinum Banks *ex* Koenig	*Syringodium filiforme* Kuetz.	*Ruppia maritima* L.
Halophila engelmannii Aschers.	*Halodule wrightii* Aschers.	
Halophila decipiens Ostenfeld		
Halophila baillonis Aschers.(?)		

BENTHIC (ROOTED OR ATTACHED) ALGAE

Chlorophyta

Acetabularia spp.	*Caulerpa sertularioides* (Gmelin) Howe
Batophora oerstedi J. Agardh	*Codium isthmocladum* Vickers
Caulerpa ashmeadii Harvey	*Codium tayori* Silva
Caulerpa cupressoides (West) C. Agardh (?)	*Halimeda incrassata* (Ellis) Lamouroux
Caulerpa fastigiata Montagne	*Penicillus capitatus* Lamarck
Caulerpa mexicana (Sonder) J. Agardh	*Penicillus dumetosus* (Lamouroux) Blainville
Caulerpa paspaloides (Bory) Greville	*Udotea conglutinata* (Ellis et Solander) Lamx
Caulerpa prolifera (Forsskael) Lamouroux	*Udotea flabellum* (Ellis et Solander) Lamx

Phaeophyta

Padina vickersiae Hoyt	*Sargassum pteropleuron* Grunow
Sargassum filipendula C. Agardh	

DRIFT MACROALGAE

Rhodophyta

Agardhiella tenera (J. Agardh) Schmitz	*Halymenia floridana* J. Agardh
Centroceras clavulatum (C.Ag.) Montagne	*Halymenia floresia* (Clemente) C. Agardh
Ceramium elegans (Ducluzeau) C. Agardh	*Hypnea cervicornis* J. Agardh
Ceramium fastigiatum (Roth) Harvey	*Hypnea cornuta* (Lamouroux) J. Agardh
Ceramium subtile J. Agardh	*Hypnea musciformis* (Wulfen) Lamx.
Ceramium tenuissimum (Lyngbye) J.Agardh	*Jania pumila* Lamouroux
Champia parvula (C.Agardh) Harvey	*Laurencia intricata* Lamouroux
Chondria cnicophylla (Melville) De Toni	*Laurencia obtusa* (Hudson) Lamx.
Chondria dasyphylla (Woodward) C. Agardh	*Laurencia papillosa* (Forsskal) Greville
Chondria littoralis Harvey	*Laurencia poitei* (Lamx.) Howe
Chondria sedifolia Harvey	*Lomentaria baileyana* (Harvey) Farlow
Chondria tenuissima (Good. et Wood.) C. Ag.	*Neoagardhiella ramosissima* (Harv. et Kutz.) Wynne & Taylor
Dasya pedicellata (C. Agardh) C. Agardh	*Polysiphonia echinata* Harvey
Dasya ramossissima Harvey	*Polysiphonia harveyi* J. Bailey
Digenia simplex (Wulfen) C. Agardh	*Polysiphonia macrocarpa* Harvey
Eucheuma nudum J. Agardh	*Polysiphonia ramentacea* Harvey
Gracilaria cervicornis (Turner) J. Agardh	*Polysiphonia subtilissima* Montagne
Gracilaria debilis (Forsskal) Borgesen	*Pterocladia americana* Taylor
Gracilaria tikvahiae McLachlan	*Spyridia filamentosa* (Wulfen) Harvey
Gracilaria sjoestedtii (Kylin) Dawson	*Wurdemannia miniata* (Draparnaud) Feldmann et Hamel
Gracilaria verrucosa (Hudson) Papenfuss	
Griffithsia globulifera Harvey	

TABLE 20.1 *(continued)*
Seagrasses and common macroalgal taxa reported to occur in the Florida Big Bend. (?) = questionable record of occurrence.

Phaeophyta	Chlorophyta
Dictyota divaricata Lamouroux	*Anadyomene stellata* (Wulfen) C. Agardh
Ectocarpus elachistaeformis Heydrich	*Chaetomorpha brachygona* Harvey
Ectocarpus intermedius Kutzing	*Chaetomorpha linum* (Muller) Kutzing
Giffordia mitchelliae (Harvey) Hamel	*Cladophora gracilis* (Griff. ex Harv.) Kutz.
Hummia onusta (Kutzing) J. Fiore	*Cladophoropsis membranacea* (C. Ag.) Borg.
Rosenvingea intricata (J. Ag.)Borgs.	*Enteromorpha compressa* (L.) Greville
Rosenvingea sanctate-crucis Borgs.	*Enteromorpha crinita* Lamark
	Enteromorpha intestinalis (L.) Link
	Enteromorpha prolifera (Muller) J. Agardh

vickersiae. S. pteropleuron generally is found in areas of lower salinity, while the other two brown algae are found in higher salinity areas.

Numerous species of drift algae, which often begin life as epiphytes, are also components of the grassbeds, with 58 species reported from the Big Bend (Table 20.1). These are mostly red algae (42 taxa), with a few species of brown (7 taxa) and green (9 taxa) algae as well. Mattson (1995) reported 35 species of red, green, and brown drift algae from a survey off the coasts of Levy, Citrus, and Hernando counties. Studies of seagrass epiphytes in the Big Bend are limited. Ballantine (1972) reported 65 taxa of epiphytic algae in seagrass beds off the Anclote River. Mattson et al. (1988) reported 50 taxa of algal epiphytes on seagrasses off the Withlacoochee and Crystal rivers. Overall, few comprehensive floristic surveys have been conducted in this region (e.g., Phillips 1960a; Earle 1969 and references therein). Additional descriptive surveys of the algal flora would be beneficial in order to obtain more complete lists of the taxa found in the region.

MAPPING STUDIES

Several mapping studies of submerged aquatic vegetation (SAV) coverage, including seagrasses and associated algae, have been conducted in the region. Areal estimates vary widely (Table 20.2),

TABLE 20.2
Summary of submerged aquatic vegetation mapping studies conducted in the Big Bend and the coverage estimates from those studies.

	Acres	km²
McNulty et al. 1972 (inshore)	132,478	537
Northern Big Bend	65,664	266
Southern Big Bend	66,814	271
Continental Shelf Assoc. and Martel 1985	2,497,31 7	10,11 4
Iverson and Bittaker 1986	740,741	3,000
Livingston 1993 (Apalachee Bay w/in 3 km of shore)	14,595	59
Sargent et al. 1995	797,140	3,228
(Jefferson — Levy counties)	416,890	1,688
(Citrus — Pasco counties)	380,250	1,540

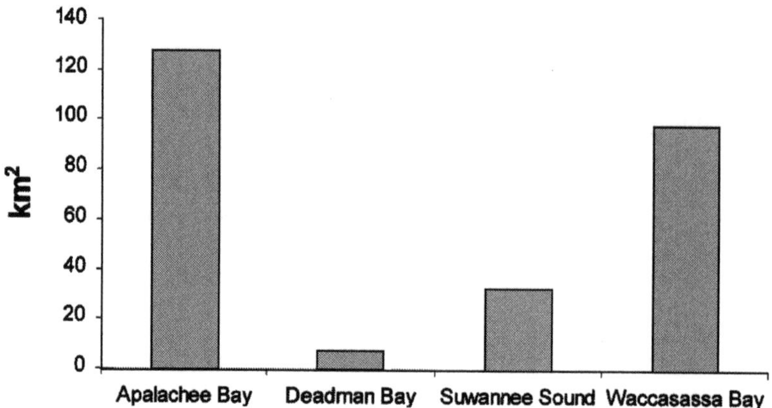

FIGURE 20.2 Graph of submerged aquatic vegetation coverage in the northern Big Bend subregions (data from McNulty et al. 1972).

in part due to the spatial boundaries of each mapping effort. Mapping conducted by McNulty et al. (1972) and more recently by Livingston (1993) was limited to inshore, shallow water regions, with Livingston's study conducted only in Apalachee Bay. The mapping study by Continental Shelf Associates and Martel Laboratories (1985) resulted in the highest estimate of SAV coverage (Table 20.2), over 10,000 km^2. However, this study appeared to misclassify inshore SAV coverage in some areas. For instance, it shows an area of "dense seagrass" in much of Suwannee Sound, which does not currently support any SAV (author's personal observation) and portions of which may never have (Moore 1963). This study was significant, however, in that it was the first, and only, attempt to map the deeper offshore SAV beds. The mapping studies of Iverson and Bittaker (1986) and Sargent et al. (1995) yield consistent estimates and appear to provide the most reasonable approximation of nearshore SAV coverage in the region (about 3000 km^2).

The northern Big Bend (from Cedar Key to Apalachee Bay) can be divided into four subregions (Figure 20.1). McNulty et al. (1972) gave estimates of inshore seagrass coverage in each region (Figure 20.2). They underestimated coverage in Deadman Bay (the estuary of the Steinhatchee River) since they did not appear to map the extensive SAV beds which occur in the southern portion of the bay in the vicinity of Pepperfish Keys. They also appear to have overestimated SAV coverage in Waccasassa Bay, based on current conditions of limited SAV coverage in that region.

SEAGRASS COMMUNITY CHARACTERISTICS

Limited amounts of field data on seagrasses have been collected in the Big Bend. The following sections summarize the existing data.

Distribution and Zonation

The most comprehensive region-wide study of seagrasses in the Big Bend was that of Iverson and Bittaker (1986). They sampled seagrasses at about 300 sites throughout the region (Figure 20.3). The five most commonly observed seagrasses were *T. testudinum*, *S. filiforme*, *Halodule wrightii*, *Halophila engelmannii* and *R. maritima*. *T. testudinum*, *S. filiforme*, *H. wrightii*, and *H. engelmannii* occurred throughout the region. *Halophila decipiens* was only observed at a few offshore, deepwater sites (Figure 20.3), while *R. maritima* was only observed at inshore sites influenced by freshwater inflow. *H. engelmannii* has classically been described as an offshore, deepwater plant which favors higher salinities (Humm 1973), yet as indicated above, it is more widespread than traditionally believed. At numerous estuaries throughout the Big Bend, including the Econfina, Steinhatchee, Withlacoochee and Crystal rivers, this plant is found at inshore sites, sometimes within 1–2 km of

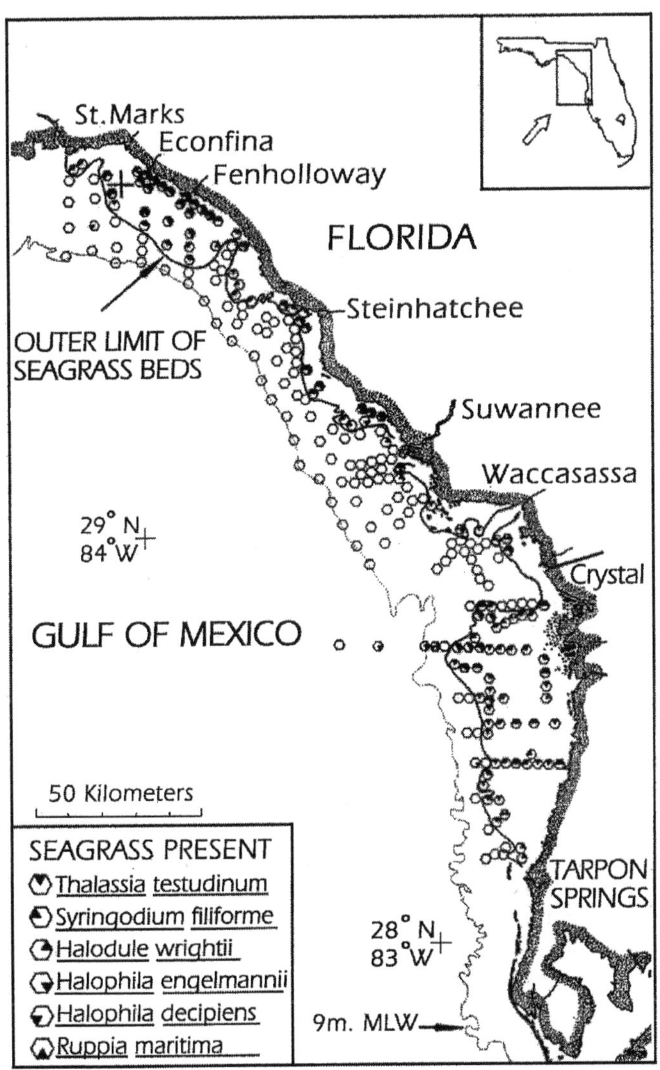

FIGURE 20.3 Map showing Iverson and Bittaker's sampling sites in the Big Bend and seagrass species occurrence. (From Iverson, R. L. and H. F. Bittaker. 1986. *Estuarine, Coastal and Shelf Science* 22:579. With permission.)

the river mouth in salinities which may drop to < 5 ppt for portions of the year (Zimmerman and Livingston 1976a; Mattson 1995 and personal observation). This plant's ability to tolerate lower light conditions (which occur around the river mouths in more highly colored water) may give it a competitive advantage in these inshore areas, even though salinities may not be optimum.

As one way of classifying or describing seagrass communities, Strawn (1961) and McNulty et al. (1972) recognized a zonation of seagrasses in Florida west coast grassbeds (Figure 20.4), with *H. wrightii*/*R. maritma* occurring inshore, then zones of *T. testudinum* and *S. filiforme* (mixed or pure stands), and finally *Halophila* spp. or *H. wrightii* offshore. Strawn (1961) studied seagrass zonation in the Cedar Key area and believed it was related to the frequency of exposure of the grassbeds at low tide combined with the morphology of the plants. *H. wrightii* was able to tolerate the most amount of low tide exposure (hence being found inshore) due to its thin, flexible leaf blades which were able to lie at right angles to the short shoots. *T. testudinum* was able to tolerate brief periods of exposure during low spring tides but also occurred in deeper water. *S. filiforme*

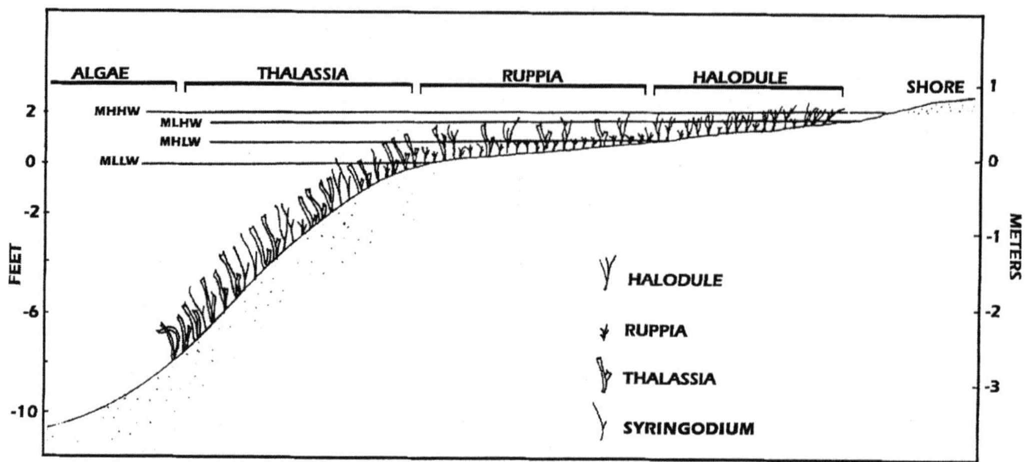

FIGURE 20.4 Diagram of an example of inshore-offshore seagrass zonation on the Florida west coast (data from McNulty et al. 1972).

appeared to be least tolerant of exposure and had thicker, brittle blades which could be easily broken off when bent at right angles to the short shoots, which could account for its occurrence only in deeper offshore zones. Iverson and Bittaker (1986) reported similar zonation in the Alligator Harbour area, adjacent to north Apalachee Bay.

Lewis et al. (1985) recognized this zonation in Tampa Bay, but in addition introduced a different way of classifying seagrass communities by identifying five structural types of seagrass meadows associated with various geomorphic features (e.g., shoals or fringing shoreline zones) and levels of environmental stress. Four of that study's five types occur in the Big Bend:

1. Healthy Fringe Perennial (Big Bend variant) — the most common community type in the Big Bend found throughout the region. All five of the common seagrasses may be found in this community type. The offshore bar associated with this type in Tampa Bay does not exist in the Big Bend seagrass meadows, hence the "variant" designation.
2. Mid-Bay Shoal Perennial — found around Cedar Key and off the Suwannee River mouth (Hedemon and Red Bank reefs). The five common seagrass taxa may occur in this community type.
3. Stressed Fringe Perennial — found near the river mouths in the region. The stresses may be natural, associated with seasonal changes in river flow and salinity, or due to human activities, as around the mouth of the Fenholloway River, where discharge of pulp mill effluent upstream creates stress on seagrasses. *H. wrightii* or *R. maritima* are more characteristic in this community type.
4. Ephemeral — found near the mouth of the Suwannee River. These are beds of *R. maritima* mixed with salt-tolerant freshwater macrophytes such as *Zannichellia palustris* and *Vallisneria americana* which may flourish during periods of higher river flows (hence lower salinities) but diminish in coverage during low flows.

Standing Crop and Biomass

Seagrass standing crop data (dry weight mass per unit area) from the Big Bend region are summarized in Table 20.3. Zimmerman and Livingston (1976a) found that *T. testudinum* and *S. filiforme* dominated standing crop in Apalachee Bay (Figure 20.5), which would be expected because of their larger, more robust thalli. Iverson and Bittaker (1986) reported that these two seagrasses accounted for most of the seagrass leaf biomass in the Big Bend, and together with *H. wrightii*

TABLE 20.3
Summary of seagrass standing crop data (g dry weight m^{-2}) from the Big Bend. Unless indicated otherwise, all data are above-ground standing crop.

	Location	Source
Thalassia testudinum		
0.05–305.85 (ab. & bel. ground)	Apalachee Bay	Zimmerman and Livingston 1979
0–409 (ab. & bel. ground)	Cedar Key	Dawes et al. 1985
0.58–10.65	Withlacoochee estuary	Mattson et al. 1988
2.37–72.0	Crystal Bay	Mattson et al. 1988
< 0.01–56.96	Crystal Bay	Mattson 1995
147–632 (ab. & bel. ground)	Homosassa Bay	Dawes et al. 1985
2.24–305.44	Weeki Wachee estuary	Mattson 1995
51–187 (ab. & bel. ground)	Weeki Wachee estuary	Dawes et al. 1985
29–342 (ab. & bel. ground)	Anclote estuary	Dawes et al. 1985
Syringodium filiforme		
< 0.01–75.3 (ab. & bel. ground)	Apalachee Bay	Zimmerman and Livingston 1979
0.31–3.20	Withlacoochee estuary	Mattson et al. 1988
1.12–31.84	Withlacoochee estuary	Mattson 1995
3.90–39.84	Crystal Bay	Mattson et al. 1988
0.94–75.36	Crystal Bay	Mattson 1995
Halodule wrightii		
0.06–4.49 (ab. & bel. ground)	Apalachee Bay	Zimmerman and Livingston 1979
0.37–3.46	Withlacoochee estuary	Mattson et al. 1988
1.12–10.56	Withlacoochee estuary	Mattson 1995
0.89–32.32	Crystal Bay	Mattson et al. 1988
< 0.01–11.36	Crystal Bay	Mattson 1995
2.24–28.16	Weeki Wachee estuary	Mattson 1995
Ruppia maritima		
0.45 (ab. & bel. ground)	Apalachee Bay	Zimmerman and Livingston 1979
0.16–61.28	Withlacoochee estuary	Mattson 1995
0.32–135.68	Crystal Bay	Mattson 1995
3.04–461.12	Weeki Wachee estuary	Mattson 1995
Halophila engelmannii		
0.07–0.69 (ab. & bel. ground)	Apalachee Bay	Zimmerman and Livingston 1979
< 0.01–3.04	Withlacoochee estuary	Mattson 1995
< 0.01–7.04	Crystal Bay	Mattson 1995

made up most of the standing crop in this grassbed ecosystem. At some locations, standing crop of the benthic and/or drift algae may exceed that of seagrasses (Zimmerman and Livingston 1979; Mattson 1995).

While a distinct inshore-offshore zonation of the different seagrass species may be observed in some areas of the Big Bend, it is also characteristic to find 3–5 species of seagrasses at a particular sampling site, with the inshore-offshore trend being a shift in relative abundance among the five species. Zimmerman and Livingston (1979) found *R. maritima* dominated standing crop at an inshore site off the Econfina River (Figure 20.6), with *T. testudinum* and/or *S. filiforme* predominant at two offshore sites, but four seagrass species were present at each of these sites (Figure 20.6). Similar results were reported by Mattson (1995) off the Withlacoochee, Crystal, and Weeki Wachee

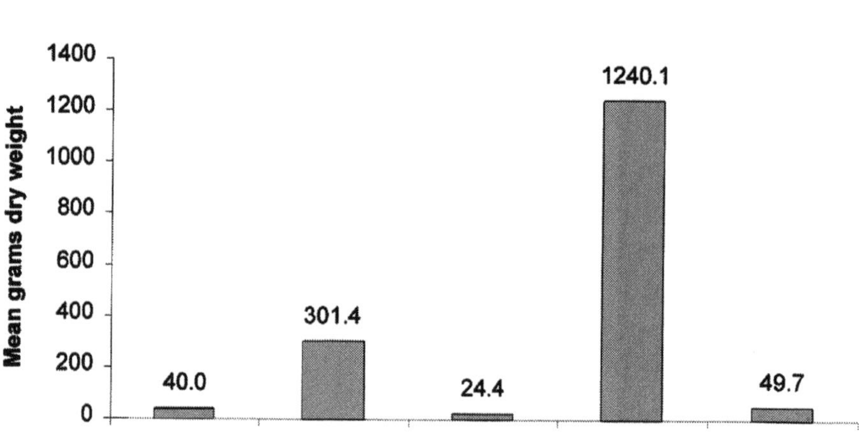

FIGURE 20.5 Comparison of total standing crop (above- and below-ground combined) among the five common seagrass taxa in Apalachee Bay (data from Zimmerman and Livingston 1976a).

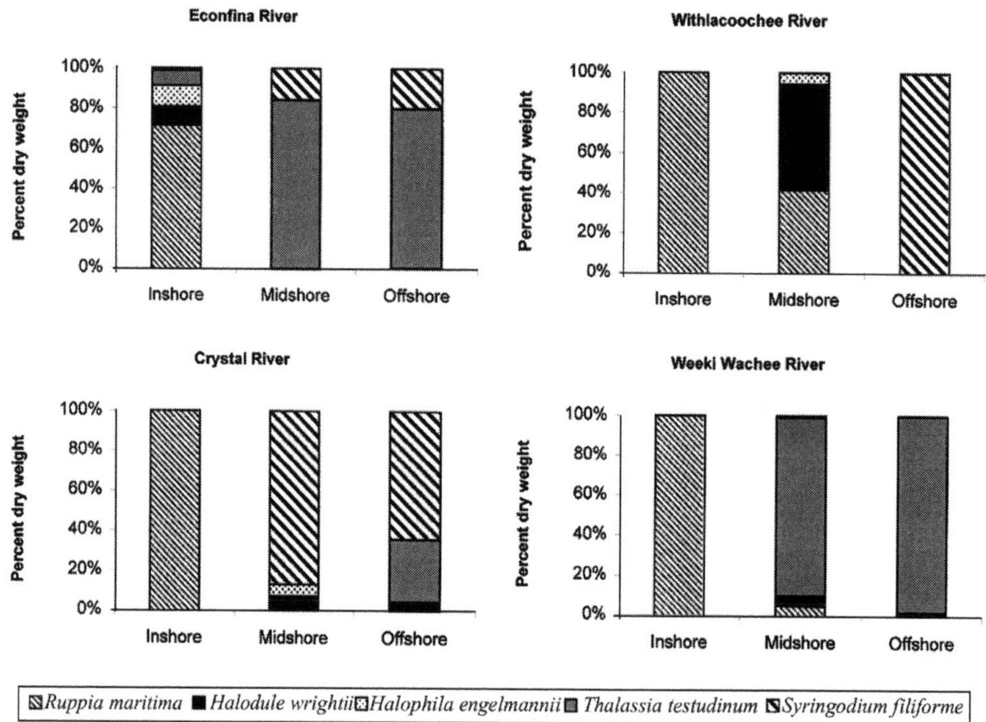

FIGURE 20.6 Comparison of seagrass relative abundance (percent standing crop by species) at inshore, midshore and offshore sampling sites from the Econfina River (data from Zimmerman and Livingston 1979) and Withlacoochee, Crystal and Weeki Wachee rivers (data from Mattson 1995).

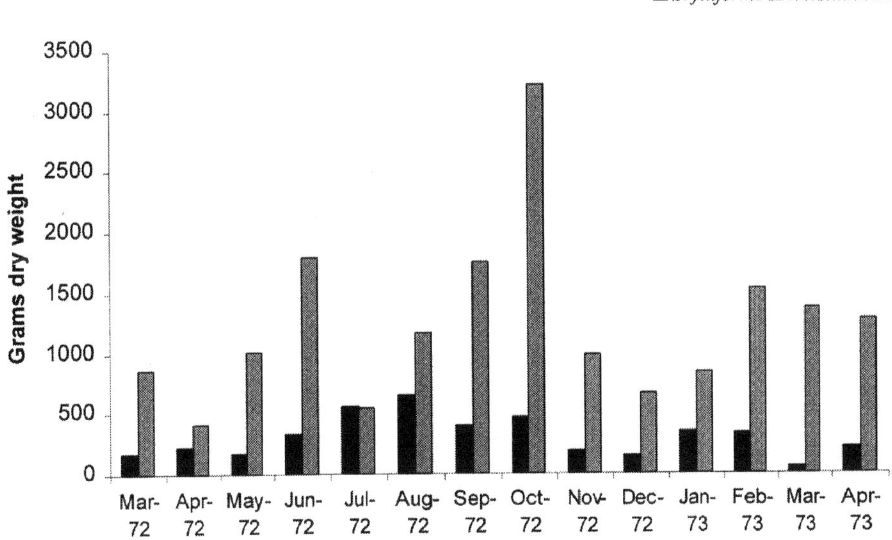

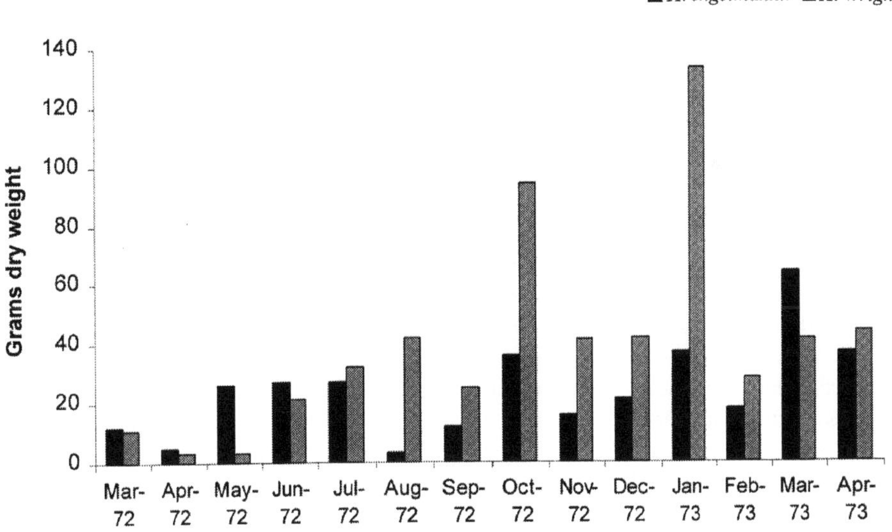

FIGURE 20.7 Monthly variation in seagrass standing crop in Apalachee Bay (data from Zimmerman and Livingston 1976a).

rivers (Figure 20.6); *R. maritima* and/or *H. wrightii* exhibited higher standing crops at inshore sites while *T. testudinum* or *S. filiforme* were higher at offshore sites.

Zimmerman and Livingston (1976a) observed peak standing crops of *T. testudinum* and *S. filiforme* generally occurred in the late spring/summer in Apalachee Bay (Figure 20.7), although *T. testudinum* also exhibited a fall or winter peak in standing crop (Figure 20.7). *H. wrightii* and *H. engelmannii* exhibited peaks in standing crop in fall or winter. Mattson et al. (1988) and Mattson (1995) found peak leaf standing crop occurred in summer or early fall off the Withlacoochee, Crystal, and Weeki Wachee rivers.

TABLE 20.4
Short shoot densities (Number per m²) of *T. testudinum, S. filiforme,* and *H. wrightii* in the Big Bend.

	Location	Source
Thalassia testudinum		
50–800	Region-wide	Iverson and Bittaker 1986
96–688	Deadman Bay	SRWMD unpublished data
300–588	Withlacoochee estuary	Mattson et al. 1988
463–963	Crystal Bay	Mattson et al. 1988
125–625	Anclote estuary	Ballantine 1972
Syringodium filiforme		
100–5000	Region-wide	Iverson and Bittaker 1986
816–1488	Deadman Bay	SRWMD unpublished data
363–813	Withlacoochee estuary	Mattson et al. 1988
738–1450	Crystal Bay	Mattson et al. 1988
700–1000	Anclote estuary	Ballantine 1972
Halodule wrightii		
100–1900	Region-wide	Iverson and Bittaker 1986
192–1248	Deadman Bay	SRWMD unpublished data
400–2486	Withlacoochee estuary	Mattson et al. 1988
629–2571	Crystal Bay	Mattson et al. 1988
425–3600	Anclote estuary	Ballantine 1972

Shoot Density and Blade Characteristics

Short shoot densities of the three major species of seagrasses in the Big Bend are shown in Table 20.4. Shoot densities measured in the Big Bend grassbeds tend to be lower than those reported for seagrass ecosystems in southern Florida (Iverson and Bittaker 1986; Zieman and Zieman 1989). A trend toward higher shoot densities toward the southern portion of the region (Withlacoochee to Anclote estuaries) does appear to be evident in the Big Bend (Table 20.4). Mattson et al. (1988) observed peak shoot densities in the summer or fall off the Crystal and Withlacoochee rivers.

Seagrass blade characteristics differ with depth in conjunction with the zonation described above. Different seagrass taxa exhibited peak leaf biomass (Figure 20.8) at different depths in Alligator Harbour (Bittaker and Iverson 1976). Leaf biomass of *H. wrightii* was highest in shallow inshore regions, biomass of *T. testudinum* was higher in areas of moderate depth, while *S. filiforme* leaf biomass was highest at deeper depths. Strawn (1961) found that leaf lengths for *H. wrightii* and *T. testudinum* increased with depth (Figure 20.8), while *S. filiforme* leaf length was similar at different depths, although this plant was only found in deeper areas (Figure 20.8).

Productivity

Few studies of seagrass productivity have been conducted in the Big Bend. Bittaker and Iverson (1976) compared two methods for measuring production of *T. testudinum* in the field in Alligator Harbour. The leaf stapling method developed by Zieman yielded an estimate of 564–705 (mean 641.5) mg C m^{-2} d^{-1}. The ^{14}C uptake method yielded estimates of 391–445 (mean 419.5) mg C m^{-2} d^{-1}. When the results of the ^{14}C method were corrected upward for factors due to the experimental setup (e.g., light extinction caused by the glass walls of the incubation chamber), both methods yielded identical production estimates. Table 20.5 summarizes the available data on

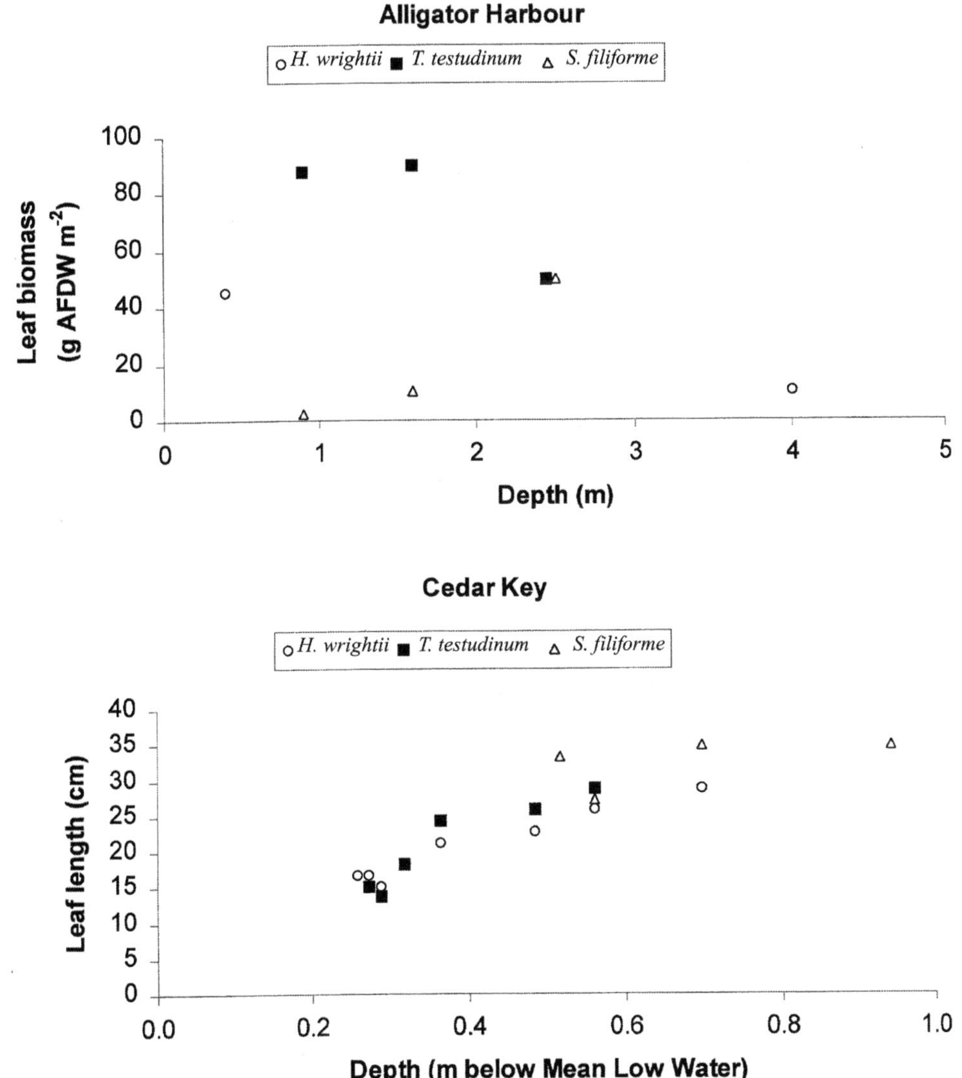

FIGURE 20.8 Comparison of leaf ash-free dry weight (data from Iverson and Bittaker 1986) and seagrass leaf length (data from Strawn 1961) with depth in the Big Bend region.

seagrass productivity in the Big Bend. The data are not entirely comparable, as some studies report production as mg C area^{-1} time^{-1} while others report it as g dry weight area^{-1} time^{-1}. Mattson et al. (1988) found highest productivity (as measured by leaf regrowth) in the summer off the Withlacoochee and Crystal rivers.

Flowering Phenology and Seed Production

Very few observations or studies are available on flowering phenologies or seed production in Big Bend seagrasses, although this deficit is reflective of the state of knowledge as a whole of seagrass reproductive ecology in the southeastern U.S. (Moffler and Durako 1987). All of the seagrasses generally are reported to flower during spring/summer (Moffler and Durako 1987). Zimmerman and Livingston (1976a) found *S. filiforme* flowering in May and *R. maritima* flowering in May-June in Apalachee Bay. *S. filiforme* was found flowering in Deadman Bay in May, *R. maritima* was

TABLE 20.5
Summary of seagrass productivity data from the Big Bend. All data are from leaf growth-based methods of measurement.

	Location	Source
Thalassia testudinum		
564–705 mg C m^{-2} d^{-1}	Alligator Harbour	Bittaker and Iverson 1976
0.04–0.35 g dry wt m^{-2} d^{-1}	Withlacoochee estuary	Mattson et al. 1988
0.07–0.98 g dry wt m^{-2} d^{-1}	Crystal Bay	Mattson et al. 1988
48–360 mg C m^{-2} d^{-1}	Anclote estuary	Zieman and Zieman 1989
Syringodium filiforme		
0.03–0.18 g dry wt m^{-2} d^{-1}	Withlacoochee estuary	Mattson et al. 1988
0.06–1.55 g dry wt m^{-2} d^{-1}	Crystal Bay	Mattson et al. 1988
48–888 mg C m^{-2} d^{-1}	Anclote estuary	Zieman and Zieman 1989
Halodule wrightii		
< 0.01–0.26 g dry wt m^{-2} d^{-1}	Withlacoochee estuary	Mattson et al. 1988
0.04–0.59 g dry wt m^{-2} d^{-1}	Crystal Bay	Mattson et al. 1988
22–34 mg C m^{-2} d^{-1}	Anclote estuary	Zieman and Zieman 1989

found flowering off the Withlacoochee, Crystal, and Weeki Wachee rivers in spring/summer, and *T. testudinum* was found flowering off the Weeki Wachee River in May (personal observation). Phillips (1960b) reported finding *T. testudinum* flowering in May near Anclote Key and *R. maritima* flowering in April, June, and in February off the Crystal River. That report found two germinating seedlings of *T. testudinum* near Anclote Key and suggested that this area may be the northernmost limit of flowering for this seagrass, although there are indications that flowering does occur further north than this (personal observation).

ENVIRONMENTAL FORCING FUNCTIONS

LIGHT

As has been observed elsewhere in coastal Florida, light appears to be the primary limiting factor influencing the distribution of seagrasses in the Big Bend (Livingston 1993). Light limitation may come from two sources: natural or human-caused. The nearshore regions adjacent to river mouths experience reduced light from high color levels during river flooding. Most of the Big Bend rivers, particularly from the Withlacoochee River northward, are highly colored during flood flows, but exhibit very low color at base flows. For instance, color levels off the Steinhatchee River mouth vary from < 10 PCU at low flows to 250 PCU at high flow (SRWMD, unpublished data). This appears to be a primary factor accounting for limited coverage of seagrasses within 1 km of the mouths of the Steinhatchee and other rivers in the Big Bend.

Human-derived light stresses also exist in the region. One of the better known estuaries in Florida is that of the Fenholloway River, which receives wastewater from a bleached kraft pulp mill. The mill effluent is highly colored with high levels of sulfate, BOD, suspended solids and nutrients. The impacts of this effluent discharge affect not only the river itself, but also the estuary and adjacent nearshore coastal waters. Zimmerman and Livingston (1976b) showed that benthic macrophyte species composition was altered and standing crop and species richness were reduced in nearshore coastal areas affected by the kraft mill effluent. In subsequent work, Livingston (1993) and McGlynn and Livingston (1994) showed how the effluent affected the overall quantity and spectral quality of light penetrating through the water column and quantified the effects of these alterations on seagrass community characteristics.

Salinity

Salinity is generally acknowledged to be an important environmental influence on seagrasses (Zieman and Zieman 1989). In the case of the estuarine areas of the Big Bend, salinity and water color (which is related to light as described above) are both intimately related to river flow, and it is difficult to separate their effects based on field data. Mattson (1995) found positive correlations between conductivity (an indicator of salinity) and above-ground standing crop for most seagrass species except *R. maritima*, which was negatively correlated with conductivity. Changes in benthic macrophyte community composition along gradients of salinity have been observed (Zimmerman and Livingston 1976a; Mattson 1995).

Increases in salinity associated with freshwater inflow reductions due to human water use is a management issue in the region. While increased salinity may not be harmful to seagrasses, except perhaps for *R. maritima*, Browder's (1991) concept of an estuary as a region of overlap of "dynamic habitat" (the area of low, variable salinities created by freshwater inflow) and "stationary habitat" (the physical structure provided by seagrasses, coastal marshes, etc.) indicates that increases in mean or minimum salinities, or alteration in salinity ranges, will likely affect the habitat values of the inshore grassbeds. This is particularly crucial for economically important fishery species. Estevez (Chapter 22, this volume) discusses this in more detail.

Climate

This environmental factor has received the least amount of research as it relates to distribution and abundance of seagrasses in the Big Bend. Yet it may be a significant influence on seagrasses here, since they reside at or close to their northernmost limits of geographic distribution and thus are under natural climatic stresses. Strawn (1961) observed that at Cedar Key, the lowest tides of the year occur in the winter (generally February), usually in the morning, when the effects of hard freezes on seagrasses might be most severe. Livingston (1987) suggested that limited recovery of seagrasses off the Fenholloway River in the late 1970s, following implementation of a pollution abatement program for the pulp mill discharge, could have been due to species such as *T. testudinum* being at or near the northern limit of their range. The natural climatic stresses experienced by the subtropical seagrasses and benthic algae in the Big Bend have management implications. The ability of the Big Bend grassbeds to recover from human impacts could be limited by the harsher winters experienced there relative to south Florida. A management approach which seeks to preserve the existing seagrass coverage may be more prudent, since recovery or restoration attempts may have a lower chance of success in the Big Bend. Clearly, this is an area where more research is needed.

MANAGEMENT ISSUES

Mapping

A basic information need for the understanding and management of the Big Bend seagrass ecosystem is the development of accurate maps of seagrass coverage in the region. While several mapping studies have been conducted, in some of these, the data are confined to a limited area (e.g., McNulty et al. 1972; Livingston 1993). Other problems include inaccuracies which do not reflect current conditions (McNulty et al. 1972; Continental Shelf Associates and Martel Laboratories 1985), seagrass coverage data not published as a map (Iverson and Bittaker 1986), or mapping based on dated photography (Sargent et al. 1995).

While mapping may be conducted easily with conventional photointerpretation methods in many inshore areas of the Big Bend, the periodic high color found near the river mouths and the depth offshore to which grasses grow (both of which limit use of aerial photography) make the technical challenges of accurate mapping much greater here than in other coastal areas in Florida. Some of the new technologies described in this symposium (J. Burns, personal communication; R. Chamberlain, personal communication) offer promise and should be evaluated for use in the Big Bend region.

Seagrass Habitat Loss

Every estuarine area in Florida has experienced some amount of loss of seagrass habitat (Livingston 1987), usually due to some combination of direct destruction and reduced water clarity due to pollution. The issue of loss of seagrass coverage is usually not associated with the Big Bend. Livingston (1993) has conducted the only quantitative effort to document seagrass habitat loss in the region in Apalachee Bay. By comparing seagrass coverage in the Fenholloway estuary with that in adjacent undisturbed drainages (the Econfina and Aucilla rivers), he determined that about 23.3 km^2 of seagrasses have been lost in the Fenholloway due to the reduced light from pulp mill effluent discharge. The only other evidence for possible changes in seagrass cover is anecdotal; for example, Putnam (1967) described extensive beds of benthic vegetation (described as "attached algae and seagrasses") in Waccasassa Bay, which do not currently exist today. Long-time fishermen from the Town of Suwannee have noted (personal communication) that seagrass cover offshore of the Suwannee River (Hedemon and Red Bank reefs and the "Spotty Bottom" area) was historically (30–40 years ago) more extensive. In the same region, Moore (1963) describes "abundant *S. filiforme*" offshore and "dense beds of turtle grass" inshore in southern Suwannee Sound near Cedar Key. Much of this area is unvegetated today. In combination with the other management issues presented in this section, documentation of seagrass loss in the region becomes important.

Development of Protective Water Quality Criteria

The concept of "resource-based restoration targets" is introduced with respect to seagrasses elsewhere in this volume (e.g., Virnstein and Morris, Chapter 16; Johansson and Greening, Chapter 21). This concept involves the development of water quality restoration targets which should allow for recovery of seagrass beds historically lost due to reduced water clarity. Livingston (1993 and personal communication) has identified water quality improvements needed in the Fenholloway estuary to allow, theoretically, recovery of seagrass coverage to that of adjacent, undisturbed drainages. The inverse of this is needed for much of the rest of the Big Bend, the development of the minimum water quality criteria necessary to maintain the existing seagrass coverage.

Increasing concentrations of nitrate-nitrogen in many Big Bend rivers (Ham and Hatzell 1996; Mattson and Flannery 1997) have generated management concerns about this nutrient increasing algal growth, which may negatively affect seagrasses. Much of this nitrate is coming from human land uses, but via a different pathway than usually thought for nonpoint pollution. Leaching of nitrate to groundwater from overlying land uses and its subsequent discharge to surface waters via springs has been documented in the region (Katz et al. 1997). A better understanding of nutrient concentrations and algal growth and determination of threshold levels to which nutrients should be kept below are all needed as part of this effort to develop protective water quality criteria.

Somewhat related in this regard is the need to be aware of potential boating impacts to seagrass beds as recreational activities in the area increase. "Ecotourism" is regarded as a viable economic development activity in the area which depends on maintenance of healthy natural resources, including seagrass beds. A recent report on boat propeller damage in Florida seagrass beds (Sargent et al. 1995) indicated that damage levels are currently low in the counties of the Big Bend. The opportunity to protect shallow-water grass beds in the region through public education and adequate marking of channel areas could work effectively at this point in time, if aggressively implemented, without having to impose severe restrictions on boat use.

Monitoring

As noted throughout this section, and as noted by Zieman and Zieman (1989), the seagrass ecosystem of the Big Bend has received less research and management attention than other seagrass areas in Florida. In part this is due to its high ecological integrity, low levels of human impact, and lack of a large population of constituent groups demanding management action from government. Whether or not these factors will persist is beyond the scope of this report. However, Florida remains one of the fastest growing states in the nation, and it seems likely that at some point in the future, the Big Bend area will feel the effects of this growth. Some individuals would argue that we have learned from our past mistakes, and that the regulatory and planning tools are now in place to allow for growth and development in the region without degradation of water quality and natural resources. The validity of this argument cannot be confirmed without a program of water quality and associated biological monitoring to evaluate if in fact the regulations and planning efforts are succeeding.

The value of seagrasses as overall environmental indicators of coastal water quality has been touted repeatedly throughout these proceedings. Developing and implementing a program to monitor seagrasses at sites throughout the Big Bend as a means of assessing their condition or health is a basic need to manage and protect the area from a scientific basis. A properly designed monitoring program (developed within a hypothesis-testing framework) could yield new understanding of seagrass ecology, as well as provide the critical feedback data needed to assess the efficacy of existing or new management programs.

ACKNOWLEDGMENTS

B. Newsome prepared the regional map (Figure 20.1). B. Shafii and D. Hornsby assisted with graphics preparation. The comments of two anonymous reviewers greatly improved this chapter.

REFERENCES

Ballantine, D. L. 1972. Epiphytes of Four Florida Seagrass Species in the Anclote Anchorage, Tarpon Springs, Florida. Master's thesis, University of South Florida, Tampa, FL.

Bittaker, H. F. and R. L. Iverson. 1976. *Thalassia testudinum* productivity: a field comparison of measurement methods. *Marine Biology* 37:39–46.

Browder, J. A. 1991. Watershed management and the importance of freshwater inflow to estuaries. In *Proceedings, Tampa Bay Area Scientific Information Symposium 2,* S. F. Treat and P. A. Clark (eds.). Tampa Bay Regional Planning Council, St. Petersburg, FL, pp. 7–22.

Clewell, A. F. 1985. *Guide to the Vascular Plants of the Florida Panhandle.* Florida State University Press, Tallahassee, FL.

Continental Shelf Associates, Inc. and Martel Laboratories, Inc. 1985. Florida Big Bend Seagrass Habitat Study Narrative Report. Final report by Continental Shelf Associates, Inc. submitted to the Minerals Management Service, Metairie, LA, Contract No. 14-12-0001-30188.

Davis, R. A. 1997. Geology of the Florida Coast. In *The Geology of Florida,* A. F. Randazzo and D. S. Jones (eds.). University Presses of Florida, Gainesville, FL, pp. 155–168.

Dawes, C. J., M. O. Hall, and R. K. Riechert. 1985. Seasonal biomass and energy content in seagrass communities on the west coast of Florida. *Journal of Coastal Research* 1(3):255–262.

Earle, S. A. 1969. Phaeophyta of the eastern Gulf of Mexico. *Phycologia* 7(2):71–254.

Eiseman, N. J. 1980. An Illustrated Guide to the Sea Grasses of the Indian River Region of Florida. Technical Report No. 31, Harbour Branch Foundation, Fort Pierce, FL.

Godfrey, R. K. and J. W. Wooten. 1979. *Aquatic and Wetland Plants of the Southeastern United States.* Monocotyledons. University of Georgia Press, Athens, GA.

Ham, L. K. and H. H. Hatzell. 1996. Analysis of Nutrients in the Surface Waters of the Georgia-Florida Coastal Plain Study Unit, 1970-91. U.S. Geological Survey Water-Resources Investigation Report 96-4037.

Humm, H. J. 1973. Seagrasses. In *A Summary of Knowledge of the Eastern Gulf of Mexico.* State University System of Florida, Institute of Oceanography, St. Petersburg, FL, pp. IIIC-1 to IIIC-10.

Iverson, R. L. and H. F. Bittaker. 1986. Seagrass distribution and abundance in eastern Gulf of Mexico coastal waters. *Estuarine Coastal Shelf Science* 22:577–602.

Katz, B. G., R. S. DeHan, J. J. Hirten, and J. S. Catches. 1997. Interactions between ground water and surface water in the Suwannee River basin, Florida. *Journal of the American Water Resources Association* 33(6):1237–1254.

Levinton, J. S. 1982. *Marine Ecology.* Prentice-Hall, Englewood Cliffs, NJ.

Livingston, R. J. 1987. Historic trends of human impacts on seagrass meadows of Florida. In *Proceedings of the Symposium on Subtropical-Tropical Seagrasses of the Southeastern United States,* M. J. Durako, R. C. Phillips, and R. R. Lewis, III (eds.). Florida Marine Research Publications No. 42, pp. 139–152.

Livingston, R. J. 1990. Inshore Marine Habitats. In *Ecosystems of Florida,* R. L. Meyers and J. J. Ewel (eds.). University of Central Florida Press, Orlando, FL, pp. 549–573.

Livingston, R. J. 1993. River/Gulf Study. Chapter III. Light Transmission Characteristics and Submerged Aquatic Vegetation. Report prepared by Environmental Planning and Analysis for the Buckeye Kraft L.P., Perry, FL.

Lewis, R. R., III, M. J. Durako, M. D. Moffler, and R. C. Phillips. 1985. Seagrass meadows of Tampa Bay: a review. In *Proceedings, Tampa Bay Area Scientific Information Symposium,* S. F. Treat, J. L. Simon, R. R. Lewis, III, and R. L. Whitman (eds.). Florida Sea Grant Report No. 65, Florida Sea Grant College, Gainesville, FL, pp. 210–246.

Mattson, R. A. 1995. Submerged Macrophyte Communities. Vol. 6 in a series: A Data Collection Program for Selected Coastal Estuaries in Hernando, Citrus and Levy Counties, Florida. Report submitted to Southwest Florida Water Management District, Brooksville, FL.

Mattson, R. A. 1997. An overview of the Big Bend ecosystem. In *Florida Big Bend Coastal Research Workshop. Toward a Scientific Basis for Ecosystem Management,* W. J. Lindberg (ed.). Florida Sea Grant Technical Paper No. 88, University of Florida, Gainesville, FL, pp. 1–3.

Mattson, R. A. and M. S. Flannery. 1997. River drainages and their characteristics. In *Florida Big Bend Coastal Research Workshop. Toward a Scientific Basis for Ecosystem Management,* W. J. Lindberg (ed.). Florida Sea Grant Technical Paper No. 88, University of Florida, Gainesville, FL, pp. 35–37.

Mattson, R. A., J. A. Derrenbacker, Jr., and R. R. Lewis, III. 1988. Effects of thermal addition from the Crystal River generating complex on the submergent macrophyte communities in Crystal Bay, Florida. In *Proceedings, Southeastern Workshop on Aquatic Ecological Effects of Power Generation,* K. Mahadevan et al. (eds.). Mote Marine Laboratory Report No. 124, Sarasota, FL, pp. 11–67.

McGlynn, S. E. and R. J. Livingston (Abstract). 1994. Light as an organizing feature of coastal benthic assemblages: experimental analysis of light transmission characteristics in polluted and unpolluted waters. *Bulletin of the North American Benthological Society* 11(1):120.

McNulty, J. K., W. N. Lindall, Jr., and J. E. Sykes. 1972. Cooperative Gulf of Mexico Estuarine Inventory and Study, Florida: Phase I, Area Description. NOAA Technical Report NMFS CIRC-368, U.S. Department of Commerce, Washington, D.C.

Moffler, M. D. and M. J. Durako. 1987. Reproductive biology of the tropical-subtropical seagrasses of the southeastern United States. In *Proceedings of the Symposium on Subtropical-Tropical Seagrasses of the Southeastern United States,* M. J. Durako, R. C. Phillips, and R. R. Lewis, III (eds.). Florida Marine Research Publication No. 42, pp. 77–88.

Moore, D. R. 1963. Distribution of the seagrass, *Thalassia,* in the United States. *Bulletin of Marine Science of the Gulf and Caribbean* 13(2):329–342.

Phillips, R. C. 1960a. The ecology of marine plants of Crystal Bay, Florida. *Quarterly Journal of the Florida Academy of Sciences* 23(4):328–337.

Phillips, R. C. 1960b. Observations on the ecology and distribution of the Florida seagrasses. Florida State Board of Conservation Professional Papers Series No. 2.

Putnam, H. D. 1967. Limiting factors for primary production in a west coast Florida estuary. *Advances in Water Pollution Research* 3:121–142.

Sargent, F. J., T. J. Leary, D. W. Crewz, and C. R. Kruer. 1995. Scarring of Florida's Seagrasses: Assessment and Management Options. Florida Department of Environmental Protection Marine Research Institute, FMRI Technical Report TR-1, St. Petersburg, FL.

Strawn, K. 1961. Factors influencing the zonation of submerged monocotyledons at Cedar Key, Florida. *Journal of Wildlife Management* 25(2):178–189.

Virnstein, R. W. 1987. Seagrass-associated invertebrate communities of the southeastern U.S.A.: a review. In *Proceedings of the Symposium on Subtropical-Tropical Seagrasses of the Southeastern United States,* M. J. Durako, R. C. Phillips, and R. R. Lewis, III (eds.). Florida Marine Research Publication No. 42, pp. 89–116.

Zieman, J. C. and R. T. Zieman. 1989. The ecology of the seagrass meadows of the west coast of Florida: a community profile. U. S. Fish and Wildlife Service Biological Report 85(7.25).

Zimmerman, M. S. and R. J. Livingston. 1976a. Seasonality and physico-chemical ranges of benthic macrophytes from a north Florida estuary. *Contributions in Marine Science* 20:33–45

Zimmerman, M. S. and R. J. Livingston. 1976b. Effects of kraft-mill effluents on benthic macrophyte assemblages in a shallow-bay system (Apalachee Bay, North Florida, USA). *Marine Biology* 34:297–312.

Zimmerman, M. S. and R. J. Livingston. 1979. Dominance and distribution of benthic macrophyte assemblages in a north Florida estuary (Apalachee Bay, Florida). *Bulletin of Marine Science* 29(1):27–40.

21 Seagrass Restoration in Tampa Bay: A Resource-Based Approach to Estuarine Management

J. O. Roger Johansson and Holly S. Greening

Abstract Historical (pre-1930s) seagrass meadows in Tampa Bay are believed to have covered 31,000 ha of the shallow bay bottom. However, impacts to the bay from increasing population and industrial development of the Tampa Bay area resulted in large seagrass reductions. By 1982, approximately 8,800 ha of seagrass remained. Recently, Tampa Bay seagrass monitoring programs have shown that the trend of seagrass loss has been reversed. The bay-wide seagrass cover in 1997, was estimated at 10,930 ha. In Hillsborough Bay, the bay segment that historically has had the poorest water quality, seagrass increased from near 0 ha in 1984 to about 57 ha in 1998. The Tampa Bay seagrass expansion apparently started in response to water quality improvements that occurred from the late 1970s to the mid-1980s, which included reductions in phytoplankton biomass and water column light attenuation. These improvements followed a nearly 50% reduction in external nitrogen loading from domestic and industrial point-sources, primarily discharging to Hillsborough Bay, that occurred in the early 1980s. However, most recently, high rainfall amounts during the years 1995, 1996, and the 1997–1998 El Niño event increased nitrogen loading to the bay. Both phytoplankton biomass and light attenuation increased in all major bay segments during this period of high rainfall. These influences on seagrass growth are often detrimental and appear to have reduced the rate of seagrass expansion during the last few years.

The reduced rate of expansion is most evident in the Hillsborough Bay section of Tampa Bay for which the most detailed and current seagrass information is available. Recognizing the link between nitrogen loading, water quality, and seagrass protection, local, state, and federal partners working cooperatively through the Tampa Bay National Estuary Program (TBNEP) have agreed to adopt nitrogen loading targets for Tampa Bay based on the light requirements of *Thalassia testudinum*. A long-term goal has been adopted to achieve 15,400 ha of seagrass coverage, or 95% of the seagrass estimated for 1950. Reaching the goal will require preservation of the approximately 10,400 ha of seagrass present in the bay in 1992 and restoration of an additional 5,000 ha.

Field measurements in Tampa Bay indicate that 20–25% of surface irradiance is required for sustained growth of *T. testudinum*. Two independent water quality models were used to estimate nitrogen loading rates and associated water column chlorophyll *a* concentrations required to maintain irradiance levels at the apparent maximum depth of seagrass growth in 1950 for each major bay segment. Based on monitoring data, it appears that light levels can be maintained at these depths in most bay segments with the existing nitrogen loading rates. However, to achieve the long-term seagrass restoration goal, increases in nitrogen loading associated with a projected increase in the watershed's human population must be offset.

0-8493-2045-3/99/$0.00+$.50
© 2000 by CRC Press LLC

TBNEP partners have, therefore, identified and committed to specific nitrogen load reduction projects to ensure that nitrogen management targets are met. An interagency bay-wide seagrass monitoring program has been established to document the progress of Tampa Bay seagrass protection and restoration. Citizen-based educational and voluntary programs are underway to help reduce propeller scarring of seagrass meadows and to plant seagrass in areas which currently lack vegetation.

INTRODUCTION

Seagrass is a vital natural resource present in many shallow estuaries and coastal areas of the world, including Tampa Bay, Florida. This habitat supports coastal food-chains through high levels of primary and secondary production and serves as a nursery for numerous animal species, including many of recreational and commercial importance. Seagrass also reduces impacts from waves and currents, helps stabilize sediments, and is an integral component of the shallow water nutrient cycling process.

Seagrass is an important indicator of estuarine health primarily because of its sensitivity to changes in submarine light availability. Seagrass losses caused by light reductions generally result from a build-up of algae populations (i.e., phytoplankton, epiphytic algae and macro-algae) following excessive inputs of man-made nutrients (eutrophication) or by an increase in turbidity from suspended solids often associated with dredging activities and stormwater discharges.

Seagrass has been selected as a central component in many estuarine management efforts that aim to restore a natural balance between primary producers (e.g., seagrass and phytoplankton) by reducing excessive nutrient inputs. This management approach has been attempted for Tampa Bay as well. The Tampa Bay National Estuary Program (TBNEP) has adopted a seagrass restoration and protection goal to be reached through the reduction of external nitrogen loadings to the bay.

To place the ongoing Tampa Bay seagrass trends and management efforts in a historical perspective, it is necessary to understand the history of both seagrass coverage and water quality, including the results of efforts to improve bay water quality. Therefore, this report will discuss the major changes that have occurred in seagrass abundance since the earliest estimate of Tampa Bay seagrass coverage and some of the causes which most probably have contributed to those changes.

In addition to the historical perspective, an up-to-date account of the current water quality and seagrass trends in the bay will also be given. Attention will be focused on Hillsborough Bay, because this bay segment has the most extensive water quality record and the most detailed and up-to-date information on the recent seagrass recolonization.

Finally, the historical and current management actions that have been developed to protect and restore Tampa Bay seagrass will be discussed. This section will primarily address the important role seagrass plays in the Comprehensive Conservation and Management Plan (CCMP), which was agreed upon by the TBNEP members in 1996. This section will also briefly discuss a newly established seagrass monitoring program that will be used, in combination with ongoing seagrass mapping efforts and water quality monitoring, to periodically evaluate the success of the protection and restoration plan established for Tampa Bay. Finally, citizen-based educational and voluntary programs will be discussed that are underway to help protect and restore Tampa Bay seagrass meadows.

HISTORICAL TRENDS

In 1982, at the first Tampa Bay Area Scientific Symposium (BASIS) conference, Lewis et al. (1985) presented a comprehensive paper on Tampa Bay seagrass. Lewis et al. estimated that approximately 31,000 ha of seagrass were present in Tampa Bay during the late 1800s (Figure 21.1). The estimate was mainly based on bathymetry rather than actual seagrass coverage, and therefore, the estimate represents the potential Tampa Bay seagrass coverage at a time when man's influence on the bay was limited.

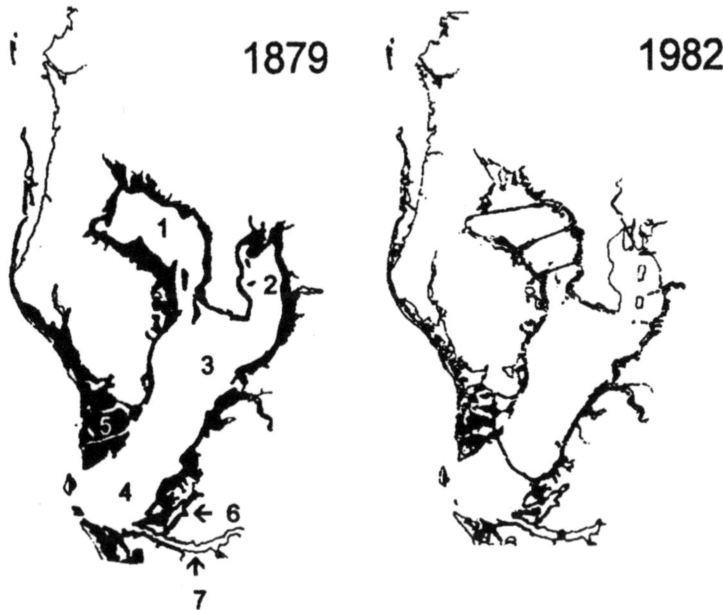

FIGURE 21.1 Seagrass coverage in Tampa Bay in 1879 and 1982 (After Lewis et al. 1985). Black areas denote seagrass meadows. Area 1 = Old Tampa Bay; Area 2 = Hillsborough Bay; Area 3 = Middle Tampa Bay; Area 4 = Lower Tampa Bay; Area 5 = Boca Ciega Bay; Area 6 = Terra Ceia Bay; Area 7 = Manatee River.

At the second BASIS meeting in 1991, Lewis et al. (1991) presented the distribution of the 1950 and 1982 Tampa Bay seagrass cover. These estimates were based on state-of-the art analysis of aerial photographs and showed that the 1950 coverage of 16,500 ha was reduced to some 8,800 ha by 1982 (Figure 21.1). The seagrass had apparently receded in all segments of the bay with major losses in the upper bay portions. Specifically, all seagrass appears to have been lost in Hillsborough Bay by 1982. Large reductions had also occurred in Boca Ciega Bay as a result of extensive residential dredge-and-fill activities.

The cause of these large seagrass losses, perhaps as much as 70% of the historical Tampa Bay seagrass cover, is undoubtedly related to man's impact on the bay. One major impact was eutrophication. This impact is directly related to the population growth of the bay area and the associated increase in commercial activities. Eutrophication as indicated by water column chlorophyll *a* concentrations (Figure 21.2A) peaked in the late 1970s and early 1980s (Johansson 1991; Johansson and Lewis 1992; Boler 1995). Another leading cause of large seagrass loss was various dredging operations and shoreline developments. These included in-bay shell dredging, port construction, ship channel expansion, causeway construction, and residential dredge-and-fill projects. Impacts from these activities culminated during the 1950s, 1960s, and 1970s.

Researchers generally agree that eutrophication and dredging operations were the major reasons for the large seagrass loss, although it is unclear which of these impacts was most serious. Also, more detailed questions remain about the process of seagrass loss. For example, did dredging operations mainly cause losses through direct physical destruction of the seagrass meadows, or through more indirect impacts such as increased turbidity of the water column and increased sediment deposition on the meadows? Details also remain unclear about the effects of eutrophication. Tampa Bay seagrass losses attributed to eutrophication are generally assumed to have resulted from a decrease in light availability, which in turn was caused by an increase in phytoplankton and epiphyte biomass. It is also well known, particularly in Hillsborough Bay (FWPCA 1969; Kelly 1995; Avery 1997) and Old Tampa Bay (J.O.R. Johansson, personal observations), that the increased

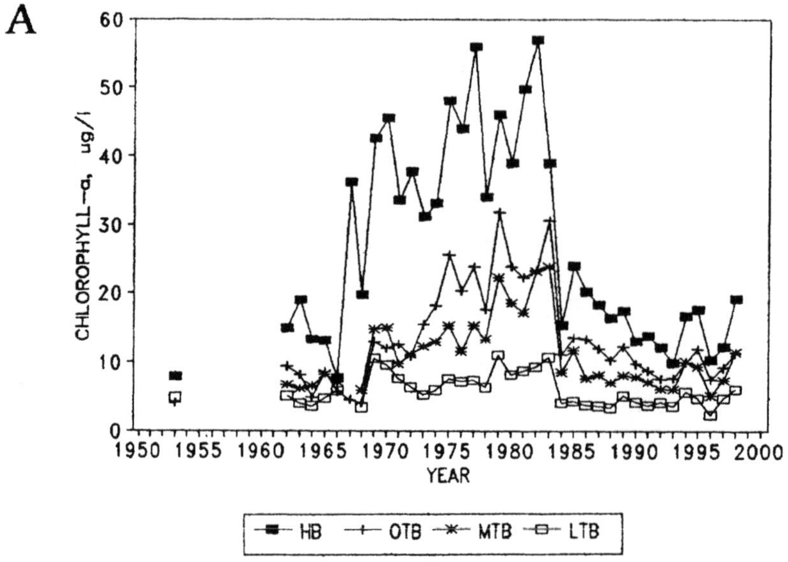

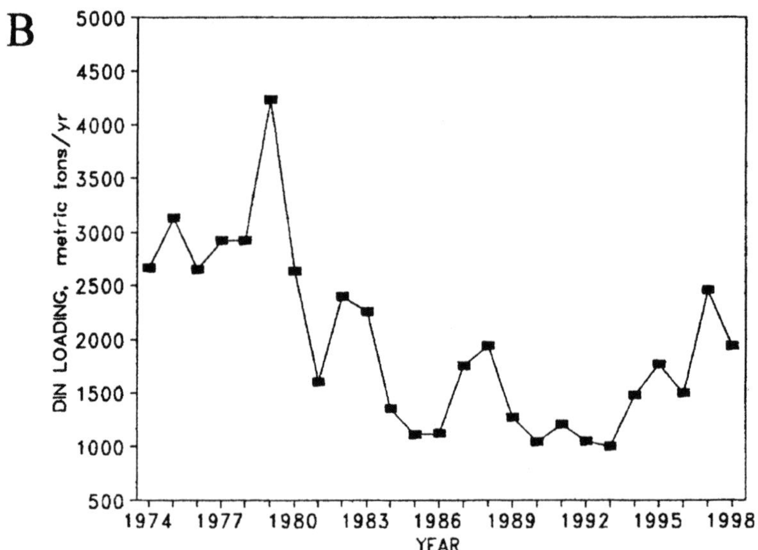

FIGURE 21.2 (A) Annual average chlorophyll *a* concentrations for the major segments of Tampa Bay, 1953–1998 (sources include National Marine Fisheries Service, Hillsborough County Environmental Protection Commission, and City of Tampa). HB = Hillsborough Bay; OTB = Old Tampa Bay; MTB = Middle Tampa Bay; LTB = Lower Tampa Bay. (B) Dissolved inorganic nitrogen loading to Hillsborough Bay from major external sources, 1974–1998. Loadings in 1997 and 1998 are estimated from rainfall amounts. Flow and concentration data were not available for these years.

nutrient loading stimulated the growth of large amounts of drift macro-algae. Dense mats of macro-algae that often accumulated in the shallow areas may have limited seagrass colonization through shading, abrasion, and hypoxia.

A large reduction in nitrogen loading to Tampa Bay occurred between the late 1970s and the early 1980s (Figure 21.2B). This reduction was primarily caused by improved wastewater treatment from domestic and industrial point-sources that discharged to Hillsborough Bay (Johansson 1991; Johansson and Lewis 1992; Zarbock et al. 1994). A large decrease in phytoplankton biomass (chlorophyll *a*) soon followed the nitrogen reduction. By 1984, chlorophyll *a* concentrations were about half of the levels found only a few years earlier (see Figure 21.2A). At about the same time that chlorophyll *a* concentrations declined in Hillsborough Bay, small isolated patches of *Halodule wrightii* (shoal grass) were observed in a shallow area of southeastern Hillsborough Bay that lacked seagrass vegetation a few years earlier (R.R. Lewis, personal communication).

SEAGRASS MONITORING PROGRAMS

After many years of severe eutrophication, one of the first tangible signs of water quality improvement in Tampa Bay was the return of seagrass to barren areas that historically had seagrass cover. Recognizing the importance of closely monitoring the progress of seagrass growth in the bay, the City of Tampa (COT) Bay Study Group and the Surface Water Improvement Management (SWIM) program of the Southwest Florida Water Management District (SWFWMD) created two independent seagrass monitoring programs in 1986 and 1988, respectively. Later, an interagency-coordinated field seagrass monitoring program, under the auspices of the TBNEP, was initiated in fall of 1998. This program is discussed in greater detail by Avery (Chapter 10 of this book) and Kurz et al. (Chapter 12 of this book), and briefly in the nitrogen load and seagrass management section later in this report.

The COT program primarily monitors seagrass in Hillsborough Bay through the use of extensive ground measurements and aerial photography. This program generates detailed information on seagrass species, condition (including short-shoot density, blade length, and epithytic cover), and composition, as well as an annual seagrass coverage estimate for this major section of Tampa Bay. The monitoring program has been modified several times in response to the rapid seagrass expansion; however, the basic survey program has remained intact. Aerial and ground surveys are conducted in September or October in order to minimize effects of seasonal growth patterns in the yearly estimates. Presently only two seagrass species, *H. wrightii* and *Ruppia maritima,* are found in Hillsborough Bay. Although the presence of *R. maritima* is recorded, only *H. wrightii* coverage is included in the annual estimates. *R. maritima* appears to be ephemeral and it is uncertain if, or how, its highly variable year-to-year growth pattern is related to water quality.

The SWIM program conducted its initial seagrass survey in 1988. This was the first comprehensive bay-wide mapping of Tampa Bay seagrass since the Florida Department of Environmental Regulation sponsored a mapping project in 1982 (Haddad 1989). SWIM has used high altitude, true color, photographs taken about every 2 years, usually during the winter, to generate groundtruthed geographical information system (GIS) maps which show Tampa Bay seagrass distribution and density information.

RECENT SEAGRASS AND WATER QUALITY TRENDS

Results from the two monitoring programs will be discussed briefly; however, a more detailed discussion of methods and findings of these programs can be found in Ries and Avery (1996), Avery (Chapter 10 of this book), and Kurz et al. (Chapter 12 of this book). Both the SWIM and the COT have found that the Tampa Bay seagrass cover has been increasing since the start of the programs. The SWIM program has now completed five bay-wide surveys over the period 1988 through 1997 (November–December 1988, December 1990, February 1993, January 1995, and May 1997) and has found that Tampa Bay seagrass coverage has increased from 9,590 ha in 1988 to 10,930 ha in 1997 (Figure 21.3A). Boca Ciega Bay, Lower Tampa Bay, and Old Tampa Bay are the bay segments that have seen the greatest increase in the number of hectares revegetated since 1988. Hillsborough Bay, however, has had by far the greatest percentage increase in seagrass cover. The COT program shows a similar trend of rapidly expanding seagrass growth in Hillsborough

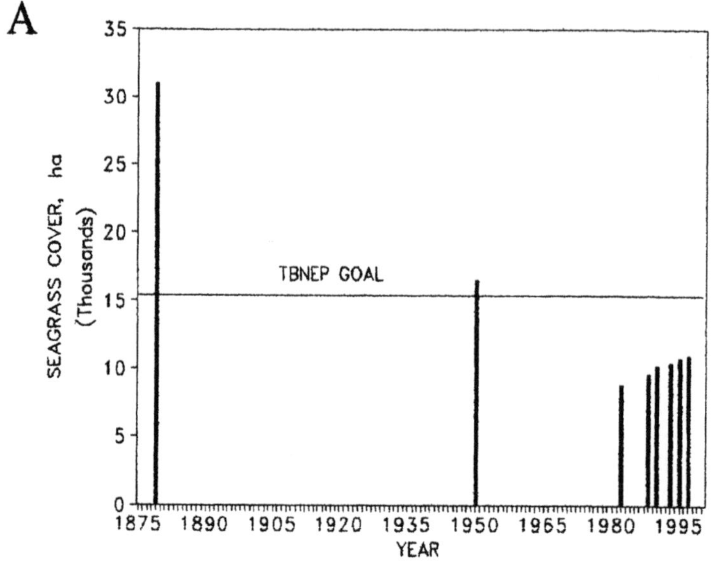

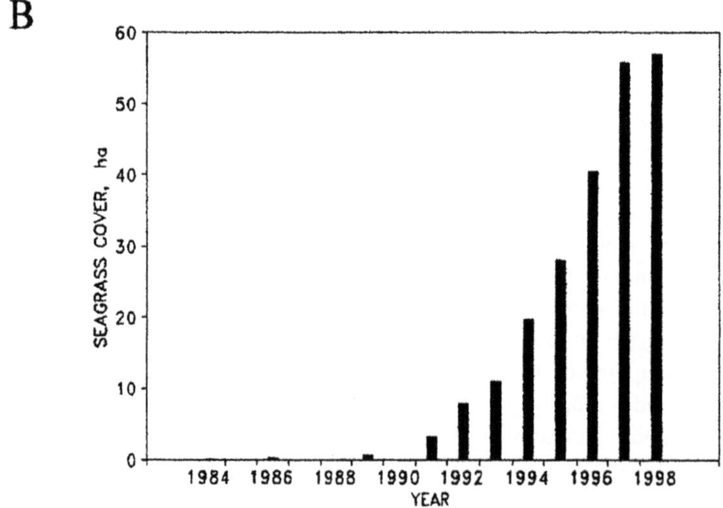

FIGURE 21.3 (A) Long-term trend of Tampa Bay seagrass coverage, 1879–1997 (sources include Lewis et al. 1985; Lewis et al. 1991; Johansson and Ries 1997; Kurz et al., Chapter 12 of this book). The horizontal line marked TBNEP GOAL denotes the Tampa Bay National Estuary Program's seagrass restoration and protection goal. (B) Hillsborough Bay seagrass coverage estimated by the COT, 1984–1998. The estimate for 1984 is exclusively based on interpretation of aerial photographs. No estimates were performed for 1985, 1987, 1988, and 1990.

Bay. In 1984, the Hillsborough Bay seagrass coverage was limited to a few isolated patches, comprising a total of less than 0.02 ha. Since then, each successive COT survey, until 1998, has shown a substantial increase in *H. wrightii* seagrass cover (Figure 21.3B). The 1998 Hillsborough Bay coverage was estimated at about 57 ha.

There are differences in the estimates of seagrass coverage between the two programs. In 1988 the SWIM program reported 6 ha in Hillsborough Bay, while the COT program in 1989 reported less than 1 ha. In 1997, the SWIM program found 81 ha, but the COT program reported only 56 ha.

In addition, the SWIM program reported a substantial loss of seagrass in Hillsborough Bay between 1990 and 1993. Although the COT program did not measure seagrass cover in 1990, it did find a large increase in coverage between 1989 and 1993. These discrepancies are probably due to differences in seagrass survey techniques and the methods used to calculate coverage. The SWIM estimates, for example, include both *H. wrightii* and the ephemeral seagrass *R. maritima*, while the COT estimates only include the more temporally and spatially stable *H. wrightii*.

The detailed seagrass information collected by the COT in Hillsborough Bay can be used to search for potential relationships between seagrass expansion and water quality trends. Year-to-year variations in Hillsborough Bay chlorophyll *a* concentrations, which indicate changes in the trophic state of the bay and also the amount of nitrogen being discharged to the bay, are not reflected in the annual trend of Hillsborough Bay seagrass expansion (Figure 21.4A). Such short-term relationships between the relatively slow process of seagrass areal expansion and more variable water quality parameters should not be expected. Instead, the ongoing expansion of seagrass in Hillsborough Bay, and in other sections of Tampa Bay as well, most probably resulted from the large decrease in eutrophic state that occurred in the early 1980s and which is reflected in the Hillsborough Bay chlorophyll *a* and Secchi depths records (Figure 21.4A and 4B). However, most recently, high rainfall amounts during the years 1995, 1996, and the 1997–1998 El Niño event (Figure 21.4B) have increased nitrogen loading to the bay. Both phytoplankton biomass and light attenuation have increased in Hillsborough Bay and similar increases have also occurred in all major bay segments during this period of high rainfall. These often detrimental influences on seagrass growth appear to have reduced the rate of seagrass expansion during the last few years, at least in the Hillsborough Bay section of Tampa Bay for which the most detailed and up-to-date seagrass information is available. The 1998 Hillsborough Bay survey was the first survey in several years that showed only minimal seagrass expansion over the previous survey. However, it is encouraging that seagrass coverage did not decline in Hillsborough Bay in 1998 despite the increase in light attenuation and phytoplankton biomass.

As noted above, most bay segments surveyed by the SWIM program show a net increase in seagrass coverage over the 9-year monitoring period. However, the most recent surveys in 1995 and 1997 detected a general decrease in density of the seagrass meadows in all bay segments except Hillsborough Bay. This apparent thinning of the meadows may be a result of the recent increase in light attenuation that has occurred in all major bay segments since 1994.

SEAGRASS PROTECTION AND RESTORATION GOALS

The concept of resource-based management has been under development in the Tampa Bay area since at least 1982, when a local group of scientists and resource managers formed the Tampa Bay Management Study Committee and identified the loss of seagrass meadows as a priority problem facing bay managers (Lewis et al. 1985). In response to perceived degradation of water quality and natural systems, the Florida Legislature designated Tampa Bay a priority water body in 1987. Acting on the advice of the Agency on Bay Management of the Tampa Bay Regional Planning Council, the Southwest Florida Water Management District subsequently adopted a Tampa Bay water quality assessment strategy which sought to identify links between nitrogen loading, bay water quality, and the management of seagrass systems (SWFWMD 1988; Spaulding et al. 1989; Ries 1993).

In 1990, Tampa Bay was accepted into the U.S. Environmental Protection Agency's National Estuary Program. The TBNEP, a partnership that includes three regulatory agencies and six local governments, has built on the resource-based approach initiated by earlier bay management efforts. Further, it has developed water quality models (Zarbock et al. 1994; Janicki and Wade 1996; Martin et al. 1996; Zarbock et al. 1996; Morrison et al. 1997) to quantify linkages between nitrogen loadings and bay water quality and models that link loadings and water quality to seagrass goals (Janicki and Wade 1996; Greening et al. 1996).

A

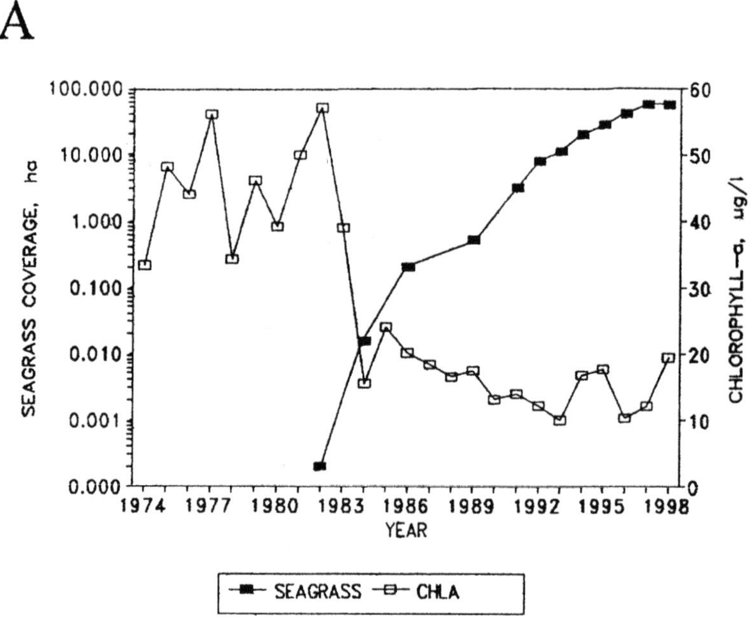

B

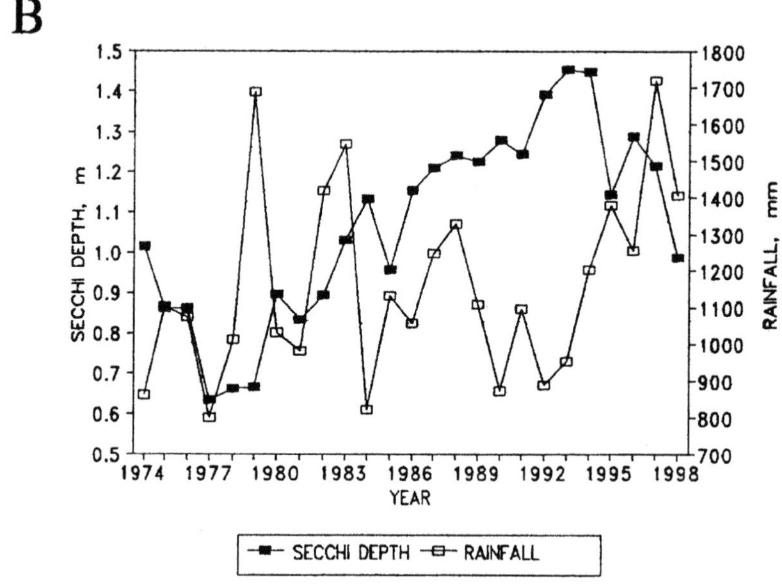

FIGURE 21.4 (A) Hillsborough Bay annual average chlorophyll *a* concentrations and seagrass coverage. Showing the start of seagrass recovery in 1984 following the rapid decline in chlorophyll from 1982 to 1984. Chlorophyll *a* concentrations measured by the Hillsborough County Environmental Protection Commission and the City of Tampa. Seagrass coverage measured by the City of Tampa. (B) Hillsborough Bay annual average Secchi depths measured by the Hillsborough County Environmental Protection Commission and rainfall at the Tampa International Airport measured by the National Oceanographic and Atmospheric Administration, National Climatic Data Center.

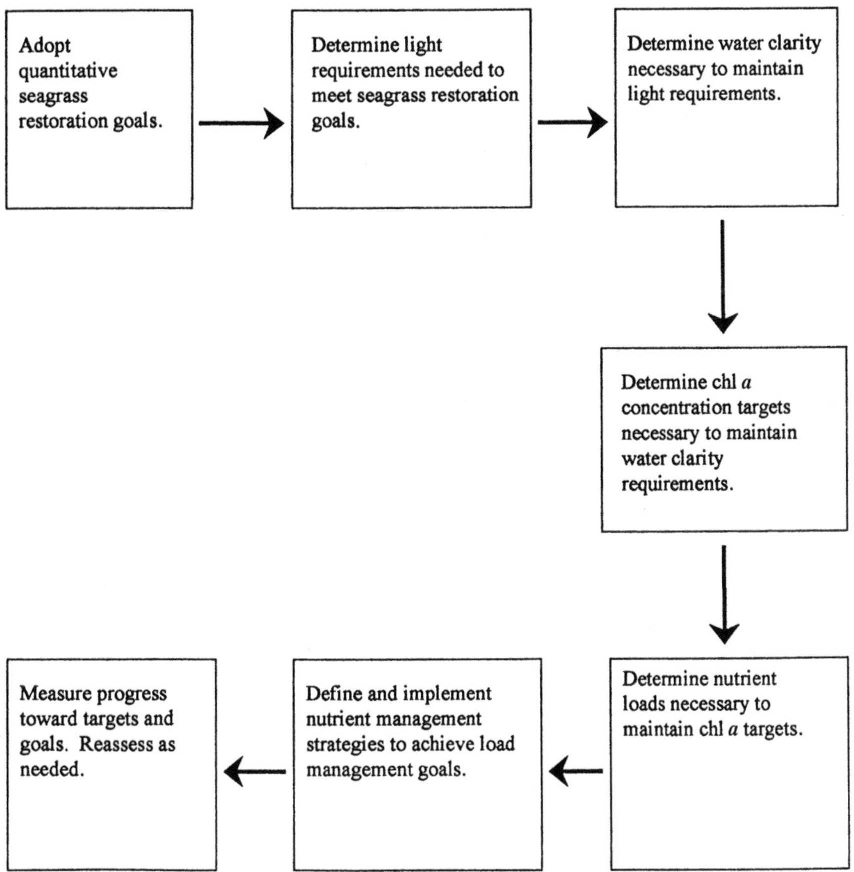

FIGURE 21.5 The approach used by the Tampa Bay National Estuary Program partners to develop and implement the seagrass protection and restoration management program for Tampa Bay.

A simplified approach was used by the TBNEP partnership to develop and implement a seagrass protection and restoration management program for Tampa Bay (Figure 21.5). The major steps were:

1. Set specific, quantitative seagrass coverage goals for each bay segment.
2. Determine seagrass water quality requirements and appropriate nitrogen loading targets.
3. Define and implement the nitrogen management strategies needed to achieve the load management targets.

Step 1. Set Quantitative Resource Management Goals

The establishment of clearly defined and measurable goals is crucial for a successful resource management effort. The TBNEP Management Conference adopted the initial goal to increase the current Tampa Bay seagrass cover to that estimated for 1950.

Based on digitized aerial photographic images, it was estimated that approximately 16,500 ha of seagrass existed in Tampa Bay in 1950 (Lewis et al. 1991). At that time, the seagrass apparently grew to depths of 1.5 to 2 m in most areas of the bay. By 1992, approximately 10,400 ha of seagrass remained in Tampa Bay (Janicki and Wade 1996), a loss of more than 35% since the 1950 benchmark period. Some of the observed loss (about 160 ha) occurred as the result of direct habitat destruction associated with the construction of navigation channels and other dredging and filling projects within existing seagrass meadows and is assumed to be non-restorable through water quality management actions.

In 1996, the TBNEP adopted a bay-wide minimum seagrass goal of 15,400 ha (see Figure 21.3A). This goal represented 95% of the estimated 1950 seagrass cover, and includes the protection of the existing 10,400 ha plus the restoration of an additional 5,000 ha (TBNEP 1996).

STEP 2. DETERMINE SEAGRASS WATER QUALITY REQUIREMENT AND APPROPRIATE NITROGEN LOADING RATES

Once the seagrass restoration and protection goal was established by the participants, the next steps established the environmental requirements necessary to meet the agreed-upon goal and subsequent management actions necessary to meet those requirements. Elements of this process included the following:

- *Determine environmental requirements needed to meet the seagrass restoration goal.* Recent research indicates that the deep edges of *Thalassia testudinum* meadows, the primary seagrass species for which nitrogen loading targets are being set, correspond to the depth at which 20.5% of subsurface irradiance (the light that penetrates the water surface) reaches the bay bottom on an annual average basis (Dixon and Leverone 1995). The long-term seagrass coverage goal can thus be restated as a water clarity and light penetration target. Therefore, in order to restore seagrass to near 1950 levels in a given bay segment, water clarity in that segment should be restored to the point that allows 20.5% of subsurface irradiance to reach the same depths that were reached in 1950.
- *Determine water clarity necessary to allow adequate light to penetrate to the 1950 seagrass depths.*
  Water clarity and light penetration in Tampa Bay are affected by a number of factors, such as phytoplankton biomass, non-phytoplankton turbidity, and water color. Janicki and Wade (1996) used regression analyses, based on long-term monitoring data collected monthly by the Environmental Protection Commission of Hillsborough County, to develop an empirical model describing water clarity (Secchi depth) variations in the four largest bay segments (Old Tampa Bay, Hillsborough Bay, Middle Tampa Bay, and Lower Tampa Bay) as a function of chlorophyll *a* and turbidity. Water color may be an important cause of light attenuation in some bay segments; however, including color in the regression model did not produce a significant improvement in the predictive ability. Results of the modeling effort indicate that, on a bay-wide basis, variation in chlorophyll *a* concentration is the major factor affecting variation in water clarity.
- *Determine chlorophyll a concentration targets necessary to maintain water clarity needed to meet the seagrass light requirement.*
  The empirical regression model was used to estimate chlorophyll *a* concentrations necessary to maintain water clarity needed for seagrass growth for each major bay segment. Annual average chlorophyll *a* targets (8.5 μg/l for Old Tampa Bay, 13.2 μg/l for Hillsborough Bay, 7.4 μg/l for Middle Tampa Bay, and 4.6 μg/l for Lower Tampa Bay) were adopted as intermediate measures for assessing success in maintaining adequate water quality to meet the long-term seagrass goal.
- *Determine nutrient loadings necessary to achieve and maintain the chlorophyll* a *targets.*
  Water quality conditions in 1992–1994 appear to allow an annual average of more than 20.5% of subsurface irradiance to reach target depths (i.e., the depths to which seagrass grew in 1950) in three of the four largest bay segments (Hillsborough Bay, Old Tampa Bay, and Lower Tampa Bay). Water quality in the Middle Tampa Bay segment currently allows slightly less than 20.5% to target depth. Thus, a management strategy based on "holding the line" at 1992–1994 nitrogen loading rates should be adequate to achieve

> the seagrass restoration goals in three of the four segments. This "hold the line" approach, combined with careful monitoring of water quality and seagrass extent, was adopted by the TBNEP partnership in 1996 as its initial nitrogen load management strategy.

However, a successful adherence to the "hold the line" nitrogen loading strategy may be hindered by the projected population growth in the watershed. A 20% increase in population, and a 7% increase in annual nitrogen load, are anticipated by the year 2010 (Zarbock et al. 1996). Therefore, if the projected loading increase (a total of 17 U.S. tons per year) is not prevented or precluded by watershed management actions, the "hold the line" load management strategy may not be achieved.

STEP 3. DEFINE AND IMPLEMENT NITROGEN MANAGEMENT STRATEGIES NEEDED TO ACHIEVE LOAD MANAGEMENT GOALS

Local government and agency partners in the TBNEP signed an Intergovernmental Agreement in 1998 pledging to carry out specific actions needed to "hold the line" on nitrogen loading. The agreement lists the responsibility of each partner for meeting the nitrogen management goals and a timetable for achieving them. How those goals are reached will be left up to the individual communities as defined by them in their action plans (TBNEP 1998a).

To maintain nitrogen loadings at 1992–1994 levels, local government action plans address the portion of the nitrogen target which relates to non-agricultural stormwater runoff and municipal point sources within their jurisdictions, a total of 6 U.S. tons of nitrogen per year through the year 2010.

A Nitrogen Management Consortium of a local electric utility, industries, and agricultural interests, as well as the local governments and regulatory agency representatives in the TBNEP, was established to address the remaining 11 U.S. tons per year of the projected nitrogen loading increase. The consortium developed a nitrogen management action plan that combines for each bay segment all local government, agency, and industry projects that will contribute to meeting the 5-year nitrogen management goal (TBNEP 1998b).

Table 21.1 summarizes expected reductions from those projects which are or will be completed by the end of 1999. Old Tampa Bay, Hillsborough Bay, and Middle Tampa Bay are expected to meet (and exceed) the year 2000 nitrogen management goal with completed and ongoing projects alone. Lower Tampa Bay and Boca Ciega Bay are expected to meet (and exceed) the load reduction

TABLE 21.1
Tampa Bay Nitrogen Management Consortium summary of goals and expected reductions expressed as tons of total nitrogen load reduced or precluded for each year until 1999.

Bay Segment	1995–1999 Nitrogen Reduction Goal	Expected Reduction: Completed or Ongoing Projects[1]	Expected Reduction: Pending Projects[2]
Old Tampa Bay	2.1	3.6	1.5
Hillsborough Bay	41.5	62.0	3.9
Middle Tampa Bay	11.1	14.8	6.3
Lower Tampa Bay	25.4	25.0	11.2
Boca Ciega Bay	3.9	0.8	4.8
Total	84.0	106.2	27.7

[1] Projects that have been completed or are under construction. These summaries do not include reductions expected from atmospheric deposition reductions.

[2] Projects that have funding available and are in the planning or permitting stages. These summaries do not include reductions expected from atmospheric deposition reductions.

goal with ongoing and pending projects. Estimates of nitrogen loadings to the bay from all sources will be updated in 1999 and the effectiveness of the proposed projects will be evaluated.

The approach advocated by the TBNEP stresses cooperative solutions and flexible strategies to meet the nitrogen management goals. This approach does not prescribe the specific types of projects that must be included in the action plan; consortium partners have been encouraged to pursue the most cost-effective options to achieve the agreed-upon goals. The goals will be reviewed and revised every 5 years, or more often if needed.

MEASURING PROGRESS TOWARDS TARGETS AND GOALS

Two important elements of the resource-based management approach are the measurement of progress toward the resource goal (seagrass coverage) and tracking implementation of the nitrogen management strategies needed to reach the long-term seagrass restoration goal.

In addition to the seagrass areal and density information from the SWIM mapping effort, an interagency seagrass field monitoring program was initiated in 1998 to document the progress of seagrass restoration. The field program includes several TBNEP partner agencies, as well as other organizations involved with bay issues. The program consists of annual determinations of seagrass species, condition, and distribution at approximately 57 permanently placed transects along the periphery of the bay (Avery, Chapter 10 of this book; Kurz et al., Chapter 12 of this book).

Further, measured bay segment-specific chlorophyll *a* concentrations will be compared annually to the adopted chlorophyll *a* targets. This comparison is an intermediate measure for assessing the success in maintaining water quality requirements necessary to meet the long-term seagrass goal. Between 1992 and 1997, annual chlorophyll *a* concentrations in all bay segments have fluctuated around the adopted targets, indicating that careful adherence to the "hold the line" nitrogen management strategy will be necessary to maintain adequate water quality conditions to meet the long-term seagrass restoration goal (Figure 21.3A).

CITIZEN-BASED SEAGRASS PROTECTION AND RESTORATION EFFORTS

In addition to the governmental and private industry initiatives to manage nitrogen loadings to Tampa Bay, several citizen-based efforts have contributed to the restoration and protection of seagrass. For example, when Hillsborough County proposed a boat closure area and other restrictions in the Cockroach Bay Aquatic Preserve due to extensive propeller scarring of seagrass in that area, local citizens and fishermen organized in 1995 and proposed alternative, non-regulatory ways for protecting the seagrass from further propeller scarring. Hillsborough County agreed to allow a 3-year "test period" for a strong educational and voluntary slow speed protection approach. Projects have included development and distribution of educational material to encourage the boating public to take responsibility for its actions. Educational signs have been posted at public boat ramps in Cockroach Bay and 25 seagrass area marker buoys have been placed at the deep edge of the seagrass meadows. Further, a unique boating stop-light tidal gauge has been developed and placed to alert boaters of the tidal water level in relation to the seagrass. Recent monitoring of the Cockroach Bay area has indicated that several previously heavily scarred seagrass areas are recovering.

Citizen volunteers have also been involved with several seagrass plantings projects that have been coordinated and directed by Tampa BayWatch, a local non-profit organization focused on Tampa Bay resource restoration. These projects are aimed at involving the public in bay issues. They also help restore seagrass to areas of the bay which currently lack seagrass vegetation, but now appear to have adequate water quality to promote seagrass growth.

CONCLUSION

Lewis et al. (1991) and others have questioned whether the expansion of Tampa Bay seagrass meadows between 1982 and 1988 was caused by the large reduction in anthropogenic nitrogen loading that occurred in the late 1970s and early 1980s, or was related to a period of relatively low rainfall in the mid-1980s. The authors cautioned that increased inputs of both nutrients and water with a high color content during future periods of high rainfall might reverse the recent water quality improvements and negatively impact the ongoing seagrass expansion by increasing water column light attenuation.

A decade of additional information is now available since 1988. The 1990s have included an extended period of high rainfall that started in 1994 and lasted through the 1997–1998 El Niño event. Increased nitrogen loading during this period appears to have reduced the water quality of the bay, specifically by increasing chlorophyll *a* concentrations and reducing water clarity. The reductions in water clarity, as indicated by relatively shallow Secchi depths, may have reduced the rate of seagrass expansion during the last few years. The rate reduction appears to have occurred in Hillsborough Bay for which the most up-to-date seagrass information is available. The scenario of increased rainfall and reductions in seagrass expansion supports the early caution by Lewis et al. (1991) and others. Water quality impacts related to the recent high rainfall period may not yet have climaxed. However, water quality has remained excellent when compared to conditions found when nitrogen loadings were high, prior to the large anthropogenic nitrogen loading reductions that occurred in the late 1970s and early 1980s. Further, it is encouraging that overall seagrass coverage has not declined in Hillsborough Bay despite the increase in light attenuation and phytoplankton biomass.

The Tampa Bay management community has agreed that the protection and restoration of the Tampa Bay living resources are of primary importance. Nitrogen loading targets were adopted for Tampa Bay in 1991 and these targets were based on the water quality requirements of the seagrass *Thalassia testudinum.* A long-term goal has been adopted to achieve 15,400 ha of seagrass in Tampa Bay, or 95% of that observed in 1950. To reach the long-term seagrass restoration goal, a projected 7% increase in nitrogen loading associated with an increase in the watershed's human population over the next 20 years must be offset. Partners in the TBNEP and organizations participating in the Nitrogen Management Consortium have, therefore, identified and committed to specific nitrogen load reduction projects to ensure that the water quality conditions necessary to meet the long-term living resource restoration goals for Tampa Bay are achieved.

ACKNOWLEDGMENTS

We would like to thank S. Bortone, R. Virnstein, A. Squires, W. Avery, and one anonymous reviewer for critical reading of the manuscript and valuable suggestions.

REFERENCES

Avery, W. 1997. Distribution and abundance of macroalgae and seagrass in Hillsborough Bay, Florida, from 1986 to 1995. In *Proceedings, Tampa Bay Area Scientific Information Symposium 3,* S. F. Treat (ed.). Tampa, FL, pp. 151–165.

Avery, W. 2000. Monitoring submerged aquatic vegetation in Hillsborough Bay, Florida. In *Seagrasses: Monitoring, Ecology, Physiology, and Management,* S. A. Bortone (ed.). CRC Press, Boca Raton, FL, pp. 137–145.

Boler, R. 1995. Surface water quality 1992–1994, Hillsborough County, Florida. Hillsborough County Environmental Protection Commission, Tampa, FL.

Dixon, L. K. and J. R. Leverone. 1995. Light requirements of *Thalassia testudinum* in Tampa Bay, Florida. Final report to the Surface Water Improvement and Management Department, Southwest Florida Water Management District, Tampa, FL.

FWPCA. 1969. Problems and management of water quality in Hillsborough Bay, Florida. Hillsborough Bay Technical Assistance Project, Technical Programs, Southeast Region, Federal Water Pollution Control Administration, Tampa, FL.

Greening, H. S., G. Morrison, R. M. Eckenrod, and M. J. Perry. 1996. The Tampa Bay resource-based management approach. In *Proceedings, Tampa Bay Area Scientific Information Symposium 3,* S. F. Treat (ed.). Tampa, FL, pp. 349–355.

Haddad, K. D. 1989. Habitat trends and fisheries in Tampa and Sarasota Bays. In *Tampa and Sarasota Bays: Issues, Resources, Status, and Management,* E. D. Estevez (ed.). NOAA Estuary-of-the-Month Seminar Series No. 11, National Oceanic and Atmospheric Administration, Washington, D.C., pp. 113–128.

Janicki, A. and D. Wade. 1996. Estimating critical nitrogen loads for the Tampa Bay estuary: an empirically based approach to setting management targets. Tampa Bay National Estuary Program, Technical Publication 06-96, Coastal Environmental, Inc., St. Petersburg, FL.

Johansson, J. O. R. 1991. Long-term trends of nitrogen loading, water quality and biological indicators in Hillsborough Bay, Florida. In *Proceedings, Tampa Bay Area Scientific Information Symposium 2,* S. F. Treat and P. A. Clark (eds.). Tampa, FL, pp. 157–176.

Johansson, J. O. R. and R. R. Lewis. 1992. Recent improvements in water quality and biological indicators in Hillsborough Bay, a highly impacted subdivision of Tampa Bay, Florida, U.S.A. *Science of the Total Environment* Supplement 1992:1199–1215.

Johansson, J. O. R. and T. Ries. 1997. Seagrass in Tampa Bay: historic trends and future expectations. In *Proceedings, Tampa Bay Area Scientific Information Symposium 3*, S. F. Treat (ed.). Tampa, FL, pp. 139–150.

Kelly, B. O. 1995. Long-term trends of macroalgae in Hillsborough Bay. *Florida Scientist* 58:179–192.

Kurz, R. C., D. A. Tomasko, D. Burdick, T. F. Ries, S. K. Patterson, and R. Finck. 2000. Recent trends in seagrass distributions in southwest Florida coastal waters. In *Seagrasses: Monitoring, Ecology, Physiology, and Management,* S. A. Bortone (ed.). CRC Press, Boca Raton, FL, pp. 157–166.

Lewis, R. R., M. J. Durako, M. D. Moffler, and R. C. Phillips. 1985. Seagrass meadows of Tampa Bay: a review. In *Proceedings, Tampa Bay Area Scientific Information Symposium,* S. F. Treat, J. L. Simon, R. R. Lewis, and R. L. Whitman, Jr. (eds.). Burgess Publishing Company, Minneapolis, MN, pp. 210–246.

Lewis, R. R., K. D. Haddad, and J. O. R. Johansson. 1991. Recent areal expansion of seagrass meadows in Tampa Bay: real bay improvement or drought-induced? In *Proceedings, Tampa Bay Area Scientific Information Symposium 2*, S. F. Treat and P. A. Clark (eds.). Tampa, FL, pp. 189–192.

Martin, J. L., P. F. Wang, T. Wool, and G. Morrison. 1996. A mechanistic management-oriented water quality model for Tampa Bay. Final report to the Surface Water Improvement and Management Department, Southwest Florida Water Management District, Tampa, FL.

Morrison, G., A. J. Janicki, D. L. Wade, J. L. Martin, G. Vargo, and J. O. R. Johansson. 1997. Estimated nitrogen fluxes and nitrogen-chlorophyll relationships in Tampa Bay, 1985–1994. In *Proceedings, Tampa Bay Area Scientific Information Symposium 3*, S. F. Treat (ed.). Tampa, FL, pp. 249–268.

Ries, T. 1993. The Tampa Bay experience. In *Proceedings and Conclusions of Workshops on Submerged Aquatic Vegetation and Photosynthetically Active Radiation,* L. J. Morris and D. A. Tomasko (eds.). St. Johns River Water Management District, Special publication SJ93-SP13, Palatka, FL, pp. 19–24.

Ries, T. and W. Avery. 1996. Seagrass coverage. In *Tampa Bay Environmental Monitoring Report,* 1992–1993, A. P. Squires, A. J. Janicki, and H. Greening (eds.). Tampa Bay National Estuary Program, Technical Publication No. 15-96, St. Petersburg, FL, pp. 6-1 to 6-5.

Spaulding, M. L., D. P. French, and H. Rines. 1989. A review of the Tampa Bay water quality model system. Final report to the Southwest Florida Water Management District, Tampa, FL.

SWFWMD. 1988. Tampa Bay Surface Water Improvement and Management Plan. Southwest Florida Water Management District, Tampa, FL.

TBNEP. 1996. Charting the Course for Tampa Bay: Final Comprehensive Conservation and Management Plan. Tampa Bay National Estuary Program, St. Petersburg, FL.

TBNEP. 1998a. Tampa Bay National Estuary Program 1995–1999: collated action plans. Tampa Bay National Estuary Program, St. Petersburg, FL.

TBNEP. 1998b. Partnership for Progress: The Tampa Bay Nitrogen Management Consortium Action Plan 1995–1999. Tampa Bay National Estuary Program, St. Petersburg, FL.

Zarbock, H. W., A. J. Janicki, D. L. Wade, D. Heimbuch, and H. Wilson. 1994. Estimates of total nitrogen, total phosphorus, and total suspended solids loadings to Tampa Bay, Florida. Tampa Bay National Estuary Program, Technical Publication 04-94, Coastal Environmental, Inc., St. Petersburg, FL.

Zarbock, H. W., A. J. Janicki, D. L. Wade, and R. J. Pribble. 1996. Model-based estimates of total nitrogen loading to Tampa Bay. Tampa Bay National Estuary Program, Technical Publication 05-96, Coastal Environmental, Inc., St. Petersburg, FL.

22 Matching Salinity Metrics to Estuarine Seagrasses for Freshwater Inflow Management

Ernest D. Estevez

Abstract Changes in the amount, timing, or location of freshwater inflow are recognized as primary stressors to estuarine and marine seagrass species and communities. Progress has been made assessing impacts of too little or too much freshwater inflow, by using salinity as a first-order stressor per se, and as an indicator of associated, second-order stressors. Seagrass species can be associated with typical mean values of salinity and also with extreme values of salinity where species persist or perish. But a growing body of anecdotal, observational, and experimental data suggests that patterns of salinity variation also have significant effects on seagrass presence, persistence, and condition.

To set the stage for a discussion on whether metrics of salinity variation can be identified that are meaningful in terms of seagrass biology, examples of salinity impacts to seagrasses are reviewed. Next, a simple theoretical model is proposed for mean salinity and salinity variation across an estuarine gradient, and implications are illustrated for interpreting seagrass data and selecting salinity metrics. Existing published data on the salinity range of wigeongrass (*Ruppia maritima*) are investigated to illustrate how tolerable salinity ranges can be interpreted in the context of the model (and misinterpreted by water managers). Types of salinity data are evaluated. Principal among these are historic data, records of continuous, *in situ* instruments, and outputs available from hydrodynamic and salinity models. A coefficient of salinity variation is proposed for seagrass studies based on the use of tidal datum planes (mean high water, mean lower low water, etc.) to describe continuous changes of water level in estuaries.

INTRODUCTION

Pritchard (1967) defined an estuary as "a semi-enclosed coastal body of water which has a free connection with the open sea and within which sea water is measurably diluted with fresh water derived from land drainage." The definition remains meaningful today, albeit with recognition of some natural exceptions and variations. Measurable dilution of sea water with fresh water establishes salinity variation as one of the estuary's dominant physico-chemical characteristics. Salinity variation has spatial and temporal aspects which give structure and regularity to estuaries. Spatial gradients act as signposts, and temporal trends act as time-givers, for a number of ecological features and processes (Snedaker et al. 1977). Although annual salinity variation across the length of an estuary may be quite large, patterns of salinity variation in undisturbed estuaries tend to be more constrained in space and predictable through time.

Freshwater inflow and the estuarine salinity fields created by such inflows regulate estuarine productivity at many levels and by many mechanisms (Livingston et al. 1997). Ample evidence documents the effects of salinity from the molecular to ecosystem levels of biological organization in estuaries (Kinne 1967). Biological responses to salinity are evident in the physiology, morphology,

0-8493-2045-3/99/$0.00+$.50
© 2000 by CRC Press LLC

growth, reproduction, behavior, and ecology of estuarine biota. At the level of the coastal landscape, estuaries of similar geomorphic type exhibit analogous and regular patterns in the distribution, abundance, composition, and condition of habitats and communities (Odum et al. 1974).

Estuarine productivity ultimately is controlled by freshwater inflow, salinity fields, and other important co-variants of inflow such as nutrient supply. The exact nature of this relationship has been difficult to evaluate empirically, but an increasing number of studies have made progress in this direction. Three useful insights are emerging. The first is that freshwater inflows can be related to measures of estuarine productivity with increased significance when the interaction of freshwater and estuarine habitat are considered (Deegan et al. 1986). The second is that "goodness of fit" between inflows and associated productivity measures (such as fishery landings or catch per unit effort) is improved when production values are lagged so as to account for recruitment delays (Longley 1994). Third, primary and secondary productivity seem to have meaningful relationships to patterns of salinity variation through space (Jassby et al. 1995) and time (Montague and Ley 1993).

Anthropogenic changes to inflows and salinity fields are known to affect estuaries, often deleteriously (Aleem 1972; Gunter et al. 1973; Mahmud 1985). Although inflows to the estuary may be increased, as in the case of flood control, they more commonly are decreased by such projects as impoundments and diversions of major tributaries. The location or timing of inflows also may be changed. Such changes to the drainage of fresh water from land cause changes first, and most noticeably, to estuarine salinity fields closest to land. Small inflow changes may cause large (relative) salinity changes in tidal river reaches, brackish backwater bays, and other low salinity environments (Sklar and Browder 1998).

Other changes in estuaries affect their salinity fields because a particular estuary's salinity structure is the result of a unique interaction among inflows, tides, and the system's physical geometry. Changes to inlets and other sea connections affect high salinity areas of an estuary, often independent of inflow changes. Physical changes, such as channelization, in middle estuary areas affect middle-range salinities through salinization of channel beds or freshening of the estuary's shallow flanks. Changes to inflow, tidal action, or system geometry may also change the estuary's vertical salinity characteristics (Richards and Grant 1986).

Ecological impairments of estuaries resulting from anthropogenic inflow and salinity changes are known from many case studies (Cross and Williams 1981; Halim 1990). The majority of reports concern adverse impacts to fauna. Pertaining to estuarine plants, several accounts of phytoplankton impacts are reported. As a result of inflow and salinity change, phytoplankton may be extirpated, entrained by reverse flows, killed, or changed in terms of community composition or productivity (Estevez and Marshall 1993). Riparian vegetation is affected directly and in several ways by inflow change when inundation patterns are changed. Salinity effects are less well evaluated, but riparian vegetation: (1) is often dispersed along estuarine salinity gradients; (2) is broadly adaptable to salinity changes over wide ranges; and (3) can be affected by natural or anthropogenic changes in the timing of inflow and salinity variation (McMillan 1974; Smalley and Thien 1976; Beare and Zedler 1987).

This chapter reviews and interprets selected reports on salinity and seagrasses to identify what is known, and not known, concerning the ecology and management of seagrasses in estuaries affected by freshwater inflow alteration. Reference will be made to submerged aquatic vegetation (SAV) other than true seagrasses. This chapter also points out that salinity-seagrass relationships are not understood well enough for estuarine management, that effects of salinity variation on seagrasses hold considerable promise for new research and management studies, and that certain ways of calculating salinity variation deserve more formal analysis.

SALINITY AND SEAGRASSES

Seagrasses occur over a wide range of salinities. When SAV associated with upper estuaries is also considered, benthic angiosperms may be said to grow across the extended range of salinity from true fresh (tidal) water to hypersaline conditions. Several authors have tabulated extreme salinities, salinity ranges, and average or optimal salinities for a number of seagrass and SAV species. Individual species are found in salinity ranges narrower than the full range of salinities occupied by benthic angiosperms, and their presence is taken as indicative of general salinity conditions where they occur (Dennison et al. 1993). In Florida waters and other lower-latitude estuaries, for example, "The general conclusions drawn are that shoal grass (*Halodule wrightii*) is broadly euryhaline; turtle grass (*Thalassia testudinum*) thrives only in intermediate salinity ranges (roughly 20‰ to 40‰); and manatee grass (*Syringodium filiforme*) tolerates only a narrow range of salinity near sea-strength" (Montague and Ley 1993: page 703).

Other species are associated with lower salinities. Tidal river and upper estuary SAV such as water nymph (*Najas guadalupensis*), tapegrass (*Vallisneria americana*), pondweeds (*Potamogeton* spp.), and introduced species including Eurasian water-milfoil (*Myriophyllum spicatum*) and hydrilla (*Hydrilla verticillata*) tolerate low salinity and some species survive variable exposures to about one-third sea salinity (Twilley and Barko 1990). Wigeongrass (*Ruppia maritima*), discussed in more detail below, may occupy a range of salinities, often replacing *H. wrightii* where salinity decreases or *V. americana* where salinities increase. So long as other factors are not limiting, it seems that species of benthic angiosperms are available to occupy every point along the typical gradient of mean salinities present in estuaries.

Why don't many species of seagrass co-occur in a given estuary, and why don't they overlap broadly in their spatial distributions? To investigate these questions insofar as the singular role of salinity regulation is concerned, this study examines (1) the responses of seagrasses to salinity changes caused by humans; (2) the salinity structures of natural and altered estuaries; and (3) the salinity range of *Ruppia maritima*.

1. Seagrass Response to Altered Salinity

Limited literature is available on the effects of salinity change on seagrasses, although some reports present a consistent pattern of impact. The Etang de Berre near Marseilles, France was a high salinity lagoonal basin connected to the Mediterranean Sea. Typical salinities ranged from 31 to 34 ppt. It originally contained extensive beds of the seagrass, *Zostera*, and the mussel, *Mytilus* (Bellan 1972). Since 1966, a canal from the Durance River and hydropower plant has pulsed large quantities of water (to 125 cubic meters per second) into the lagoon on a highly erratic schedule. By 1970 the system had changed significantly. Vast seagrass areas had disappeared and mussel beds contracted to areas farthest from the canal (Stora and Arnoux 1983).

In Florida, large quantities of water have been pulsed on erratic schedules to the Sebastian and St. Lucie rivers (east coast), northeastern Florida Bay, the Ten Thousand Islands (south peninsula), and the Caloosahatchee and Manatee rivers (west coast). In the Indian River Lagoon, *Halodule wrightii,* which once was abundant near the mouth of the Sebastian River, has retreated about 2 km, over the 30-year life of a flood control canal that augments river flow (Estevez and Marshall 1993). Freshwater macrophytes once were abundant in the St. Lucie River. In 1924, a canal connecting the river to Lake Okeechobee was completed for flood relief and navigation. By the 1960s, estuarine seagrasses were present, but sparse, and today, 75 years after the initial canal, the St. Lucie estuary is almost barren of seagrass. In both the Sebastian and St. Lucie rivers, stabilization and enlargement of nearby tidal inlets have been concurrent with canal improvements. Faka Union Bay (Ten Thousand Islands) receives pulses of flood water from the Faka Union Canal, constructed in 1969. Seagrass meadows were present in the bay during the early 1970s, but by 1986, almost no seagrasses persisted except for small aggregations of star grass (*Halophila engelmanii*). Seagrass beds are thought to be more extensive in nearby Pumpkin Bay, unaffected by freshets from the Faka Union Canal (Browder et al. 1989).

Along the northern Gulf of Mexico, "The greatest anthropogenic impact on seagrasses in Mississippi Sound is the release of fresh water from the Mississippi River during periodic openings of the Bonnet Carré Spillway." (Eleuterius 1987). Changes to the volume, location, and timing of freshwater inflows extirpated *Halophila engelmanii*, caused catastrophic declines of *Syringodium filiforme*, and detrimentally affected *Thalassia testudinum*. *Ruppia maritima* flourished under prolonged low-salinity conditions but erratic inflows harmed all benthic angiosperms. Seasonal growth patterns of seagrasses also were disrupted.

The Laguna Madre of Texas is a hypersaline lagoon, not an estuary, that has experienced salinity changes caused by channel dredging, increased rainfall, and agricultural drainage. Once dominant *Halodule wrightii* beds have receded extensively, replaced by bare bottom or *Syringodium filiforme* and *Thalassia testudinum* admixtures over a period of 50 years. Biological changes continue decades after major hydrological changes, including shifts in species ranges by some unexplained mechanism of patch saltation (Quammen and Onuf 1993).

These case studies and less documented accounts of seagrass response to salinity change suggest the following findings.

- Salinity changes affect seagrass distribution and community structure; changes in morphology, physiology, and productivity also occur but studies of such phenomena risk missing the larger impacts occurring at landscape levels.
- Seagrass changes include extirpation, decline in species diversity, shifts in location or relative size, "halo" effects, shortened seasons for species with annual cycles, and alternation of species within particular estuarine reaches.
- Seagrass distribution and community structure may change soon, after only a few years, following large salinity changes — but seagrass changes can continue for long periods after the onset of inflow alterations.
- Seagrass changes caused by inflow and salinity changes are amplified by changes to estuary geometry or connections to the sea — and sometimes changes are beneficial.
- Seagrass changes are caused as much by changes in salinity variation as by changes in average salinity conditions — most case studies involve erratic pulses of large volumes of fresh water.
- Seagrasses may never adapt to salinity conditions driven by erratic freshwater inflows, this is also true of benthic faunal communities that are simply replaced by assemblages of eurytopic species.

2. Salinity Structure of Natural and Altered Estuaries

Estuaries may be classified by many schemes. The general salinity structure of an estuary allows it, or the reaches of very large estuaries such as Chesapeake Bay, to be classified as tidal fresh (< 0.5‰), oligohaline (0.5–5‰), mesohaline (> 5–18‰), or polyhaline (> 18‰) for research and management purposes. Other salinity intervals and ranges also are in use (Bulger et al. 1990). Specific isohalines are regarded as average values and the ranges captured by pairs of isohalines describe the salinity variation between them. There are presently no known classification systems for estuaries that are based on descriptions of salinity variation more detailed than ranges (personal observation).

Mean salinity down the length of a perfect estuary increases from pure fresh water to pure sea water values, as illustrated by line 1 in Figure 22.1 (upper graph). For discussion sake, line 1 may be interpreted as applying to mean salinity in a vertically homogeneous estuary, although it could also be interpreted as surface or bottom salinity in a stratified system. Continuous variation of salinity between these boundary conditions allows for the dilution-curve analysis of non-conservative constituent concentrations used in estuarine nutrient and pollution studies.

Mean salinity rarely follows the form of line 1 down real estuaries because of the integration of inflowing fresh water and tidally driven sea water, within a physical system of irregular geometry. Lines 2 and 3 in Figure 22.1 illustrate two of many other possibilities. Line 2 does not originate

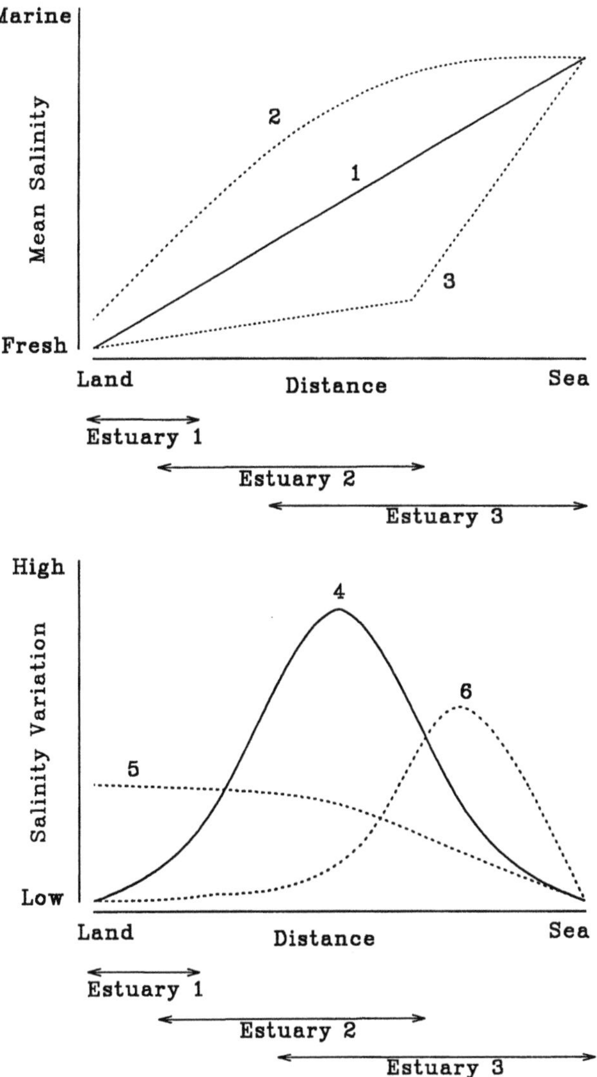

FIGURE 22.1 Patterns of spatial variation in (upper graph) mean salinity and (lower graph) salinity variation across estuaries. Specific estuaries experience mean salinities and ranges in salinity variation, particular to their situation in the coastal landscape.

in fresh water, for example, to depict a common salinity consequence of instream structures built in tidal river reaches. Rotation of line 3 illustrates a common effect of physical control on salinity, such as a sill or constricted tidal inlet. Natural variation, anthropogenic changes to inflow, tidal action, or system geometry cause some salinity gradients to be dominated by low values; others are dominated by high values.

We tend to think of estuaries as "semi-enclosed coastal bodies of water," although there are cases where estuaries are not enclosed (for example, Florida's Big Bend coast). For this reason we tend not to think of estuaries as extending over the full range of measurable sea water dilution. For purely geological reasons, some semi-enclosed coastal bodies of water encompass a wide range of mean salinities whereas other estuaries are situated on narrower segments of the full dilution curve (Figure 22.1). In terms of mean salinity fields, different estuaries can present different conditions for seagrasses and other biota.

Salinity variation down an estuary can be evaluated similarly. Salinity variation must be negligible near an estuary's boundary conditions for salinity. By definition fresh water is always fresh and the sea is usually salty. Between these two locations, salinity variation will be measurable and some reach of the estuary will have higher salinity variation than other reaches (Figure 22.1, bottom graph). In a symmetrical example (line 4), the largest salinity variation occurs at intermediate values of mean salinity (line 1). Highest salinity variations may extend across larger geographic areas and values of mean salinity (line 5) or be shifted seaward (line 6). As line 5 also illustrates, instream structures in tidal rivers can create and maintain high salinity variations where mean salinity and salinity variation normally would be low. Some estuaries encompass a wide range of salinity variance whereas other estuaries do not. In terms of salinity variances, different estuaries can present different conditions for seagrasses and other biota.

Specific estuaries comprise particular combinations of mean salinity conditions and salinity variances. To venture a generalization, every estuary has diagnostic fields of mean salinity and salinity variance; the fields usually do not capture the complete range of possibilities, and no two estuaries have identical fields. True estuarine organisms have the evolved capacity to adapt to a salinity field, including a field of salinity variation they encounter in the specific estuary they inhabit. However, these organisms cannot adapt to any combination of mean salinity and salinity variation or to the full range of combinations possible across all estuarine landscapes. Many studies have shown that estuarine plants, including seagrasses, and animals have the capacity to cope with low, high, stable, and fluctuating salinities, but that limits to such capacity also exist.

3. Local and Global Salinity Ranges of *Ruppia maritima*

Adapting Kantrud (1991), wigeongrass (*Ruppia maritima*):

- is often found with the seagrasses but is not a true marine plant; it is considered a freshwater species with a pronounced salinity tolerance;
- behaves as an annual in habitats subject to drought, lethal increases in salinity, or other extremes; or as a perennial in deeper, more stable environments;
- has specialized features enabling survival under varying salinities and high temperature beyond those tolerated by other submersed angiosperms;
- usually occurs at low intertidal elevations in estuaries, but mixes with true seagrasses up to at least 1.5 km offshore in large oceanic bays.

R. maritima has been reported in waters of salinity ranging from fresh to hypersaline concentrations, greatly exceeding the tolerable salinity range of other seagrass species presently under discussion. Kantrud (1991) tabulated salinity data from 128 records of *R. maritima* occurrence, including 109 records where the dominant cation was chloride. Of these, 68 records provide ranges of salinity in which *R. maritima* was found. Figure 22.2 illustrates the distribution of reported ranges at 68 locations, relative to sea water salinity (lower panel), and relative to *R. maritima*'s highest known salinity tolerance (upper panel).

R. maritima grows in salinities higher than those of sea water, but most records of its occurrence show that the range of salinity tolerated at a particular place is considerably lower than sea salinity. The median range is approximately 40% sea salinity, signifying that half of the local salinity ranges for *R. maritima* are less than 16‰. When expressed as percentages of its maximum salinity (Figure 22.2, upper panel), all but 8 records of local salinity ranges tolerated by *R. maritima* fall below 10%. By this analysis the very wide range of salinity over which *R. maritima* has been found clearly is not the range in which it actually grows within a particular water body. There seems to be a limit to the plant's adaptability to salinity variation, probably involving the rate of salinity change (Kantrud 1991), which may prevent it from occurring in natural waters where the limit is exceeded. Such a limit would exist irrespective of the estuary's mean salinity.

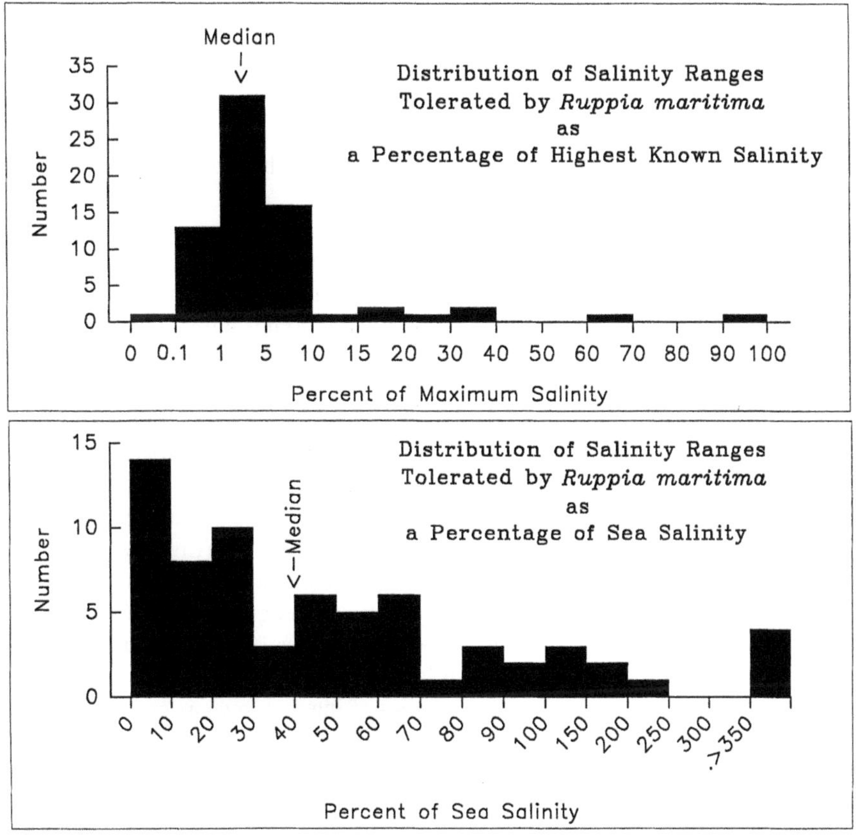

FIGURE 22.2 Observed salinity ranges of *Ruppia maritima* in waters dominated by chlorides (N = 68 sites), adapted from Kantrud (1991). Local salinity ranges are expressed as percentages of sea salinity (lower panel) and percentages of the highest salinity in which *R. maritima* is known to occur (upper panel).

To recapitulate, estuarine seagrasses have been harmed by salinity changes resulting from alterations made to freshwater inflows. Case studies involve instances where irregular, large volumes of fresh water are pulsed to estuaries, suggesting that salinity variation may be one factor responsible for seagrass responses observed at landscape levels. Although the estuarine landscape is characterized by variable salinities, each estuary has a particular pattern of mean salinity and salinity variation resulting from the interplay of its shape, tides, and water supply. An analysis of salinity ranges tolerated by the cosmopolitan species, *R. maritima*, suggests that this adaptable species usually tolerates local salinity ranges that are much smaller than its global salinity range. Other species of seagrasses may have more or less narrow working ranges of tolerance to salinity variation (personal observation). Improved analyses of salinity patterns in estuaries can increase our understanding of seagrass ecology and the management of inflows and salinity in altered estuaries.

SALINITY METRICS

Seagrass attributes are measured through time with frequencies that differ from the measurement of salinities in an estuary. A trend has developed for seagrass physiology and productivity studies to employ daily, weekly, or continuous records of independent variables such as light. These records can be extended mathematically to represent the entire annual light climate of seagrasses in a bed (see Dixon, Chapter 2 of this book). Although average annual salinity of an estuary can be described empirically by routine measurement, such mean annual conditions are not comparable to annualized light climates that account for daily and seasonal solar variations.

Estuary managers interested in salinity optimization for seagrasses use the landscape-level attributes of presence and absence, centers and end-points of dispersion, overlaps, and temporal features, such as seasonal alternation or extirpation of species, as dependent variables. Condition measures such as shoot density or growth rates can be used within perennial beds for added insight to salinity effects.

How should salinity be measured and described to best enable seagrass scientists and managers to recognize and attribute the variance of such seagrass data to salinity variance? Routine monitoring is a common source of salinity data in estuaries. Monitoring occurs opportunistically or on a routine calendar basis. Evidence is sought for general water quality trends through time and under-samples shallows. Because routine monitoring is designed with statistical considerations other than seagrass in mind, salinity fields near seagrasses must often be extrapolated from distant data. Some monitoring employs only surface or mid-depth sampling and measurement and consequently misses conditions experienced by seagrasses. Results of many monitoring programs, including those employing continuous recorders of *in situ* data, are reported in terms of central tendencies (mean or median salinity) and ranges. An estuary's salinity regime can take widely different forms depending on how continuous or repetitive salinity data are queried (Livingston 1987).

Mathematical models provide a second source of salinity data for estuaries. Steady-state models are basically useless for questions of inflow, salinity, or seagrass management, and why they continue to be used is a mystery. Dynamic models capture system geometry and are driven by boundary conditions of freshwater inflow and tidal action. Curvilinear models can be fit to irregular shorelines, shallow waters, and small or patchy areas of interest, such as grassbeds. Contemporary models of circulation and salinity can be calibrated and verified with precision as well as accuracy, and are also able to incorporate wind effects. Models simulate conditions over small intervals of space and time, meaning that tidal variations and surface to bottom differences in salinity can be resolved. Physical modelers have the ability to output salinity data in a variety of forms useful to estuarine ecologists and managers, although this rarely occurs.

To the extent that seagrasses are affected by salinity variation as well as average salinity conditions, we need to design and utilize monitoring programs and estuarine models so as to improve what we know about salinity variation. We must do better than calculating the simple range and mean of salinity values over time, better than using depth-averaged and tide-averaged descriptions of salinity over the length of an estuary, and better than trying to interpret the conventional engineering analysis of steady-state salinity at the 10th, 50th, and 90th percentiles of freshwater inflow. We seek descriptions of salinity that seagrasses actually experience, and measures of salinity variation that can be evaluated with respect to seagrass dispersion, abundance, or condition.

Several tools are available but untested. Estuarine managers are concerned that the rate of salinity change may harm seagrasses, but we have yet to experiment with the first or second differentials of salinity with respect to time, as independent variables. Geographic information systems could depict salinity differences between natural conditions and those resulting from a proposed inflow change, and then depict the spatial relationship of seagrass attributes to such changes. Comparative analyses of well-studied estuaries could seek regional patterns in the relationship of salinity variation and seagrasses. And, as explained previously, the distribution and abundance of benthic angiosperms could be evaluated over the full range of salinity variation from land to sea, rather than being constrained to the physical boundaries of the semi-enclosed estuary.

In the short term, the most promising avenue for seagrass research and management would be fuller appreciation, use, and analysis of tidal variations in salinity. Tidal variation is an integral part of the estuarine environment and must be considered in the design of certain types of sampling and measurement (Hutchinson and Sklar 1993). As continuous harmonic variations in water level caused by astronomical forces, tidal stands of practical significance have long been expressed in terms of "mean high water," "mean lower low water," "mean tide level," and the like. Some monitoring programs are intentionally biased to capture high-tide or low-tide conditions. Data analysis is sometimes improved when all but slack-tide salinity data are censored from the record.

Fits of tidally filtered salinity data to estuarine features are suggestive. The location of the 1–2‰ high-tide, mean surface salinity in tidal rivers on the Florida west coast corresponds to the location of a transition from alluvial to tidal soils (Estevez et al. 1991a). Tidal wetland species in estuaries have up-estuary limits that correspond to low-tide surface salinities, and tidal freshwater wetland species have down-river limits that correspond to high-tide surface salinities (Estevez et al. 1991b). In the Loxahatchee River, Florida, oysters are largest in the estuarine reach with the highest salinity variation, whereas benthic angiosperms are fewest. In the tidal Manatee River, Florida, where salinity fields are disrupted by an instream dam and reservoir, benthic angiosperms are absent in reaches where the coefficient of salinity variation is greater than about 0.35 (Estevez and Marshall 1997).

Salinity variation can be expressed as the variance about a mean value; the standard deviation about the mean, or the coefficient of variation (or coefficient of variability — the standard deviation divided by the mean). The coefficient of variation (CV) holds promise as a descriptor of salinity variation in the study of seagrasses. Intrinsically, the CV always is positive, is normalized by the mean of the sample population, and can be represented as a percentage. In application to salinity studies, the CV can be calculated from historic, current, or modeled data (few metadata are necessary); it can be interpolated between two locations for which sample data are available; and it can be mapped. It follows from previous remarks that coefficients of salinity variation can be computed from salinities corresponding to particular tidal conditions — mean higher high and mean lower low waters, for example.

Work in progress in several tidal rivers along the southwest Florida coast suggests that down-estuary ranges of benthic plants and animals can be interpreted with meaning relative to salinity variation, and that tidally filtered coefficients of salinity variation are useful as analytical tools in such applications. The extent to which tidally filtered coefficients of variation improve our understanding of seagrass community structure, distribution, and condition remains to be seen. Undertaking such simple analyses is bound to advance our knowledge of salinity-mediated consequences of altering freshwater inflows to estuaries.

DISCUSSION

Many factors other than salinity affect seagrasses and the effect of some factors is profound. These have been deliberately set aside to highlight the role of salinity in regulating seagrasses in estuaries beset by problems of freshwater inflow. Seagrasses are regulated by factors other than salinity, but which co-vary with salinity, such as current speed, temperature, and anion–cation balances. Seagrass response to salinity variation can be studied in isolation (Fears 1993), but the singular effect of salinity on seagrasses found in nature will be difficult to separate from such factors. Other factors important to seagrasses, such as nutrient concentrations, light climates, and oxygen stress, co-vary with salinity in regular ways. However, they also may vary independently of salinity — their singular effects, and interactions with salinity, will be less difficult to evaluate. In the real world, seagrasses simultaneously face these and other regulatory forces, such as epiphytism, herbivory, and disease. But in the real world of estuarine science and management, the issues of freshwater inflows and salinity impacts continue to receive less attention than they deserve (Kennish 1992).

The idea that salinity variation can harm seagrasses is not new. In a rule-based ecological model of an estuarine lake, Starfield et al. (1989) concluded that the abundance of underwater plant biomass was sensitive to the rate of salinity change rather than the salinity level per se, but this inference was left unverified by empirical testing. In a study of salinity variation and benthic vegetation of northeastern Florida Bay Montague and Ley (1993) found that mean plant biomass decreased as salinity variation increased, even though biomass and mean salinity were positively associated. Montague and Ley (1993: page 715) reached three significant conclusions:

1. "The negative effect of salinity fluctuation may be greater at stations where the lowest salinities approach fresh water … because most marine species do not tolerate fresh water for long. For example, fluctuation between 0‰ and 10‰ may be more damaging than fluctuation between 20‰ and 30‰, where the same change occurs well above fresh water."
2. "If the proposed salinity-fluctuation hypothesis is correct, the regression of plant biomass may predict the impact of human-induced changes in the salinity regime of northeastern Florida Bay. Roughly, for every 3‰ increase in *standard deviation* of salinity, total benthic plant biomass could decline by an order of magnitude."
3. "Previously, some researchers have discounted the importance of water-project-induced salinity changes because of relatively small predicted changes in the *average* salinities and a general belief that seagrasses are tolerant of salinity change... such a conclusion seems unwarranted." [Emphasis added by original authors.]

In fact, the idea has been proffered that salinity variation, at least human-caused variation in excess of the variation expected from natural causes, adversely affects multiple resources and processes in estuaries, and not just seagrasses. Rozengurt (1992) opined that inflow alterations (and consequent salinity variation) that exceed the natural variation of an estuary will damage it, and that the coefficient of variation of freshwater inflow corresponds to the natural limit of tolerable inflow alteration. No mechanistic cause for this claim has come forward, and the coefficient of variation of freshwater inflow has been computed only for a relatively few estuaries. But the generality of the claim marks how far we have advanced from the idea that any salinity variation is tolerable by estuarine biota as long as it occurs within the estuary.

Of the many factors known to regulate or harm seagrasses in estuaries, salinity is possibly the easiest to diagnose and correct. Salinity problems arise when estuaries receive too much or too little fresh water, or water at improper times (Odum 1970). Stormwater runoff from urbanized areas can contribute to such problems and is a complex and expensive problem to fix. More often, though, large salinity problems can be traced to the presence, design, or operation of one or a few water management structures, many of which are public works built before legal and social prescriptions to consider estuarine impacts. Dams, reservoirs, navigation locks, flood control structures, and canals can become inflexible, single-mission structures guarded jealously by local governments, water suppliers, and other utilities. By the same token, though, it often takes the decision of only a single person, board, or agency to modify a structure, or alter its operation. It will be necessary to change such structures and their operations in order to protect seagrasses and other estuarine resources. Doing so will require scientific, resource-based data that, in the words of one conference participant, "are simple, understandable, and explainable."

River science and management have outpaced estuarine science and management insofar as the provision of restorative flows and protection of optimal flows. This stems in no small way from the fact that flow alterations in rivers are more conspicuous than in estuaries, which are below sea level and "always full." The science of instream flows for rivers is at least decades old. While we can speak of estuarine science's venerable age, it is difficult to say that a science exists by which the freshwater needs of estuaries can be determined. Managers of regulated rivers employ a number of tools that could be adapted for estuarine inflows. Base flows, for example, are presently not required in some Florida rivers impounded for water supply. It is a simple matter to add spillways to fixed-crest dams in such tidal streams. Likewise, ramped discharges are common practice in regulated rivers. No Florida estuary with regulated rivers presently receives ramped discharges — dams that may have been closed tight for months, or years, suddenly open and enormous volumes of fresh water pour through tidal rivers and into estuaries with scant notice but to boaters and landowners who might be harmed.

The types of data employed in riverine flow studies are surprisingly "simple, understandable, and explainable." Estuaries are arguably more complex ecosystems than rivers, but there is no reason why the methods of river science cannot be adapted for use in estuaries. It is not necessary that we know everything there is to know about an estuary before being able to answer a question about one of its components. Rather, we should emulate the river scientists who decide what values and benefits are important in a river, and work from those back through proximate controlling factors to answer practical questions concerning optimal flows of fresh water to estuaries (Chamberlain and Doering 1998a, 1998b). River biologists attach considerable importance to stage variation in setting optimal flows. Water levels in estuaries are set mostly by tidal action. By tracking the variation in salinity associated with different tidal stages, we may improve the science and management of estuarine seagrasses.

ACKNOWLEDGMENTS

I appreciate the assistance of Jay Sprinkel (graphics) and Sue Stover (library research). L. Kellie Dixon and three anonymous reviewers provided several helpful suggestions.

REFERENCES

Aleem, A. A. 1972. Effect of river outflow management on marine life. *Marine Biology* 15:200–208.

Beare, P. A. and J. B. Zedler. 1987. Cattail invasion and persistence in a coastal salt marsh: the role of salinity reduction. *Estuaries* 10(2):165–170.

Bellan, G. 1972. Effects of an artificial stream on marine communities. *Marine Pollution Bulletin* 3:74–78.

Browder, J. A., J. Tashiro, E. Coleman-Duffie, A. Rosenthal, and J. Wang. 1989. Documenting estuarine impacts of freshwater flow alterations and evaluating proposed remedies. In *Proceedings, Wetlands and River Corridor Management,* J. A. Kusler and S. Daly (eds.). Association of Wetland Managers, Berne, NY and OmniPress, Madison, WI, pp. 300–312.

Bulger, A. J., B. P. Hayden, M. G. McCormick-Ray, M. E. Monaco, and D. M. Nelson. 1990. A proposed estuarine classification: analysis of species salinity ranges. National Oceanic and Atmospheric Administration, National Ocean Survey ELMR Report Number 5, Rockville, MD, 28 pp.

Chamberlin, R. H. and P. H. Doering. 1998a. Freshwater inflow to the Caloosahatchee Estuary and the resource-based method for evaluation. In *Proceedings, Charlotte Harbor Public Conference and Technical Symposium,* 1997 March 15–16, S. F. Treat (ed.). Charlotte Harbor National Estuary Program Technical Report Number 98-02, South Florida Water Management District, West Palm Beach, FL, pp. 81–90.

Chamberlin, R. H. and P. H. Doering. 1998b. Preliminary estimate of optimum freshwater inflow to the Caloosahatchee Estuary: a resource-based approach. In *Proceedings, Charlotte Harbor Public Conference and Technical Symposium,* 1997 March 15–16, S. F. Treat (ed.). Charlotte Harbor National Estuary Program Technical Report Number 98-02, South Florida Water Management District, West Palm Beach, FL, pp. 121–130.

Cross, R. D. and D. L. Williams (eds.). 1981. *Proceedings of the National Symposium on Freshwater Inflow to Estuaries.* U.S. Fish and Wildlife Service OBS-81/04, 2 vols.

Deegan, L. A., J. W. Day, Jr., J. G. Gosselink, A. Yanez-Arencibia, G. S. Chavez and P. Sanchez-Gil. 1986. Relationships among physical characteristics, vegetation distribution and fisheries yield in Gulf of Mexico estuaries. In *Estuarine Variability,* D. A. Wolfe (ed.). Academic Press, New York, pp. 83–100.

Dennison, W. C., R. J. Orth, K. A. Moore, J. C. Stevenson, V. Carter, S. Kollar, P. W. Bergstrom, and R. A. Batiuk. 1993. Assessing water quality with submersed aquatic vegetation. *BioScience* 43(2):86–94.

Eleuterius, L. N. 1987. Seagrass ecology along the coasts of Alabama, Louisiana, and Mississippi. In *Proceedings of the Symposium on Subtropical-Tropical Seagrasses of the Southeastern United States,* M. J. Durako, R. C. Phillips and R. R. Lewis, III (eds.). Florida Department of Natural Resources, Marine Research Publication No. 42, pp. 11–24.

Estevez, E. D. and M. J. Marshall. 1993. Sebastian River salinity regime. Contract 92W-177, Report to St. Johns River Water Management District, Palatka, FL. Mote Marine Laboratory Technical Report No. 308, 171 pp.

Estevez, E. D. and M. J. Marshall. 1997. A landscape-level method to assess estuarine impacts of freshwater inflow alterations. In *Proceedings, Tampa Bay Scientific Information Symposium 3,* S. F. Treat (ed.). Tampa Bay National Estuary Program, pp. 217–236.

Estevez, E. D., R. E. Edwards, and D. M. Hayward. 1991a. An ecological overview of Tampa Bay's tidal rivers. In *Proceedings, Tampa Bay Scientific Information Symposium 2,* S. F. Treat and P. A. Clark (eds.). Tampa Bay Regional Planning Council, pp. 263–277.

Estevez, E. D., L. K. Dixon, and M. S. Flannery. 1991b. West coastal rivers of peninsular Florida. Chap. 10, In *Rivers of Florida,* R. J. Livingston (ed.). Springer-Verlag, New York, Ecological Studies 83, 289 pp.

Fears, S. 1993. The role of salinity fluctuation in determining seagrass distribution and species composition. Master's thesis, Department of Environmental Engineering, University of Florida, Gainesville, FL, 91 pp.

Gunter, G., B. S. Ballard, and A. Venkataramaiah. 1973. A review of salinity problems of organisms in United States coastal areas subject to the effect of engineering works. *Gulf Research Reports* 4(3):380–475.

Halim, Y. 1990. Manipulations of hydrological cycles. In *Technical Annexes to the Report on the State of the Marine Environment.* United Nations Environment Programme Regional Seas Reports and Studies Number 114/1, Chap. 6.

Hutchinson, S. E. and F. H. Sklar. 1993. Lunar periods as grouping variables for temporally fixed sampling regimes in a tidally dominated estuary. *Estuaries* 16(4):789–798.

Jassby, A. D., W. J. Kimmerer, S. G. Monismith, C. Armor, J. E. Cloern, T. M. Powell, J. R. Shubel, and T. J. Vendlinski. 1995. Isohaline position as a habitat indicator for estuarine populations. *Ecological Applications* 5(1):272–289.

Kantrud, H. A. 1991. Wigeongrass (*Ruppia maritima* L.): a literature review. U.S. Fish and Wildlife Service, Fish and Wildlife Research 10, 58 pp.

Kennish, M. J. 1992. *Ecology of Estuaries: Anthropogenic Effects.* CRC Press, Boca Raton, FL, 494 pp.

Kinne, O. 1967. Physiology of estuarine organisms with special reference to salinity and temperature: general aspects. In *Estuaries*, G. H. Lauff (ed.). American Association for the Advancement of Science, Publication No. 83, Washington, D.C., pp. 525–540.

Livingston, R. J. 1987. Field sampling in estuaries: the relationship of scale to variability. *Estuaries* 10:194–207.

Livingston, R. J., X. Niu, G. Lewis, III, and G. C. Woodsum. 1997. Freshwater input to a gulf estuary: long-term control of trophic organization. *Ecological Applications* 7(1):277–299.

Longley, W. L. (ed.). 1994. Freshwater inflows to Texas bays and estuaries: ecological relationships and methods for determination of needs. Texas Water Development Board and Texas Parks and Wildlife Department, Austin, TX, 386 pp.

Mahmud, S. 1985. Impacts of river flow changes on coastal ecosystems. In *Coastal Resources Management: Development Case Studies,* J. R. Clark (ed.). RPI Renewable Resources Information Series Coastal Management Publication Number 3, Columbia, SC, Chap. 7.

McMillan, C. 1974. Salt tolerance of mangroves and submerged aquatic plants. In *Ecology of Halophytes,* R. J. Reimold and W. H. Queen (eds.). Academic Press, New York, pp. 379–390.

Montague, C. L. and J. A. Ley. 1993. A possible effect of salinity fluctuation on abundance of benthic vegetation and associated fauna in northeastern Florida Bay. *Estuaries* 16(4):703–717.

Odum, W. E. 1970. Insidious alteration of estuarine environment. *Transactions American Fisheries Society* 99:836–847.

Odum, H. T., B. J. Copeland, and E. A. McMahan. 1974. Coastal ecological systems of the United States. Conservation Foundation, Washington, D.C., 4 vols.

Pritchard, D. W. 1967. What is an estuary: physical viewpoint. In *Estuaries*, G. H. Lauff (ed.). American Association for the Advancement of Science, Publication No. 83, Washington, D.C., pp. 3–5.

Quammen, M. L. and C. P. Onuf. 1993. Laguna Madre: seagrass changes continue decades after salinity reduction. *Estuaries* 16(2):302–310.

Richards, D. R. and M. A. Granat. 1986. Salinity redistributions in deepened estuaries. In *Estuarine Variability,* D. A. Wolfe (ed.). Academic Press, New York, pp. 463–482.

Rozengurt, M. A. 1992. Alteration of freshwater inflows. In *Stemming the Tide of Coastal Fish Habitat Loss*, R. H. Stroud (ed.). National Coalition for Marine Conservation, Marine Recreational Fisheries 14, Savannah, GA, pp. 73–80.

Sklar, F. H. and J. A. Browder. 1998. Coastal environmental impacts brought about by alterations to freshwater inflow in the Gulf of Mexico. *Environmental Management* 22(4):547–562.

Smalley, A. E. and L. B. Thien. 1976. Effects of environmental changes on marsh vegetation with special reference to salinity. Final report for contract NASA-9-14501 by Tulane University, New Orleans. NASA Report 147585.

Snedaker, S. C., D. P. De Sylva, and D. Cottrell. 1977. Role of freshwater in estuarine ecosystems. Southwest Florida Water Management District Planning Report Number 1977-2, Brooksville, FL. Vol. 1: Summary.

Starfield, A. M., B. P. Farm, and R. H. Taylor. 1989. A rule-based ecological model for the management of an estuarine lake. *Ecological Modelling* 46:107–119.

Stora, G. and A. Arnoux. 1983. Effects of large freshwater diversions on benthos of a Mediterranean lagoon. *Estuaries* 6(2):115–125.

Twilley, R. R. and J. W. Barko. 1990. The growth of submersed macrophytes under experimental salinity and light conditions. *Estuaries* 13(3):311–321.

Index

D

E

F

G

H

I

J

K

L

M

N

P

R

S

T

U

V

W

Y

Z

Milton Keynes UK
Ingram Content Group UK Ltd.
UKHW052019071024
449327UK00027B/2345